Structural Engineering Analysis by Finite Elements

The XB-70A, engineered by North American Aircraft operations using finite element analysis.

The World Trade Center.

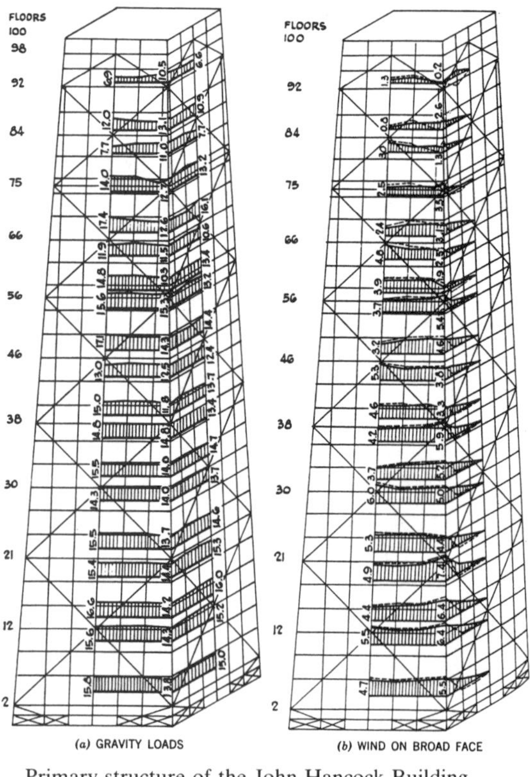

Primary structure of the John Hancock Building.

Structural Engineering Analysis by Finite Elements

Robert J. Melosh
Duke University

PRENTICE-HALL INTERNATIONAL SERIES
IN CIVIL ENGINEERING AND ENGINEERING MECHANICS

William J. Hall, Editor

PRENTICE HALL, Englewood Cliffs, New Jersey 07632

Library of Congress Cataloging-in-Publication Data

Melosh, Robert J.
Structural engineering analysis by finite elements / Robert J. Melosh.
p. cm. -- (Prentice Hall international series in civil engineering and engineering mechanics)
Includes bibliographies and index.
ISBN 0-13-855701-2
1. Structural analysis (Engineering) 2. Finite element method.
I. Title. II. Series.
TA645.M45 1990
624.1'71--dc19 89-31175
CIP

to Barbara

Editorial/production supervision and
interior design: **Kathleen Schiaparelli**
Manufacturing buyer: **Mary Noonan**
Cover design: **Lundgren Graphics, Ltd.**
Cover art: Finite element model of the NASA Ames 12-foot high
Reynold's number subsonic wind tunnel. Courtesy of Ken Hamm and Mladen Chargin.
Frontispiece art:
(a) XB-70A courtesy of North American Aircraft Operations.
(b) World Trade Center courtesy of Marc Anderson.
(c) John Hancock Building courtesy of The American Society of Civil Engineers.

Printed in the United States of America

10 9 8 7 6 5 4 3 2 1

ISBN 0-13-855701-2

Prentice-Hall International (UK) Limited, *London*
Prentice-Hall of Australia Pty. Limited, *Sydney*
Prentice-Hall Canada Inc., *Toronto*
Prentice-Hall Hispanoamericana, S.A., *Mexico*
Prentice-Hall of India Private Limited, *New Delhi*
Prentice-Hall of Japan, Inc., *Tokyo*
Simon & Schuster Asia Pte. Ltd., *Singapore*
Editora Prentice-Hall do Brasil, Ltda., *Rio de Janeiro*

Contents

PREFACE **xi**

1 FINITE ELEMENT ANALYSIS **1**

1.1 Finite Element Technology 2
1.2 Analysis Steps 9
1.3 Role of the Computer Configuration 11
1.4 Role of the Engineer in FEA 12
1.5 Homework 13
1.6 References 14

2 PROCESSING ELEMENT MODELS **16**

2.1 Assembling Element Macro Equations 16
2.2 Imposing System Boundary Conditions 18
2.3 The Finite Element Process as an Energy Process 19
2.4 The Contragradient Transformation 23
2.5 System Stiffness Matrix 24

2.6 General Use of the Finite Element Process 26
2.7 Homework 27

3 THE COMPUTER CONFIGURATION **32**

3.1 The EASE Computer Configuration 32
3.2 Problem-descriptive Data 34
3.3 Validating the Computer Configuration 36
3.4 Computerwork 38

4 LATTICE ELEMENT MODELS **42**

4.1 Types of Lattice Element Models 42
4.2 Stiffness Matrix Characteristics 46
4.3 Testing Elements for the Lattice Property 49
4.4 Finding Equivalent Loads 52
4.5 Microscopic Models 54
4.6 Unbraced and Braced Elements 57
4.7 Transforming Element Models for General Use 59
4.8 Cogradient Transformation of FEA 62
4.9 Deriving Macro Models from Test Data 65
4.10 Formulating Lattice Element Models 69
4.11 Discretization of Lattice Structures 71
4.12 Homework 71
4.13 Computerwork 76
4.14 References 80

5 SOLVING SYSTEM EQUATIONS **81**

5.1 Triangular Factorization: A Direct Method 82
5.2 Gauss-Seidel Iteration 89
5.3 *N*-Step Processes 94
5.4 Selecting the Solver 95
5.5 Validating the Solver 96

5.6 Homework 97
5.7 Computerwork 98
5.8 References 101

6 CONTROLLING ROUND-OFF ERROR 102

6.1 Number Representation 103
6.2 Arithmetic Error and Accuracy 105
6.3 Determining Maximum Errors 108
6.4 Direct Measurement of Errors 111
6.5 Determining Actual Errors 114
6.6 Illustrative Problem 115
6.7 Minimizing Errors 120
6.8 Increasing Solution Accuracy 122
6.9 Round-Off Error Control 122
6.10 Homework 125
6.11 Computerwork 126
6.12 References 129

7 CONTINUUM ELEMENT MODELS 130

7.1 Types of Continuum Element Models 131
7.2 Mathematical Basis for Continuum Models 137
7.3 Illustrative Element Models 141
7.4 Derivation of Models by Potential Energy 152
7.5 Shape and Interpolation Functions 157
7.6 Gauss Quadrature 163
7.7 Parametric Mapping 169
7.8 Stiffness Models: Minimum Requirements 171
7.9 Equivalent Loading: Minimum Requirements 172
7.10 Micro Models: Minimum Requirements 174
7.11 Methods of Deriving Models 175

7.12 Selection of Element Models 176
7.13 Homework 177
7.14 Computerwork 180
7.15 References 182

8 PREASSESSING PERFORMANCE OF FINITE ELEMENT MODELS 184

8.1 Two-Node Element Tests 185
8.2 Eigenvalue Test 187
8.3 Extended Lattice Test 189
8.4 Performance of the Beam Element Model 190
8.5 Multiple-Node Element Tests 194
8.6 Eigendata Test 195
8.7 Subdivisibility Test 196
8.8 Performance of a Membrane Element Model 199
8.9 Direct Comparison of Element Models 205
8.10 Preassessment Tests 206
8.11 Homework 207
8.12 Computerwork 210
8.13 References 213

9 SELF-QUALIFYING FINITE ELEMENT ANALYSES 214

9.1 Mathematical Basis for Torsion Analysis 215
9.2 Finite Elements for Analysis of Torsion 219
9.3 Torsional Stiffness of a Square Shaft 221
9.4 Optimum Grids 224
9.5 Discretization Accuracy Measures 227
9.6 Adaptive Local Grid Refinement 230
9.7 Adaptive Polynomial Global Refinement 236
9.8 Extrapolation Using Regular Grid Refinement 239

9.9 Designing and Refining Meshes 243
9.10 Homework 244
9.11 Computerwork 245
9.12 References 248

10 ANALYZING SCELERNOMIC SYSTEMS 249

10.1 Structures with Counters 251
10.2 Limit Analysis 257
10.3 Structures of Hyperelastic Materials 266
10.4 Linear Buckling Analysis 270
10.5 Structures with Prestresses 276
10.6 Slotted-Joint Structures 277
10.7 Features of Scelernomic Analysis 282
10.8 Homework 283
10.9 Computerwork 286
10.10 References 295

APPENDIX A CROSS-CHAPTER EQUATION NOTATION 296

APPENDIX B ANSWERS AND COMMENTS ON SELECTED HOMEWORK AND COMPUTERWORK PROBLEMS 298

INDEX 306

Preface

It is often claimed that engineering education teaches methods of thinking that are useful in problem solving. If so, an engineering textbook should emphasize the basis for thinking about a subset of engineering problems. Accordingly, this book presents information and exercises that will prepare the student for informed interpretation of finite element analysis results.

The focus is on linear and stepwise linear static analysis in structural engineering. The theory the book projects encompasses understandings that we consider essential in using finite element analysis in structural engineering practice. The book concerns implications of the choice of models and solution methods, effects of mesh design, and the insinuations of computer-induced errors.

Its exercises emphasize hands-on experiences with finite element computer codes. Computerwork experiments on problems with few elements and nodes serve to develop the engineer's deductive capabilities—so critical to circumspect use of a production code in practice. Homework analytical problems and problems supportable with a hand-held calculator serve to clarify the meanings of the theory.

This text is supported by a computer problem workbook. Reproductions of computer output for many of the Computerwork problems are included to accommodate teaching without direct use of a microcomputer.

Computer codes, capable of solving all the Computerwork problems, are also available to aid learning. These are menu-driven and input data self-prompting. They provide several levels of printout so that the student can examine models and solution steps in more detail than with a production code. They preserve problem input to accommodate later revision for parametric studies. I will gladly supply information on obtaining the codes and a site license if you let me know of your interest.

I have developed this book while teaching finite element analysis classes at Duke University, the University of California at Berkeley, Virginia Polytechnic Institute and State University, and San Jose State University. The material of Chapters 1, 2, and part of 4 has been useful in introducing undergraduates to finite element analysis. All of the material has been used with graduate students.

I have found that the material in the book is more than enough for a semester course. I recommend proceeding through the book sequentially, skipping parts of Chapter 10 if necessary due to time limitations.

More problems are included than can be solved during a semester. I have found that students benefit when assignments of parameter values and solutions of multipart problems are divided among class members and results shared. This tactic involves students in more of the problems than solving problems independently and increases the opportunities for discovery learning.

A significant amount of the material of the book has not been available elsewhere. Much of this material evolved in the process of classifying state-of-the-art models and solution methods of structural analysis. Some of it, particularly that in Chapters 8 and 10, is the result of recent research. Though these chapters may be especially vulnerable to criticism, they are consonant with the goal of the book: informed interpretation of finite element analysis results.

I would like to thank Christian Meyer of Columbia University and Subhash C. Anand of Clemson University for their thoughtful reviews of this mansucript.

Robert Melosh
Civil and Environmental Engineering Department
Duke University
Durham, NC 27706

Chapter 1

Finite Element Analysis

Finite element analysis (FEA) is the most popular means of simulating an engineering system on a digital computer. Hundreds of thousands of engineers, worldwide, have used the method for a variety of technical disciplines and equation forms. FEA has been used to predict behavior of structural, mechanical, thermal, fluid flow, electrical, and chemical systems. The analysis method has been applied to systems whose equations are linear and nonlinear. These equations may be time independent or include rate and/or acceleration terms. FEA commonly deals with systems formed of virtually any material, of regular and irregular geometry, and including geometry changes caused by spatial and time-varying boundary conditions. A measure of its popularity is that over $1 billion is spent yearly in the United States for FEA computer runs.

Thousands of engineers, scientists, and mathematicians are performing research in FEA: describing its basis, potential improvements, extensions, and new applications knowledge in more than 8000 articles in professional journals, more than 3000 Ph.D. and master's theses, and hundreds of additional professional society conference papers. More than 200 textbooks and monographs have been published on this subject.[1]

Recent research emphasizes FEA use across traditional disciplinary and career boundaries. Attention has focused on soil-structure, fluid-structure, thermoelastic, servoelastic, and magnetoelastic interactions. Besides integrity analysis, finite element technology is rapidly becoming an essential and integral part of design improvement, optimization, production process simulation, and failure assessment.

This finite element industry was stimulated by a few engineers working in Boeing's Structural Dynamic Unit under the research-conducive leadership of M. Jon Turner. Beam and extended beam theory had proved to be inadequate for predicting the stiffness of airplane wings of the triangular planform efficient for supersonic flight. Accordingly, Turner set up a research group within the unit and set the goal of establishing an improved analysis approach. The engineering literature from 1956 to 1962 chronicles the achievements of this group.[2–7]

Though claims conflict about who initiated some of the advances and established the field itself, no one disputes the fact that the Turner paper[2] is the most commonly cited reference in the literature of finite elements.

The reader of John Robinson's book *Early FEM Pioneers*[8] will be rewarded, on several levels, with accounts of finite element beginnings. These show that beginnings involved more than the Boeing cast and that evolution of the technology moved in several directions simultaneously, including both significant steps forward and some imaginative false steps.

The frontispiece shows North American's B-70, a bomber designed in the late 1950s and the first production airplane in which FEA was depended on for stiffness and stress predictions—a dramatic debut demonstrating the applicability of the method to practical hardware. The World Trade Center in New York, and the John Hancock Center in Chicago also shown, were the first buildings of more than 100 floors designed on the basis of FEA.[9,10]

From the student's viewpoint, the most important aspect of this history is that it is recent. Diversity of viewpoint still characterizes many areas of FEA. This diversity makes FEA a lively and evolving technology. An analyst who commits to a career in this subdiscipline commits to an area requiring continuing education in new concepts and understandings. Furthermore, analysts knowledgeable and circumspect in the interpretation of finite element results are at a premium.

The growing importance of FEA in engineering practice is compelling justification for including its study in an engineering curriculum. In addition, finite element analysis affords a unified and fresh view of analysis of engineering systems.

This book supports the study of FEA by presenting components of the available FEA technology in the context of linear static analysis of structures.

1.1 FINITE ELEMENT TECHNOLOGY

Finite element technology consists of a library of element models, a process for combining these models into a mathematical model of an engineering system, and a set of algorithms for numerical solution of equations. This technology is supported by computer software and hardware and by knowledge based on application experiences.

Figure 1.1 presents a numerical finite element model of a rectangular strip.

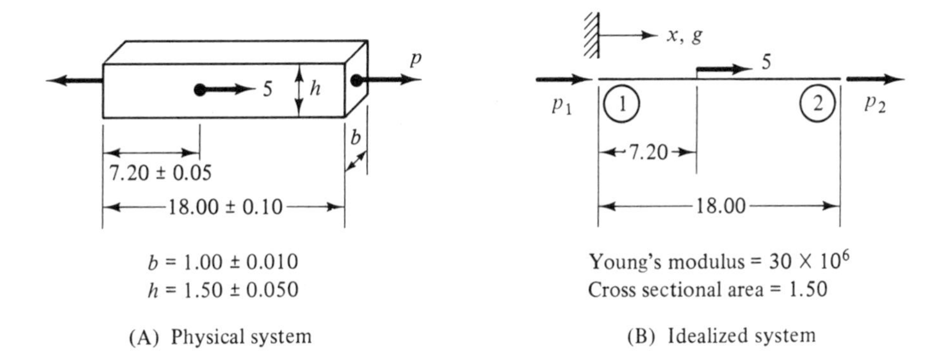

(A) Physical system

(B) Idealized system

Macro element model

$$\begin{bmatrix} 2.5 \times 10^6 & -2.5 \times 10^6 \\ -2.5 \times 10^6 & 2.5 \times 10^6 \end{bmatrix} \begin{pmatrix} g_1 \\ g_2 \end{pmatrix} - 5 \begin{pmatrix} 0.6 \\ 0.4 \end{pmatrix} = \begin{pmatrix} p_1 \\ p_2 \end{pmatrix} \qquad \text{(a)}$$

Micro element model

$$\begin{aligned} s_x &= \frac{1}{1.50}\, p_1 - \frac{0.6}{1.5} \times 5 \qquad x \leq 7.20 \\ s_x &= \frac{1}{1.50}\, p_1 - \frac{0.4}{1.5} \times 5 \qquad x > 7.20 \end{aligned} \qquad \text{(b)}$$

(C) Macro and micro mathematical models

Figure 1.1 Finite element representation of a steel strip.

Figure 1.1(A) is a sketch of the hardware. Screwed into the center of each end is an eye bolt. A small hole extends from the front surface to the back surface through the midline. The eye bolts are pulled. A wire extending through the hole is pulled to the right, each end of the wire applying a force of 2.5. Therefore, the strip has three applied forces. Figure 1.1(A) notes the dimensions and dimensional tolerances of the manufacturing specifications of this steel strip.

Figure 1.1(B) is a line drawing of an idealized mathematical model of this structure. This idealization renders the strip as a structure of uniform and perfect geometry along its nominal length. The strip is modeled by a straight line loaded at three points by stresses distributed uniformly over the cross section.

The equations of Fig. 1.1(C) define the element model for the structure. For now, we will not concern ourselves with how the numbers of the model are obtained. We will be concerned that the mathematical model has two parts, a macroscopic and a microscopic part.

The two sets of equations of the macroscopic element model include stiffness and loading equations. The stiffness equations are

$$\mathbf{K}_e \mathbf{g}_e = \mathbf{p}_e + \mathbf{q}_e \tag{1.1}$$

where

$\mathbf{K}_e$ is the symmetric element stiffness matrix; its coefficients depend on the geometry and material properties of the element;

$\mathbf{g}_e$ is the column vector of element generalized nodal displacements, (i.e., the displacements and derivatives of displacements), the independent variables of FEA;

$\mathbf{q}_e$ is the column vector of equivalent loads, representing loads applied within the element; and

$\mathbf{p}_e$ is the column vector of element loads introduced directly at the nodes.

Equations (1.1) relate the forces caused by deflections to the applied forces. The left side of these equations represents internal forces; the right, external. Equations (1.1) are the deflection-load equations. Traditionalists would call them load-deflection equations.

The loading equations are of the form

$$\mathbf{q}_e = \mathbf{L}_e\mathbf{f}_e \tag{1.2}$$

where

$\mathbf{L}_e$ is the element loading matrix, a function of element geometry and internal load distribution, and

$\mathbf{f}_e$ is the column vector defining the magnitude of internal loading.

For the macroscopic element model, Eqs. (1.1) and (1.2), provide all the structural equations needed to model the interface between an element and the rest of the structure.

This point is illustrated by Eqs. (a) of Fig. 1.1(C). Set $\mathbf{f}_e = 0$. The sum of the two equations is the statement that the element structure must be in equilibrium. The difference of the two equations is the element's load-deflection equation. If our interest is only prediction of deflections, stiffness, and element nodal forces of a structural system, the macro model represented by Eqs. (1.1) and (1.2) is enough.

The equations of the microscopic model define stresses, due to the nodal and interior loadings. The nodal loading equations are

$$\mathbf{s}_n = \mathbf{S}_n\mathbf{p}_e \tag{1.3}$$

where

$\mathbf{s}_n$ is the column vector of stress components at various points over the element and

$\mathbf{S}_n$ is the end loading stress matrix; its coefficients depend on the stress components of interest and the location of the points in the element where stress is being evaluated.

The interior loading stress equations are

$$\mathbf{s}_i = \mathbf{S}_i \mathbf{f}_e \tag{1.4}$$

where the subscript i denotes interior loading. The total stress at a point is expressed by the superposition of Eqs. (1.3) and (1.4):

$$\mathbf{s} = \mathbf{s}_n + \mathbf{s}_i \tag{1.5}$$

Equations (b) of Fig. 1.1 are of the form of Eqs. (1.5) and show that for interior point loads different stress equations of the form of Eqs. (1.4) may govern the stress, depending on the position of the stress point. This dependence characterizes the micro equations.

The finite element library can be thought of as a collection of computer subroutines, each of which can generate coefficients like those of Fig. 1.1 for a particular finite element type. In many of the illustrative models of this book, formulas are given for the coefficients. Some of the models are defined only by the procedures required to evaluate their coefficients. However, the common concept of a model is a computer logic package which is usually provided in a subroutine.

Figure 1.2 exhibits line drawings of the minimum number of finite element types that would be needed in a finite element library for structural practice. A minimal library must include two-noded elements and surface and solid elements based on triangular shapes.

The two-noded element is a straight-axis element that can simulate stretching, bending, twisting, and shearing about its axis. This element can be used to predict response of three-dimensional "stick" structures formed of members that are long compared with their width and thickness. Some of the flexibilities can be inhibited to simulate two- and one-dimensional stick structures. If a member is curved, it can be approximated as a folded line. As the number of folds increases, the true length and behavior of the curved line will be represented. This element is useful in modeling trusses, frames, and surface stiffeners.

The triangular finite element types of this minimal library are suited directly to modeling flat membranes, flat plates, and three-dimensional solids. Here we define a membrane as a surface that deforms only in biaxial stretching or shearing in its plane. Conversely, the plate only deforms in bending and shearing normal to its plane. A shell can deform as a linear combination of membrane and plate deformations. A three-dimensional solid may be stretched independently along any one of its three axes and shear independently on any of the three coordinate planes or in a linear combination of these behaviors.

With care, the flat triangular surface elements can be used for predicting behavior of curved surfaces by approximating the surfaces by flat pieces. Similarly, any three-dimensional solid can be represented by a collection of plane-surfaced tetrahedra.

Fredriksson and Mackerle[11] cataloged a professional library of more than 250

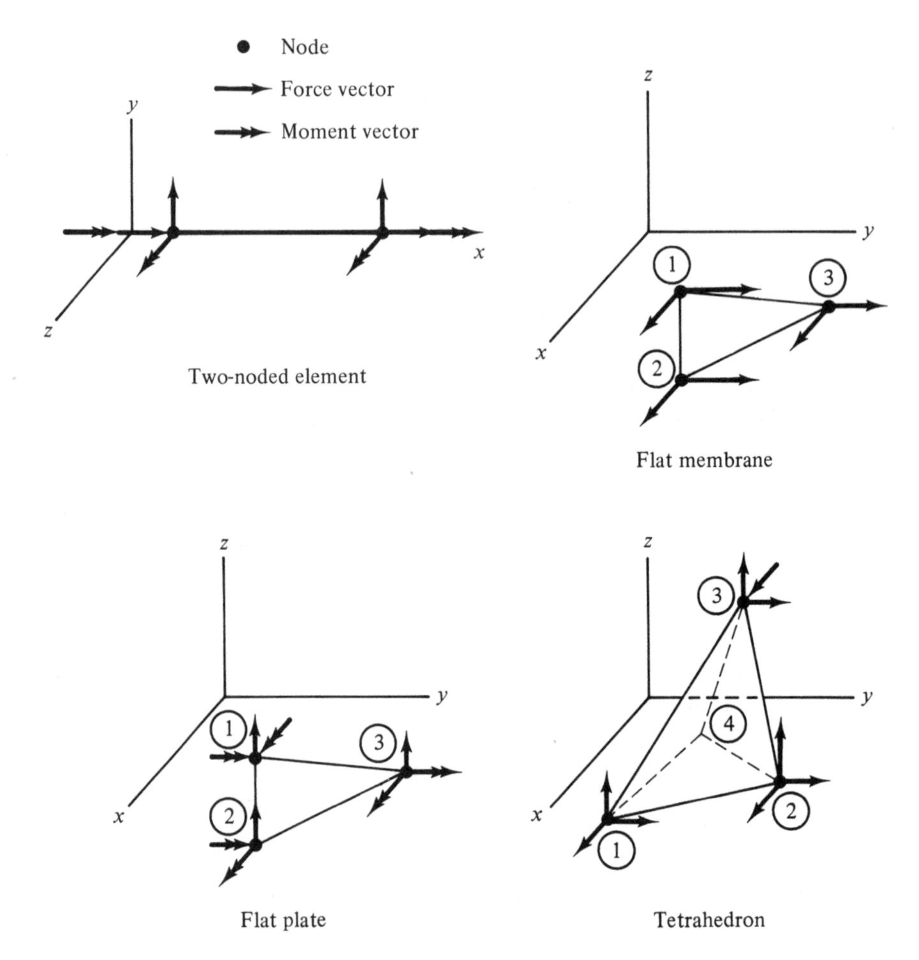

Figure 1.2 Element types of a minimal finite element library.

elements in 1978. Since providing more efficient element models is a continuing target of FEA research, the literature currently contains more than 500 element models.

Figure 1.3 shows a series spring consisting of two finite elements, each modeled by the element model of Fig. 1.1. The figure gives the macro models of each element and the system equations produced by the FE process. The form of the processed element model equations, as indicated in the last equations of Fig. 1.3, is

$$\mathbf{K}_c \mathbf{g}_c = \mathbf{p}_c \tag{1.6}$$

where

$\mathbf{K}_c$ is the constrained symmetric stiffness matrix of the structural system;

$\mathbf{g}_c$ is the column vector of unknown nodal displacements; the number of

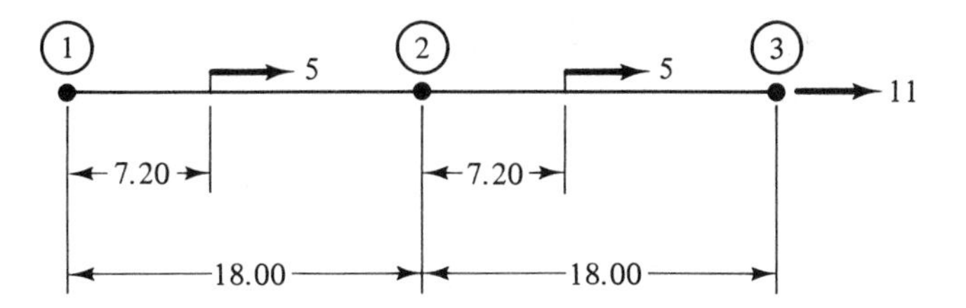

Two-element structural system

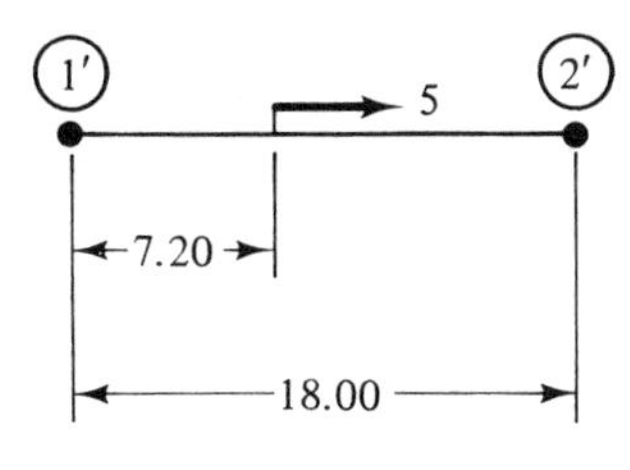

Left-side element

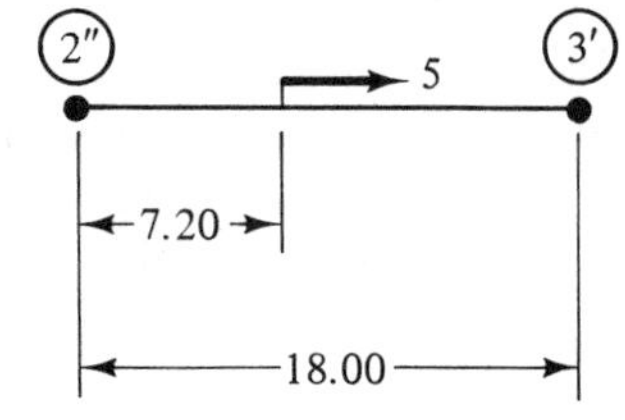

Right-side element

$$2.5 \times 10^6 \begin{bmatrix} 1 & -1 \\ -1 & 1 \end{bmatrix} \begin{pmatrix} g_{1'} \\ g_{2'} \end{pmatrix} + \begin{pmatrix} 3 \\ 2 \end{pmatrix} = \begin{pmatrix} p_{1'} \\ p_{5'} \end{pmatrix}$$

Left-side macro model

$$2.5 \times 10^6 \begin{bmatrix} 1 & -1 \\ -1 & 1 \end{bmatrix} \begin{pmatrix} g_{2''} \\ g_{3'} \end{pmatrix} + \begin{pmatrix} 3 \\ 2 \end{pmatrix} = \begin{pmatrix} p_{2''} \\ p_{3'} \end{pmatrix}$$

Right-side macro model

$$2.5 \times 10^6 \begin{bmatrix} 2 & -1 \\ -1 & 1 \end{bmatrix} \begin{pmatrix} g_2 \\ g_3 \end{pmatrix} = \begin{pmatrix} 5 \\ 13 \end{pmatrix}$$

Constrained system stiffness model

$$\therefore \begin{pmatrix} g_2 \\ g_3 \end{pmatrix} = \frac{1}{2.5 \times 10^6} \begin{pmatrix} 18 \\ 31 \end{pmatrix}$$

Solution of system equations

Figure 1.3 Results of the FE process and equation solution.

displacement components in $\mathbf{q}_c$ will be the total number of nodal displacements less those prescribed by displacement boundary conditions; and

$\mathbf{p}_c$ is the column vector of external loads, which include externally applied nodal forces and equivalent nodal loads arising from internal loads.

Note that the constrained system stiffness equations depend only on the macroscopic element models and the displacement boundary conditions.

The equation solving algorithms provide alternative ways of solving Eqs. (1.6). Formally, the solution is

$$\mathbf{g}_c = \mathbf{K}_c^{-1}\mathbf{p}_c \tag{1.7}$$

where the exponent -1 indicates the inverse of the matrix. In practice, the explicit inversion is avoided to simplify the data processing. Rather, the equations are solved for the particular loadings. The equation-solving process requires algorithms for solving a set of linear simultaneous equations. Though the solution of these equations is unique as long as the stiffness matrix is nonsingular, efficient use of computer resources and assessment of round-off error make access to more than one solving algorithm advantageous.

The elegance of this finite element technology is that only the element models change when the application discipline changes. More than that, the form of the finite element equations as represented by Eqs. (1.1) through (1.7) is unchanged.

The computer program is the key ingredient of the supporting computer configuration. The program translates requests from the engineer into directions for computer activity to produce numerical results.

Many major leasable or purchasable FEA computer codes are available to the practicing structural engineer. Each can operate on two or more types of hardware systems. Many large engineering firms have developed codes for in-house use. In addition, a number of codes reside in software depositories, copies of which are available for little more than reproduction costs.

Change characterizes the computer software environment. New finite element technology drives change. So do new computer science concepts, computer utilization methods, and computer hardware capabilities. Indeed, a revolution in computer implementation of FEA is fomenting around the parallel processor hardware evolution.

The analyst inspired to develop FEA computer software can obtain an excellent grounding in FEA methods by studying existing software and consulting some of the books on the subject.[12,13]

The focus of this book's presentation is intelligent use of an FEA computer code without detailed knowledge of its inner workings. Accordingly, we will deal with computer issues from the perspective of the applications analyst.

Recent knowledge developed in applications of FEA to structural problems appear in journals such as the *Journal of Computers and Structures* and the *Finite Elements in Analysis and Design Journal*.[14,15] Many relevant case studies are documented in the *NASTRAN User's Experiences* colloquia proceedings.[16] Meyers presents guidelines for FEA of civil engineering systems based on assimilated experiences of structural engineers.[17] Kamal and Wolf present practices and guidelines for highway vehicle structural analysis.[18]

1.2 ANALYSIS STEPS

The sequence of steps in performing a finite element analysis is as follows:

1. Idealization of the physical system
2. Discretization of the mathematical model
3. Application of the finite element process
4. Solution of the equations
5. Evaluation of stresses
6. Interpretation of numerical results

Idealization involves selection and particularization of the mathematical models of the system. This includes models for geometry, boundary conditions, material behavior, stress equilibrium, and displacement continuity.

The mathematical model of geometry is Euclidean. Within this constraint, element geometry is usually taken to be simple: uniform area, uniform thickness, and coordinate axes and neutral surfaces defined by low-order polynomials.

Displacement boundary conditions are idealized to prescribed displacement or rotation components at each node. Imposed forces are idealized as point forces, uniformly distributed forces, or forces with intensity varying linearly or quadratically with distance measured along the boundary.

Materials are idealized by Hooke's law, which, in general, takes the form:

$$\mathbf{s} = \mathbf{E}\mathbf{e} \tag{1.8}$$

where

$\mathbf{s}$ is the column vector of the six stress components of three-dimensional elasticity,

$\mathbf{E}$ is the symmetric material properties matrix, and

$\mathbf{e}$ is the column vector of strain components.

This general form is specialized to biaxial or uniaxial stress states for surface and two-noded elements, respectively.

Two classes of stress equilibrium equations are used: elasticity and structural equations. When elasticity equations are selected, the target of analysis is solution of the differential equations of the theory of elasticity. The structural equations are the equations of elasticity transformed by particular stress and strain distribution assumptions to equations for beams, plates, and shells. Mathematical models of displacement continuity across finite element boundaries range from continuity only at the nodes to continuity everywhere along the boundary.

Discretization involves dividing the system into pieces along imagined cuts in the structure. The sketches of Fig. 1.3 represent the two pieces of the structural system shown at the top of the page.

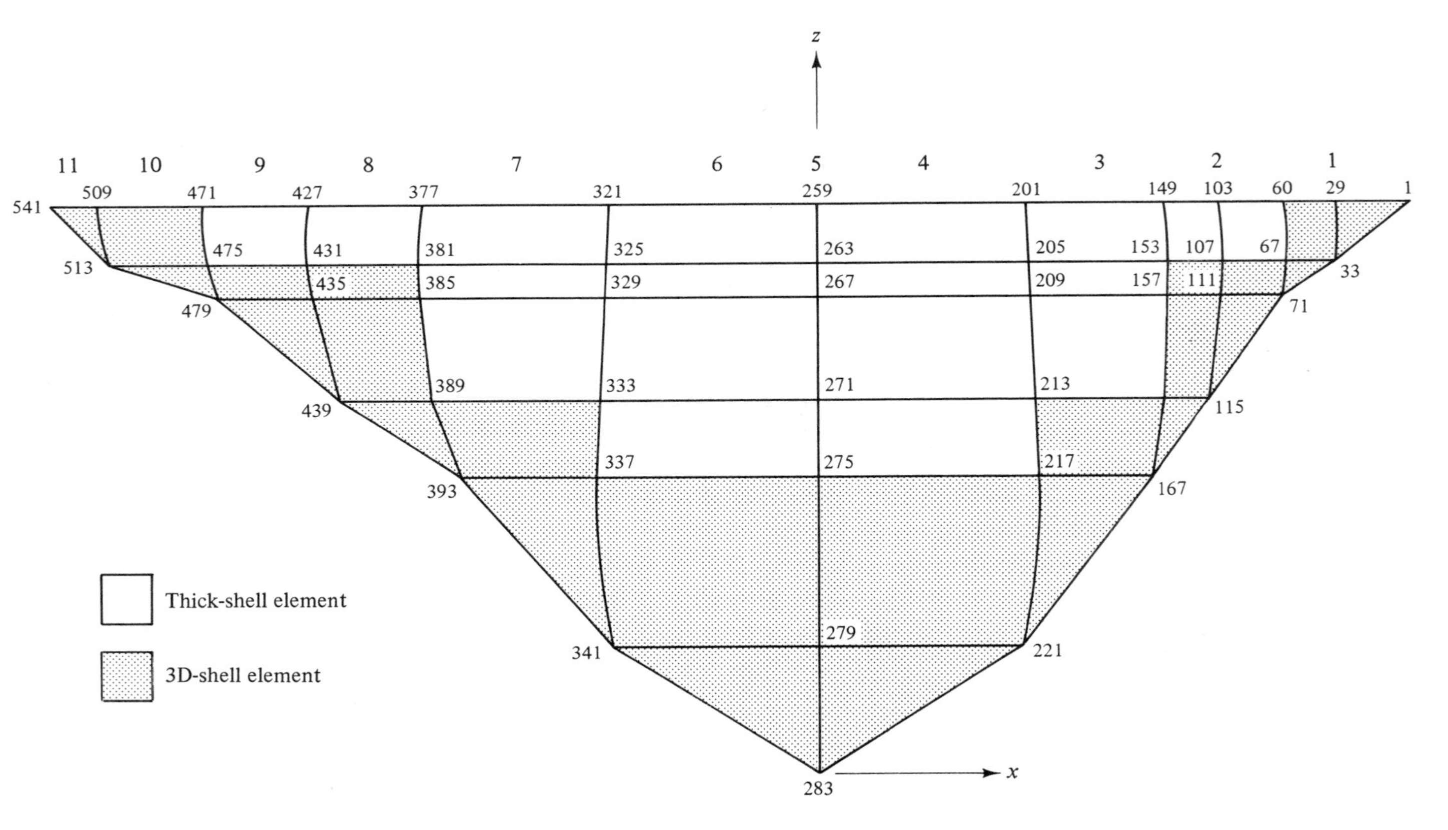

Figure 1.4 Projection of the upstream face of a dam on the xz plane.

Figure 1.4 is the plot of a discretized mathematical model of a gravity dam.[19] The lines of the figure are "nodal lines," defining boundaries of the irregular prisms that model the system. In this figure, four nodal lines define a nodal surface: the boundary surface of a prism element.

Discretization also involves associating a particular element type with each of the structural pieces. Thus it involves choosing the behavior approximations of each piece of the structure. Figure 1.4 shows that in the case of discretization of the dam the analyst chose to use two different models for the concrete prisms to improve analysis efficiency.

Discretization is completed when numerical values have been assigned to the element stiffness matrix and equivalent loading vector coefficients of all element models of the system. This process requires considerable arithmetic, usually ranging from 30 to 40 percent of FEA calculations.

Application of the finite element process transforms the element stiffness equations into the constrained stiffness equations for the system, considering continuity and balance conditions of the engineering system. It is important to understand this process, but because it directly involves no approximations, it will not affect interpretation of FEA results.

Equation solving produces the values of the unknown nodal displacements. This step is often achieved using Gauss elimination. Because the number of equations can be extremely large—one unusual analysis involved 250,000 equations—approximations are sometimes made in this step of the analysis.

Evaluation of point stresses requires developing numerical values for element stress matrices represented by the $\mathbf{s}_n$ and $\mathbf{s}_i$ matrices of Eqs. (1.3) and (1.4). When only displacements or stiffnesses are of interest, this step can be omitted from FEA. However, if peak stresses are the analysis target, the stresses at many points within an element may need to be evaluated to establish the peak value.

Interpretation of FEA results involves determining the meaning of behavior predictions of the discretized model with respect to the physical system of interest.

1.3 ROLE OF THE COMPUTER CONFIGURATION

The computer configuration consists of the computer hardware and software. The assistance provided by the configuration is at once a blessing and a bane.

It is a blessing because it checks the problem descriptive data for omissions and inconsistencies; generates detailed descriptive data for regular geometries; plots the original structural geometries; evaluates element model coefficients; does all the data processing for the finite element process, equation solving, and stress evaluation; plots the deformed geometry and contours of stress throughout the structure; prints tables of deflection and stresses; and saves intermediate results for subsequent review by the analyst. Each of these tasks would be inordinately time consuming, and tedious, if performed by a person.

The computer is a bane because, unless the engineer works with someone who knows the computer configuration well, the learning time needed to use this configuration is high. The beginner is not only confronted with too much irrelevant information but also finds frustrating omissions of information that is relevant.

Of longer-term importance is the fact that the computer introduces errors of its own. The computer program will have logic errors—even the best programmers certify codes that have an average of one logic error per 400 instructions. The computer usually introduces round-off and number truncation errors in arithmetic, and occasionally computer errors result in spurious behavior predictions. The computer program may also incorporate insidious approximations introduced without alerting the analyst.

Nevertheless, the computer is an essential feature of FEA. Some years ago, Topp described his desk calculator–assisted solution of 18 linear simultaneous equations associated with a structural analysis.[2] Results of the calculations were needed for his master's thesis. He spent three hot summer months in pursuit of the solution. The fact that these equations can be solved on a microcomputer in a few seconds emphasizes that even for academic problems, high-speed digital computation capability is a must.

1.4 ROLE OF THE ENGINEER IN FEA

The engineer's prime responsibility in FEA is to interpret numerical results. This responsibility is both professional and legal.

The professional code notes that the engineer is obliged to exercise skills with concern for "the welfare and safety of the public." The engineer is even urged to approach the firm's client directly in support of concern for public safety, should the employer act counter to this concern.

Legally, if the structure approved causes damage, the engineer can be liable for the damage. In principle, to avoid damages, the engineer will be required to defend successfully the thesis that the care of a competent practitioner of the profession has been exercised in the structural analysis.

The importance of the engineer to structural safety is reflected in the price of liability and malpractice insurance to the structural design firm: 3 to 5 percent of the design company's gross income. At this rate, the costs are higher than for a physician in general practice. One consulting engineer noted, "As structural engineers, we must protect against killing hundreds of persons per case rather than the one-at-a-time pace of a physician."

To fulfill their responsibilities, engineers must exercise a healthy skepticism about analysis results. Engineers will seek satisfaction that idealization, discretization, and computer-induced errors do not make analysis results irrelevant. Although, by the courtesy of the computer time salesman, engineers may recover computer run costs when irrelevant results are obtained, no court case has yet determined that an engineer may be absolved for failures based on computer configuration faults.

The engineer's secondary responsibility is to ensure the relevancy and accuracy of the FEA. In subsequent chapters, we will examine ideas on which engineers can take action to increase the quality assurance of the analysis.

1.5 HOMEWORK†

Assess your understanding of the concepts of this chapter by completing the following sentences. There may be more than one phrase that correctly completes the lead phrase.

***1:1.** FEA is a means of analyzing an engineering system
- **(a)** whose idealized model is represented by discrete element models.
- **(b)** by approximating a structure by the behavior of the subassemblies of which it may be fabricated.
- **(c)** by representing a structure only formed of two-noded straight axis, flat triangular, or tetrahedral discrete elements of finite size.
- **(d)** that represents pieces of a structure by numerical models, thereby mandating use of a digital computer.

1:2. FEA is a method of analysis for finding the
- **(a)** deflections and internal forces of a structure.
- **(b)** temperature and heat flow of a thermal system.
- **(c)** areas and volume of a geometric system.
- **(d)** fluid head and flow rates of a hydrodynamic system.

***1:3.** FEA has been used to predict behavior of systems using
- **(a)** linear equations only.
- **(b)** linear and nonlinear equations.
- **(c)** time-independent equations.
- **(d)** static and dynamic equations.

1:4. The history of FEA marks it as a method
- **(a)** that is used by engineers, scientists, and applied mathematicians.
- **(b)** about which there is no longer a diversity of opinion.
- **(c)** that is well accepted despite a continuing diversity of opinions about its soundness.
- **(d)** that seems to work but can easily be misused.

***1:5.** Finite element technology includes:
- **(a)** a finite element library, each model of which is suitable to all engineering disciplines.
- **(b)** a finite element process that is applicable, unchanged, to each engineering discipline.
- **(c)** equation solving algorithms that are suitable for solving problems from any engineering discipline.
- **(d)** supporting computer configurations that are suitable for FEA in any engineering discipline.

†Homework problems are labeled by chapter number and problem number separated by a colon. Therefore, 1:1 designates the first homework problem of Chapter 1.

*Denotes problems with solutions given in Appendix B.

1:6. The finite element model of a structural element includes
- **(a)** a macroscopic model that relates internal forces due to stiffness to equivalent and external nodal forces.
- **(b)** a macroscopic model that incorporates equilibrium equations and element load-deflection equations.
- **(c)** a microscopic model that establishes nodal displacements.
- **(d)** a microscopic model that relates nodal internal and external forces to estimates of stresses at a point.

***1:7.** The finite element process produces the constrained system stiffness equations from:
- **(a)** the element macro equations.
- **(b)** the element micro equations.
- **(c)** both the element macro and micro equations.
- **(d)** the interior equivalent loading equations.

1:8. For linear static analysis, the equation solving algorithm
- **(a)** works with a symmetric coefficient matrix.
- **(b)** develops the inverse of the stiffness matrix.
- **(c)** never involves approximations in the solution.
- **(d)** can be done by more than one method in many computer codes.

***1.9.** In the steps of FEA, approximations are involved in
- **(a)** idealization.
- **(b)** discretization.
- **(c)** the finite element process.
- **(d)** equation solution.
- **(e)** stress evaluation.

1:10. The computer configuration is important in FEA because it
- **(a)** has a significant effect on analysis time and cost.
- **(b)** can cause failure in completing the analysis.
- **(c)** makes practical FEA economically viable.
- **(d)** introduces its own errors in the analysis results.

***1:11.** The engineer's responsibility in FEA includes
- **(a)** finding programming bugs.
- **(b)** correcting program documentation.
- **(c)** ensuring that the computer configuration did the analysis correctly.
- **(d)** determining the relevance and accuracy of analysis results for the physical system.

1.6 REFERENCES

1. Noor, A.K., "Textbooks and Monographs on Finite Element Technology," *Journal of Finite Elements*, Feb. 1985.
2. Turner, M.J.; Clough, R.W.; Martin, H.C.; Topp, L.J., "Stiffness and Deflection Analysis of Complex Structures," *Journal of Aeronautical Sciences*, vol. 23, no. 9, Sept. 1956, pp. 805–823.
3. Melosh, R.J.; Merritt, R.G., "Evaluation of Spar Matrices for Stiffness Analyses," *Journal of Aero/Space Sciences*, vol. 25, no. 9, Sept. 1958, pp. 537–543.

*Denotes problems with solutions given in Appendix B.

4. Turner, M.J.; Dill, E.H., Martin, H.C.; Melosh, R.J., "Large Deflections of Structures Subjected to Heating and External Loads," *Journal of Aeronautical Sciences*, vol. 27, no. 2, Feb. 1960, pp. 97–106.
5. Martin, H.C.; Weikel, R.C.; Jones, R.E.; Seiler, J.A.; Greene, B.G., "Nonlinear and Thermal Effects on Elastic Vibration," U.S. Air Force Report No. ASD-TDR-62-156, Feb. 1962.
6. Melosh, R.J., "Basis for Derivation of Matrices for the Direct Stiffness Method," *AIAA Journal*, vol. 1, no. 7, July 1963, pp. 1631–1637.
7. Turner, M.J.; Martin, H.C., Weikel, R.C., "Further Development and Application of the Stiffness Method," *AGARD Structures and Materials Panel*, AGARDograph No. 72, Paris: Pergamon Press, 1964, pp. 203–266.
8. Robinson, John, *Early FEM Pioneers*, Bath, England: Pitman Press, 1985.
9. Khan, F.; Iyengar, S.; Colaco, J., "Computer Design of the 100-Story John Hancock Center," *J. Struct. Div. ASCE*, vol. 92, no. ST6, Dec. 1966, pp. 55–74.
10. Taylor, R., "Computers and the Design of the World Trade Center," *J. Struct. Div. ASCE*, vol. 92, no. ST6, Dec. 1966, pp. 75–92. (Photo courtesy of Marc Anderson.)
11. Fredriksson, B.; Mackerle, J., *Finite Element Review*, Report No. AEC-1003, Linköping, Sweden: Advanced Engineering Corporation, 1978.
12. Hinton, D; Owen, R.J., *An Introduction to Finite Element Computations*, Swansea, England: Pineridge Press, 1979.
13. Irons, B.; Ahmed, S., *Techniques of Finite Elements*, New York: Wiley, 1980.
14. Liebowitz, H., ed., *Computers and Structures*, Elmsford, N.Y.: Pergamon Press, 1971–.
15. Pilkey, W., ed., *Finite Elements in Analysis and Design*, Amsterdam: North-Holland, 1985–.
16. *NASTRAN User's Experiences*, NASTRAN User's Colloquia: First, NASA TM-X-2378, 1971; Second, NASA TM-X-2637, 1972; Third, NASA TM-X-2893, 1973; Fourth, NASA TM-X-3278, 1975; Fifth, NASA TM-X-3428, 1976; Sixth, NASA CP-2018, 1977; Seventh, NASA CP-2062, 1978; Eighth, NASA CP-2131, 1979; Ninth, NASA CP-2151, 1980; Tenth, NASA CP-2249, 1982.
17. Christian, M., *Applied Finite Element Analysis*, New York: ASCE, 1988.
18. Kamal, M.M.; Wolf, J.A., *Modern Automotive Structural Analysis*, New York: Van Nostrand Reinhold, 1982.
19. Clough, R.W.; Chang, K.T.; Chen, H.O.; Stephen, R.M., "Dynamic Response Behavior of Quan Shui Dam," Earthquake Research Center report UCB/EGRC-8, University of California, Berkeley.

Chapter 2

Processing Element Models

Given the macro element models, the finite element process assembles the models into a system stiffness model and introduces system force and displacement boundary conditions.

The process deals with the element models implying that forces are transmitted only through shared nodes. Thus the process takes macro element models, which include element equilibrium and load-deflection equations. It processes them considering only nodal variables. It enforces the requirement that all forces at a node be in equilibrium and that displacement continuity conditions be satisfied at each node.

2.1 ASSEMBLING ELEMENT MACRO EQUATIONS

Figure 2.1 illustrates the assembly process for a structure discretized as four elements acting in series to resist a given loading.

The element macro coefficients are expressed in algebraic form in Fig. 2.1 to clarify the effects of the processing steps. These models are identical to the models for the series spring case of Fig. 1.2 except that the internal loading coefficients are zero. Primes have been used on node numbers of the discretized system to distinguish these nodes from those of the complete system.

Equations (a) of Table 2.1 define the element macro equations of the dis-

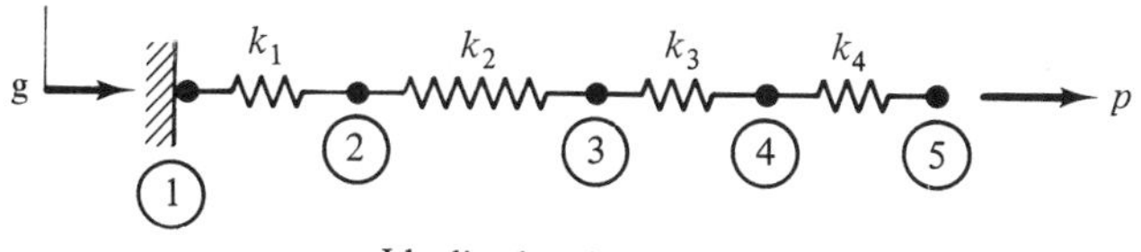

Idealized spring system

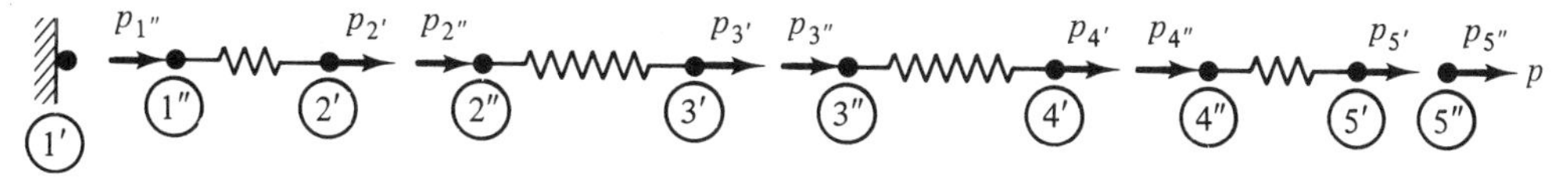

Discretized system

ELEMENT MACRO EQUATIONS, $k_e g_e = p_e$

Eq. no.

$$\begin{matrix} 1 \\ 2 \\ 3 \\ 4 \\ 5 \\ 6 \\ 7 \\ 8 \end{matrix}
\begin{bmatrix}
k_1 & -k_1 & 0 & 0 & 0 & 0 & 0 & 0 \\
-k_1 & k_1 & 0 & 0 & 0 & 0 & 0 & 0 \\
0 & 0 & k_2 & -k_2 & 0 & 0 & 0 & 0 \\
0 & 0 & -k_2 & k_2 & 0 & 0 & 0 & 0 \\
0 & 0 & 0 & 0 & k_3 & -k_3 & 0 & 0 \\
0 & 0 & 0 & 0 & -k_3 & k_3 & 0 & 0 \\
0 & 0 & 0 & 0 & 0 & 0 & k_4 & -k_4 \\
0 & 0 & 0 & 0 & 0 & 0 & -k_4 & k_4
\end{bmatrix}
\begin{Bmatrix} g_{1''} \\ g_{2'} \\ g_{2''} \\ g_{3'} \\ g_{3''} \\ g_{4'} \\ g_{4''} \\ g_{5'} \end{Bmatrix}
=
\begin{Bmatrix} p_{1''} \\ p_{2'} \\ p_{2''} \\ p_{3'} \\ p_{3''} \\ p_{4'} \\ p_{4''} \\ p_{5'} \end{Bmatrix} \qquad \text{(a)}$$

Figure 2.1 Macro models of a four-element spring.

cretized system. Because the model is fragmented, the equations for any one element have zero coupling coefficients with every other element.

As a first step in assembly, we add all the forces at each node and rewrite the element macro equations as Eqs. (b) of Table 2.1. This summing is achieved by adding pairs of Eqs. (a): The second equation of Eqs. (a) is added to the third, the fourth to the fifth, and the sixth to the seventh.

As the last step of assembly, we require that displacements of primed nodes match displacements of the corresponding system (unprimed) nodes. This reduction in the variables results in the addition of columns of Eqs. (b): columns 2 and 3, columns 4 and 5, and columns 6 and 7. This results in Eqs. (c) of Table 2.1.

The system stiffness equations, Eqs. (c), involve a symmetric stiffness matrix because the element stiffness matrices are symmetric and the row and column operations of the assembly correspond. Accordingly, diagonal coefficients in Eqs. (a) contribute only to diagonal coefficients in Eqs. (c).

With a little practice, you can write the assembled stiffness matrix directly from the separate element matrices rather than writing the matrices in the form

TABLE 2.1 ASSEMBLY OF ELEMENT MACRO EQUATIONS

1. Result of adding forces at common nodes
(adding Eqs. (3) and (2), (5) and (4), and (7) and (6) of Eqs. (a) of Fig. 2.1)

$$\begin{array}{c} \textit{Eqs.} \\ 1 \\ 2+3 \\ 4+5 \\ 6+7 \\ 8 \end{array} \begin{bmatrix} k_1 & -k_1 & 0 & 0 & 0 & 0 & 0 & 0 \\ -k_1 & k_1 & k_2 & -k_2 & 0 & 0 & 0 & 0 \\ 0 & 0 & -k_2 & k_2 & k_3 & -k_3 & 0 & 0 \\ 0 & 0 & 0 & 0 & -k_3 & k_3 & k_4 & -k_4 \\ 0 & 0 & 0 & 0 & 0 & 0 & -k_4 & k_4 \end{bmatrix} \begin{Bmatrix} g_{1''} \\ g_{2'} \\ g_{2''} \\ g_{3'} \\ g_{3''} \\ g_{4'} \\ g_{4''} \\ g_5 \end{Bmatrix} = \begin{Bmatrix} p_{1''} \\ p_{2'} + p_{2''} \\ p_{3'} + p_{3''} \\ p_{4'} + p_{4''} \\ p_{5'} \end{Bmatrix} \qquad \text{(b)}$$

2. Result of setting $g_1 = g_{1'}$, $g_2 = g_{2'} = g_{2''}$, $g_3 = g_{3'} = g_{3''}$, $g_4 = g_{4'} = g_{4''}$, $g_5 = g_{5'}$ in Eq. (b).

$$\begin{array}{c} \textit{Eqs.} \\ 1 \\ 2+3 \\ 4+5 \\ 6+7 \\ 8 \end{array} \begin{bmatrix} k_1 & -k_1 & 0 & 0 & 0 \\ -k_1 & k_1 + k_2 & -k_2 & 0 & 0 \\ 0 & -k_2 & k_2 + k_3 & -k_3 & 0 \\ 0 & 0 & -k_3 & k_3 + k_4 & -k_4 \\ 0 & 0 & 0 & -k_4 & k_4 \end{bmatrix} \begin{Bmatrix} g_1 \\ g_2 \\ g_3 \\ g_4 \\ g_5 \end{Bmatrix} = \begin{Bmatrix} p_{1''} \\ p_{2'} + p_{2''} \\ p_{3'} + p_{3''} \\ p_{3'} + p_{4''} \\ p_{5'} \end{Bmatrix} \qquad \text{(c)}$$

of Eqs. (a) and (b). You can proceed element by element or equation by equation. The number of additions and results will be the same.

No approximations are involved in the assembly process itself. It involves simply the addition of rows and columns of stiffness coefficients. Most computer implementations of the process assume that all element stiffness matrices are symmetric and operate with only a triangular representation of the matrices, forgoing the fact that the assembly process is meaningful for nonsymmetric element matrices as well as symmetric ones.

2.2 IMPOSING SYSTEM BOUNDARY CONDITIONS

Table 2.2 displays the effects on the equations of imposing the force and displacement boundary conditions of Fig. 2.1.

Force boundary conditions are imposed by setting the sum of the internal forces at each node to the external loading at the node. For this structure, this action leaves one unknown in the external loading vector: the reaction at node 1.

The displacement boundary condition of this problem is imposed by setting g_1 to zero. This condition annihilates the first column of the stiffness matrix of Eqs. (d) of Table 2.2, yielding the constrained system equations, Eqs. (e) of Table 2.2.

Fortunately, the last four equations of Eqs. (e) of Table 2.2 do not involve the unknown reactions. We can solve these equations and then use the first of

TABLE 2.2 IMPOSING SYSTEM BOUNDARY CONDITIONS

1. Result of imposing force boundary conditions on Eqs. (c) of Table 2.1 ($p_1 = p_{2'}, p_2 = p_{2'} + p_{2''} = 0, p_3 = p_{3'} + p_{3''} = 0, p_4 = p_{4'} + p_{4''} = 0, p_5 = p_{5'} = p$)

$$\begin{bmatrix} k_1 & -k_1 & 0 & 0 & 0 \\ -k_1 & k_1 + k_2 & -k_2 & 0 & 0 \\ 0 & -k_2 & k_2 + k_3 & -k_3 & 0 \\ 0 & 0 & -k_3 & k_3 + k_4 & -k_4 \\ 0 & 0 & 0 & -k_4 & k_4 \end{bmatrix} \begin{Bmatrix} g_1 \\ g_2 \\ g_3 \\ g_4 \\ g_5 \end{Bmatrix} = \begin{Bmatrix} p_1 \\ 0 \\ 0 \\ 0 \\ p \end{Bmatrix} \quad \text{(d)}$$

2. Result of imposing displacement boundary conditions on Eqs. (d) ($g_1 = 0$)

$$\begin{bmatrix} -k_1 & 0 & 0 & 0 \\ k_1 + k_2 & -k_2 & 0 & 0 \\ -k_2 & k_2 + k_3 & -k_3 & 0 \\ 0 & -k_3 & k_3 + k_4 & -k_4 \\ 0 & 0 & -k_4 & k_4 \end{bmatrix} \begin{Bmatrix} g_2 \\ g_3 \\ g_4 \\ g_5 \end{Bmatrix} = \begin{Bmatrix} p_1 \\ 0 \\ 0 \\ 0 \\ p \end{Bmatrix} \quad \text{(e)}$$

Eqs. (e) to evaluate the reaction. The system equations to be solved first involve a symmetric matrix of stiffness coefficients. Therefore, these four equations are of the form of Eqs. (1.6).

2.3 THE FINITE ELEMENT PROCESS AS AN ENERGY PROCESS

To associate the process with energy, we define the element strain energy and external work as

$$SE_e = 0.5\, \mathbf{g}_e^T \mathbf{K}_e \mathbf{g}_e; \quad EW_e = \mathbf{g}_e^T \mathbf{p}_e \tag{2.1}$$

where SE abbreviates strain energy and EW, external work.

The designation of the last of Eqs. (2.1) as external work requires that the components of $\mathbf{p}_e$ be conjugate to the components of $\mathbf{g}_e^T$. Accordingly, force and moment vector components must align with displacement and angular change components both referenced to orthogonal coordinate axes. This definition of external work complies with the conventional definition in structural theory.

We define the element energy as

$$EE_e = EW_e - SE_e \tag{2.2}$$

where EE_e denotes the element energy.

Often, the strain energy, as defined by Eq. (2.1), is the conventional strain energy of the structural theory. To make it clear that this is not always the case, we denote the difference between external work and strain energy as the energy, rather than as potential energy. These definitions ensure that if we set the differential of the element energy with respect to the nodal displacements to zero, we recover the element macro equations.

TABLE 2.3 ENERGY OF EACH SPRING ELEMENT, Eqs. (a)

Element 1-2

$$EW = \lfloor g_1\ g_2\ g_3\ g_4\ g_5 \rfloor \lfloor p_{1'}\ p_{2'}\ 0\ 0\ 0 \rfloor^T$$

$$SE = 0.5 \lfloor g_1\ g_2\ g_3\ g_4\ g_5 \rfloor \begin{bmatrix} k_1 & -k_1 & 0 & 0 & 0 \\ -k_1 & k_1 & 0 & 0 & 0 \\ 0 & 0 & 0 & 0 & 0 \\ 0 & 0 & 0 & 0 & 0 \\ 0 & 0 & 0 & 0 & 0 \end{bmatrix} \begin{Bmatrix} g_1 \\ g_2 \\ g_3 \\ g_4 \\ g_5 \end{Bmatrix}$$

Element 2-3

$$EW = \lfloor g_1\ g_2\ g_3\ g_4\ g_5 \rfloor \lfloor 0\ p_{2''}\ p_{3'}\ 0\ 0 \rfloor^T$$

$$SE = 0.5 \lfloor g_1\ g_2\ g_3\ g_4\ g_5 \rfloor \begin{bmatrix} 0 & 0 & 0 & 0 & 0 \\ 0 & k_2 & -k_2 & 0 & 0 \\ 0 & -k_2 & k_2 & 0 & 0 \\ 0 & 0 & 0 & 0 & 0 \\ 0 & 0 & 0 & 0 & 0 \end{bmatrix} \begin{Bmatrix} g_1 \\ g_2 \\ g_3 \\ g_4 \\ g_5 \end{Bmatrix}$$

Element 3-4

$$EW = \lfloor g_1\ g_2\ g_3\ g_4\ g_5 \rfloor \lfloor 0\ 0\ p_{3'}\ p_{4'}\ 0 \rfloor^T$$

$$SE = 0.5 \lfloor g_1\ g_2\ g_3\ g_4\ g_5 \rfloor \begin{bmatrix} 0 & 0 & 0 & 0 & 0 \\ 0 & 0 & 0 & 0 & 0 \\ 0 & 0 & k_3 & -k_3 & 0 \\ 0 & 0 & -k_3 & k_3 & 0 \\ 0 & 0 & 0 & 0 & 0 \end{bmatrix} \begin{Bmatrix} g_1 \\ g_2 \\ g_3 \\ g_4 \\ g_5 \end{Bmatrix}$$

Element 4-5

$$EW = \lfloor g_1\ g_2\ g_3\ g_4\ g_5 \rfloor \lfloor 0\ 0\ 0\ p_{4''}\ p_{5'} \rfloor^T$$

$$SE = 0.5 \lfloor g_1\ g_2\ g_3\ g_4\ g_5 \rfloor \begin{bmatrix} 0 & 0 & 0 & 0 & 0 \\ 0 & 0 & 0 & 0 & 0 \\ 0 & 0 & 0 & 0 & 0 \\ 0 & 0 & 0 & k_4 & -k_4 \\ 0 & 0 & 0 & -k_4 & k_4 \end{bmatrix} \begin{Bmatrix} g_1 \\ g_2 \\ g_3 \\ g_4 \\ g_5 \end{Bmatrix}$$

The equations of Table 2.3 are the energy expressions for each of the elements of the discretized spring of Fig. 2.1. These equations have been written directly in terms of displacements of the shared nodes by setting $g_i = g_{i'} = g_{i''}$, $i = 1, 2, \ldots 5$.

Equation (b) of Table 2.4 is the equation of the sum of the element energies. This equation is generated from Eqs. (a) of Table 2.3 simply by adding the element matrix stiffness coefficients. Equations (c) and (d) of Table 2.4 express the total energy after the external loads are equated to the sum of the internal and then the first node displacement set to zero.

TABLE 2.4 ASSEMBLY OF ELEMENT ENERGIES AND IMPOSITION OF BOUNDARY CONDITIONS

1. Assembly: $\sum_{e=1}^{4} EW_e = \lfloor g_1\ g_2\ g_3\ g_4\ g_5 \rfloor \lfloor p_{1''},\ p_{2'} + p_{2''},\ p_{3'} + p_{3''},\ p_{4'} + p_{4''},\ p_{5'} \rfloor^T$

$$\sum_{e=1}^{4} SE_e = 0.5 \begin{Bmatrix} g_1 \\ g_2 \\ g_3 \\ g_4 \\ g_5 \end{Bmatrix}^T \begin{bmatrix} k_1 & -k_1 & 0 & 0 & 0 \\ -k_1 & k_1 + k_2 & -k_2 & 0 & 0 \\ 0 & -k_2 & k_2 + k_3 & -k_3 & 0 \\ 0 & 0 & -k_3 & k_3 + k_4 & -k_4 \\ 0 & 0 & 0 & -k_4 & k_4 \end{bmatrix} \begin{Bmatrix} g_1 \\ g_2 \\ g_3 \\ g_4 \\ g_5 \end{Bmatrix} \qquad \text{(b)}$$

2. Imposition of force boundary conditions ($p_1 = p_{1'}$, $p_2 = p_{2'} + p_{2''} = 0$, $p_3 = p_{3'} + p_{3''} = 0$, $p_4 = p_{4'} + p_{4''} = 0$, $p_5 = p_{5'} = p$); therefore,

$$\sum EW_e = \lfloor g_1\ g_2\ g_3\ g_4\ g_5 \rfloor \lfloor p_1\ 0\ 0\ 0\ p \rfloor^T \qquad \text{(c)}$$

3. Imposition of displacement boundary conditions ($g_1 = 0$):

$$\sum_{e=1}^{4} EW_e = \lfloor g_2\ g_3\ g_4\ g_5 \rfloor \lfloor 0\ 0\ 0\ 0\ p \rfloor^T$$

$$\sum_{e=1}^{4} SE_e = 0.5 \begin{Bmatrix} g_2 \\ g_3 \\ g_4 \\ g_5 \end{Bmatrix}^T \begin{bmatrix} k_1 + k_2 & -k_2 & 0 & 0 \\ -k_2 & k_2 + k_3 & -k_3 & 0 \\ 0 & -k_3 & k_3 + k_4 & -k_4 \\ 0 & 0 & -k_4 & k_4 \end{bmatrix} \begin{Bmatrix} g_2 \\ g_3 \\ g_4 \\ g_5 \end{Bmatrix} \qquad \text{(d)}$$

4. System equations from $\dfrac{\delta}{\delta g_i}\left(\sum EW_e - \sum SE_e\right)$:

$$\begin{Bmatrix} 0 \\ 0 \\ 0 \\ p \end{Bmatrix} = \begin{bmatrix} k_1 + k_2 & -k_2 & 0 & 0 \\ -k_2 & k_2 + k_3 & -k_3 & 0 \\ 0 & -k_3 & k_3 + k_4 & -k_4 \\ 0 & 0 & -k_4 & k_4 \end{bmatrix} \begin{Bmatrix} g_2 \\ g_3 \\ g_4 \\ g_5 \end{Bmatrix} \qquad \text{(e)}$$

We get the constrained stiffness equations by differentiating the total of element energies, thereby casting Eqs. (c) and (d) into Eqs. (e) of Table 2.4. We observe that Eq. (e) is identical to Eqs. (e) of Table 2.2. We conclude that we can regard the assembly process as either adding equations and changing variables or as the addition of element energies recognizing the sharing of nodal displacements. In either interpretation, we impose force boundary conditions by requiring that internal forces at each node be in equilibrium with external.

To impose displacement boundary conditions, we use the transformation

$$\mathbf{g} = \mathbf{C}\mathbf{g}_c \qquad (2.3)$$

where

C is a constraint matrix and

c, as a subscript, denotes the constrained generalized displacements.

We note from Eq. (2.4) that the most general boundary condition the process can deal with arises when the system displacements are a general linear function of the constrained displacements.

When we interpret the process as dealing with equations, Eqs. (2.3) are substituted in the system equations as indicated in Eqs. (e) of Table 2.2.

When we view the process as dealing with energies, we substitute Eq. (2.3) in the energy form, yielding

$$TE_c = \Sigma\, EE_e = \mathbf{g}_c^T\mathbf{C}^T\mathbf{p} - 0.5\, \mathbf{g}_c^T\mathbf{C}^T\mathbf{K}\mathbf{C}\mathbf{g}_c \tag{2.4}$$

where TE_c is the total energy of the constrained system and summation is over the number of finite elements.

We obtain the macro structural equations for the system by differentiating Eq. (2.4) with respect to the constrained displacements and setting this differential to zero. This supplies

$$\mathbf{C}^T\mathbf{p} = \mathbf{p}_c = \mathbf{C}^T\mathbf{K}\mathbf{C}\mathbf{g}_c \tag{2.5}$$

Whether we view the finite element process as dealing with equations or energy, we will arrive at the same constrained equations, as we can see by comparing Eqs. (e) of Tables 2.2 and 2.4.

Table 2.5 defines the constraint equations for very general boundary conditions for the four element spring of Fig. 2.1. To provide for this case, the con-

TABLE 2.5 FOUR-ELEMENT SPRING WITH LINEAR CONSTRAINTS

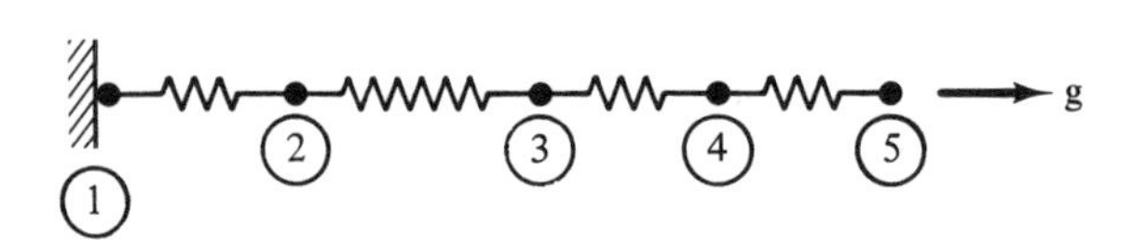

Suppose that node 1 is immovable, as before, and that

Node 2 moves 0.2 times node 4 movement
Node 3 moves 2.7 units in g direction
Node 4 moves 0.8 times the node g_4 motion plus 1.1 units

Then the constraint equations $g = Cg_c$ are

$$\begin{Bmatrix} g_2 \\ g_3 \\ g_4 \\ g_5 \end{Bmatrix} = \begin{bmatrix} 0.2 & 0 & 0 \\ 0 & 0 & 2.7 \\ 0.8 & 0 & 1.1 \\ 0 & 1 & 0 \end{bmatrix} \begin{Bmatrix} g_4 \\ g_5 \end{Bmatrix} \tag{f}$$

Substituting Eq. (f) in Eqs. (d) of Table 2.4 and deducing the system equations by differentiation gives

$$\begin{Bmatrix} o \\ p \end{Bmatrix} = \begin{bmatrix} 0.04(k_1 + k_2) + 0.64(k_3 + k_4) & -0.8k_4 \\ -0.8\,k_4 & k_4 \end{bmatrix} \begin{pmatrix} g_4 \\ g_5 \end{pmatrix} + \begin{pmatrix} -0.54k_2 - 1.32k_3 + 0.88k_4 \\ -1.1\,k_4 \end{pmatrix} \tag{g}$$

strained displacement vector includes the constant to deal with node settlement. Since the differential of this "variable" is zero, only two displacement variables appear in the constrained system, Eqs. (g) of Table 2.5.

2.4 THE CONTRAGRADIENT TRANSFORMATION

The simplest interpretation of assembly and displacement boundary condition imposition is as a contragradient transformation.

Suppose that we have written the element stiffness equations in the form used in Fig. 2.1:

$$\mathbf{K}_e'\mathbf{g}_e' = \mathbf{p}_e' \tag{2.6}$$

where the primes denote the uncoupled form of the displacement and external loading variables.

The displacement continuity conditions are a constraint of the form

$$\mathbf{g}_e' = \mathbf{C}_p\mathbf{g} \tag{2.7}$$

where $\mathbf{C}_p$ is a matrix of zeros and ones (a permutation matrix).

Then, by definition, the contragradient transformation of Eq. (2.6) is

$$\mathbf{C}_p^T\mathbf{K}_{e'}\mathbf{C}_p\mathbf{g} = \mathbf{C}_p^T\mathbf{p}_{e'} \tag{2.8}$$

which is effected by substituting Eqs. (2.7) in Eqs. (2.6) and premultiplying by $\mathbf{C}_p^T$. In comparison with the equation assembly, $\mathbf{C}_p^T$ induces the addition of internal forces at each node and $\mathbf{C}_p$ equates displacements at shared nodes. Equation (2.8) can be expressed as

$$\mathbf{K}\mathbf{g} = \mathbf{p} \tag{2.9}$$

where $\mathbf{K} = \mathbf{C}_p^T\mathbf{K}_{e'}\mathbf{C}_p$ and $\mathbf{p} = \mathbf{C}_p^T\mathbf{p}_{e'}$. $\mathbf{K}$ is the "congruent transform" of $\mathbf{K}_{e'}$. Because the kernel $\mathbf{K}_{e'}$ is symmetric, the congruent transformation produces a symmetric matrix.

From the energy interpretation, imposing displacement boundary conditions involves a second contragradient transformation of the form of Eqs. (2.3). This transforms Eqs. (2.9) to

$$\mathbf{C}^T\mathbf{C}_p^T\mathbf{K}_{e'}\mathbf{C}_p\mathbf{C}\mathbf{g}_c = \mathbf{C}^T\mathbf{C}_p^T\mathbf{p}_{e'} \tag{2.10}$$

Thus we can, in a single contragradient transformation, convert element stiffness equations to constrained system stiffness equations using the constraint

$$\mathbf{g}_c = \mathbf{C}^T\mathbf{C}_p^T\mathbf{g}_{e'} \tag{2.11}$$

The perseverance of symmetry in the stiffness matrix from element level to the system level to the constrained stiffness is guaranteed because the congruent transformation applies.

2.5 SYSTEM STIFFNESS MATRIX

We will require that all element stiffness matrices be symmetric to ensure that finite element models of structures always satisfy the Maxwell-Betti reciprocity theorem.

The Maxwell-Betti theorem is based on two premises: that the system is linear and that its behavior is conservative. We accept the linearity premise, but we would like to model unconservative as well as conservative systems. Viewing the assembly as adding equations and noting that unconservative systems may give rise to nonsymmetric matrices, we observe that the process can accommodate unconservative systems.

A matrix **K** is singular when we can find a non-null vector such that

$$\mathbf{r}^T\mathbf{K} = \mathbf{0} \tag{2.12}$$

where

r is the nontrivial column vector and

0 represents a null vector.

We observe that the unconstrained system stiffness matrix is often singular. The vector with each component equal to 1 is an **r** vector satisfying Eqs. (2.12) with respect to both the element stiffness matrices of Table 2.3 and the system matrix of Eqs. (b) of Table 2.4.

The singularity of the unconstrained system stiffness matrix is a consequence of the singularity of the element matrices. If all the elements of the structure are kinematically unstable (have singular stiffness matrices), the unconstrained system will be kinematically unstable for most collections of finite element models of structures.

To solve the constrained system stiffness equations, the stiffness matrix must be nonsingular. Nonsingularity is achieved by reducing the order of **K** by imposing enough displacement boundary conditions.

The rank of a square matrix **K** is the maximum number of columns in the constraint matrix of Eqs. (2.3) that results in a nonsingular $\mathbf{K}_c$. To ensure a nonsingular $\mathbf{K}_c$, the number of variables on the right-hand side of Eqs. (2.3) must be less than or equal to the rank of **K**.

Consider the example of Table 2.4. The order of **K** of Eq. (b) is 5, the rank 4. Therefore, the constraint equation could involve as many as four right-hand side variables. If we use the constraint equations, Eqs. (f) of Table 2.5 with the condition that $\mathbf{g}_1 = 0$, the order and rank of **K** will be reduced to 2. Thus the number of constrained displacements is small enough so that $\mathbf{K}_c$ may be nonsingular. To determine whether the particular constraints used will result in a nonsingular $\mathbf{K}_c$, we will need to know more about the element macro equations involved in **K**.

The displacement boundary conditions establish a datum to which we relate displacements. The element macro equations include the load-deflection equations—equations that relate element forces to changes between displacements of the nodes. The displacement boundary conditions must provide a basis for converting these relative displacements to absolute displacements. This is analogous to the measurement of displacements of a physical structure where our instruments measure the changes relative to some fixed reference points in space.

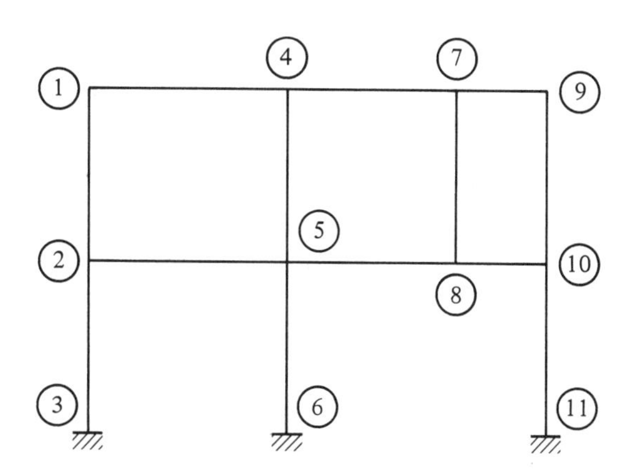

All elements have only two nodes.

Node \ Node	1	2	3	4	5	6	7	8	9	10	11
1	X										
2	X	X									
3		X	X								
4	X			X							
5				X	X						
6					X	X					
7				X			X				
8					X		X	X			
9							X		X		
10								X	X	X	
11										X	X

Symmetric (X indicates placement of possible nonzero coefficients)

Relative sparsity = $\frac{84}{121}$

Maximum nodal bandwidth = 3

Figure 2.2 Populated portion of a system stiffness matrix.

Further, we observe that the number of zero coefficients in the stiffness matrix is a significant percentage of the total number of coefficients. Let the ratio be the "sparsity ratio." The sparsity ratio for the matrix of Eqs. (b) is < 0.48.

The stiffness matrix of Eqs. (b) also displays the fact that the nonzero coefficients cluster about the matrix diagonal. This tendency can be measured by the matrix bandwidth. The bandwidth is the number of zero and nonzero coefficients that occur across a row, measured from the diagonal coefficient, until the rest of the coefficients are zero. Thus the bandwidth of each of the first three rows of the stiffness matrix of Eqs. (b) of Table 2.4 is 2.

The sparsity and bandwidth are important parameters in determining the computer resources needed for analysis. These parameters are determinable once the elements and their node numbers have been established because of the nature of the element macro equations and the finite element process.

To facilitate *ab initio* determination of bandwidth, we seek the "nodal bandwidth," that is, the bandwidth assuming only one displacement variable at a node. Figure 2.2 illustrates the relation between a structure modeled by two-noded finite elements and its system stiffness matrix. The only coefficients that can be nonzero are noted by an $\times$. In principle, to establish where the nonzeros may fall, we assume that each element stiffness matrix is fully populated and represent the assembly process without doing arithmetic.

To emphasize the pattern of nonzero coefficients in Fig. 2.2, only the population in the upper triangular part of the matrix is shown. Since this matrix must be symmetric, the coefficients of the lower half can be inferred. In any row, zeros within the nodal bandwidth are noted. All the coefficients to the right of the band are zero. These coefficient boxes are left blank in the figure.

The nodal bandwidth at any node is readily found by direct reference to the discretization sketch. Nonzeros can occur only between nodes connected by a finite element. Adding 1 to the difference of a node's number and the number of the highest node to which it is connected by an element determines the nodal bandwidth. Consider node 5 of Fig. 2.2. It is connected by elements to nodes 2, 4, 6, and 8. Thus the nodal bandwidth for node 5 is $8 - 5 + 1 = 4$.

The nodal bandwidth and maximum nodal bandwidth change with the assignment of numbers to nodes. The sparsity of the matrix is independent of this assignment.

2.6 GENERAL USE OF THE FINITE ELEMENT PROCESS

The finite element process is only slightly modified when the macro model includes an internal loading vector. This vector is treated just like the external loads vector in assembly.

The finite element process is unchanged in application to other engineering systems—physical systems that are bound together by continuity and balance conditions.

TABLE 2.6 ENGINEERING SYSTEM VARIABLES

Engineering System	Explicit Variable	Implicit Variable	Material Relation	Balance Condition
Structural	Displacement	Force	Hooke's law	Force equilibrium
Thermal	Temperature	Heat flow	Fourier's law	Heat flow in = out
Fluid	Fluid head	Fluid flow	Darcy's law	Fluid flow in = out
Electrical	Voltage	Current	Ohm's law	Current in = out
Magnetic	Magnetic potential	Magnetic flux	Ampere's law	Flux balance

Table 2.6 lists some engineering systems and the nodal variables that play the role of continuity and balance variables. The variables playing the role of displacements are explicit variables because they are measurable with nondestructive testing. The balance variables are implicit variables because their value is usually only determinable by deduction from measurements of the explicit variables.

The finite element models relate explicit and implicit variables through equations that require measurements of the properties of the element material. The finite element process combines the models into a system model, enforcing continuity and equilibrium of internal and external implicit variables at the nodes. The process is independent of the engineering system the model emulates.

2.7 HOMEWORK

2:1. Equation processing.

Given: The macro equations below for each element of a two-rod truss.

Find:

(a) The coefficients of the equations $\mathbf{K}\mathbf{g} = \mathbf{p}$

(b) The coefficients of the equations $\mathbf{K}_c\mathbf{g}_c = \mathbf{p}_c$

Method: Equation processing.

Macro Models

Element 1′, 2′	Element 2″, 3′
$$\begin{bmatrix} a & 0 & -a & 0 \\ 0 & 0 & 0 & 0 \\ -a & 0 & a & 0 \\ 0 & 0 & 0 & 0 \end{bmatrix} \begin{pmatrix} u_{1'} \\ v_{1'} \\ u_{2'} \\ v_{2'} \end{pmatrix} = \begin{pmatrix} p_{x1'} \\ p_{y1'} \\ p_{x2'} \\ p_{y2'} \end{pmatrix}$$	$$\begin{bmatrix} b & -c & -b & c \\ -c & d & c & -d \\ -b & c & b & -c \\ c & -d & -c & d \end{bmatrix} \begin{pmatrix} u_{2''} \\ v_{2''} \\ u_{3'} \\ v_{2'} \end{pmatrix} \begin{pmatrix} p_{x2''} \\ p_{y2''} \\ p_{x3'} \\ p_{y3'} \end{pmatrix}$$

2:2. Processing energies.

Given: The input for Prob. 2:1.

Find: The constrained equations of the system by following the steps of energy finite element processing.

2:3. Contragradient processing.

Given: The input for Prob. 2:1.

Find: The single contragradient transformation matrix that will produce the constrained equations of the system from the input matrices.

2:4. General linear constraints.

Given: The spring system equations, Eqs. (e) of Table 2.2, and the constraint equations (f) of Table 2.5.

Show: Equations (g) of Table 2.5 are the result of imposing the constraint equations.

***2:5.** Imposing response symmetry.

Given: The symmetric structure with the symmetric boundary conditions shown.

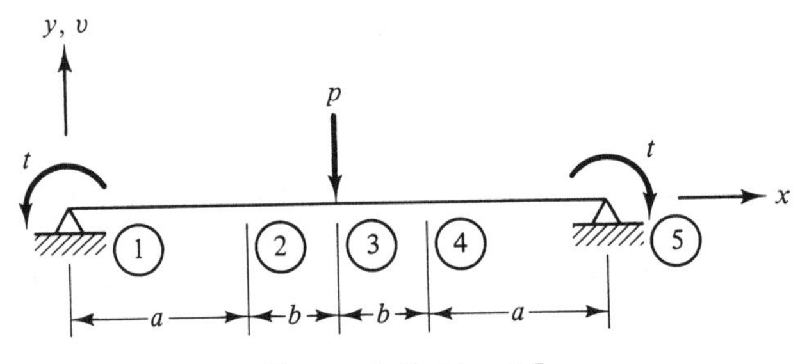

Homework Problem 2:5

Define: The constraint equations that require symmetric responses about the vertical at node 3.

2:6. Imposing response antisymmetry.

Given: The symmetric structure with the antisymmetric loading shown.

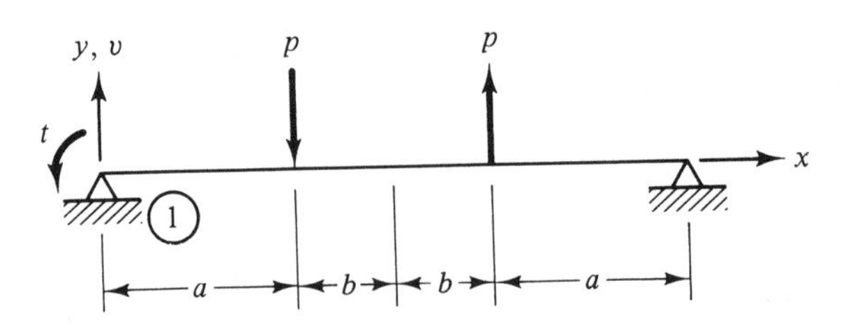

Homework Problem 2:6

Define: The constraint equations that require antisymmetric responses about the vertical at node 3.

2:7. Symmetric structure kinematic boundary conditions.

Given: Half of a structure that is symmetric and has symmetric boundary conditions as shown.

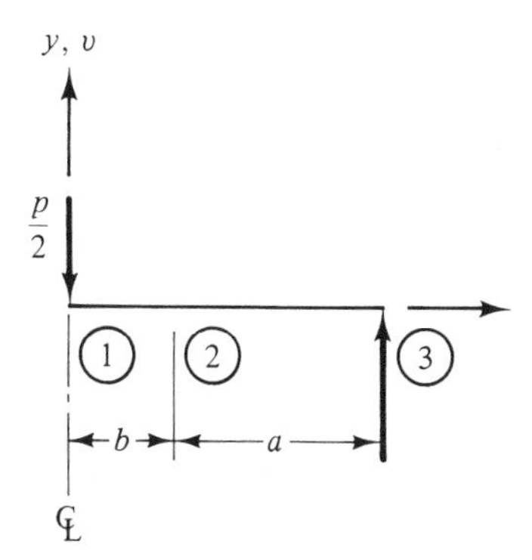

Homework Problem 2:7

Define: The kinematic boundary conditions, in terms of constraint equations, that will imply the behavior of the missing half of the structure in the analysis.

***2:8.** Antisymmetric structure kinematic boundary conditions.

Given: Half of a structure that is symmetric and has antisymmetric loading as shown.

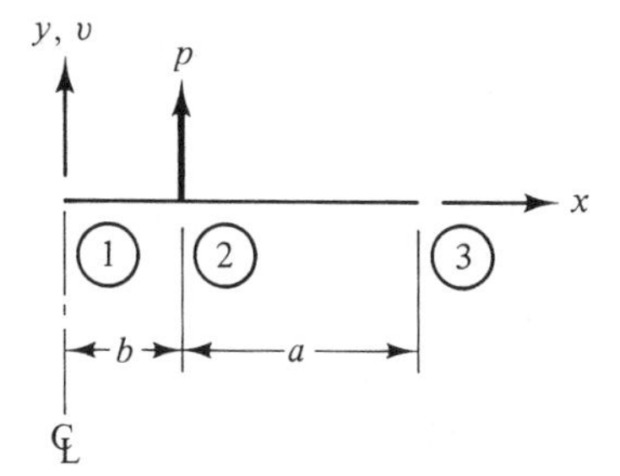

Homework Problem 2:8

Define: The kinematic boundary conditions, in terms of constraint equations, that will imply the behavior of the missing half of the structure in the analysis.

***2:9.** Representing a rigid link.

Given: The beam of the figure in the diagram is connected at node 3 to an adjacent structure. The flexible part of the beam extends from node 1 to node 2. The beam lies and deforms only in the plane of the paper.

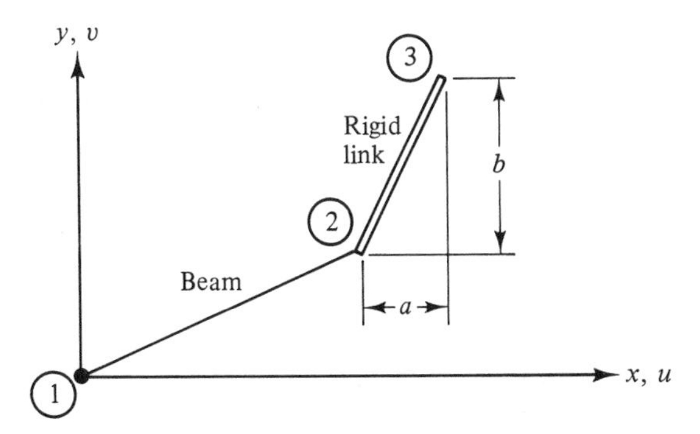

Homework Problem 2:9

Find: The constraint equations relating the generalized displacements at node 2 and node 3.

2:10. Element matrix singularity.

Given: The following element matrix.

$$\begin{bmatrix} b & -c & -b & c \\ -c & d & c & -d \\ -b & c & b & -c \\ c & -d & -c & d \end{bmatrix}$$

Find: A non-null row multiplier $\mathbf{r}^T$ of the matrix that results in a zero product.

***2:11.** System stiffness matrix sparsity.

Given: The truss structure of the figure shown and a typical set of macro equations.

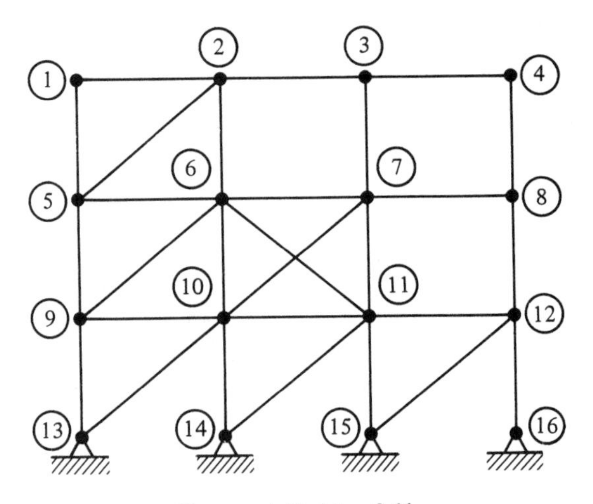

Homework Problem 2:11

Do the following:

(a) Show which coefficients of the system stiffness matrix must be zero on an $\times - \circ$ display like that of Fig. 2.2.

(b) Calculate the sparsity ratio for the matrix.

(c) Find the bandwidth of the matrix.

2:12. Energy minimization or maximization.

Given: $TE = 0.5\, \mathbf{g}^T \mathbf{K} \mathbf{g} - \mathbf{g}^T \mathbf{p}$

Show: That the partial differentiation of TE with respect to the components of the generalized displacements yields $\mathbf{K}\mathbf{g} = \mathbf{p}$.

2:13. Matrix rank.

Given: The matrix shown and three vectors:

$\mathbf{r}_1^T = (2, -1, 0)$, $\mathbf{r}_2^T = (8, -3, 1)$, $\mathbf{r}_3^T = (2, -1, 1)$.

$$\begin{bmatrix} 1 & 2 & -2 \\ 2 & 6 & 2 \\ -2 & 2 & 22 \end{bmatrix}$$

Define: What you can say about the relation between matrix order and rank based on the results obtained by independently premultiplying the matrix by each vector.

2:14. Stiffness matrix rank.

Given: The element stiffness matrix of Prob. 2:13.

Determine: The rank of the matrix.

2:15. Contragradient transformation and matrix rank.

Given: The nonsingular stiffness matrix $\mathbf{K}$ is constrained by a contragradient transformation, $\mathbf{C}^T\mathbf{K}\mathbf{C}$.

Show: That the result of the transformation is singular whenever the number of columns of $\mathbf{C}$ is greater than the number of rows.

Chapter 3

The Computer Configuration

The computer configuration is the automated data processing capability the analyst can bring to bear on an FEA problem. It includes both computer hardware and software. Knowledge of the configuration is valuable in planning the analysis scope, checking that the computer program interprets the problem description as intended, and interpreting analysis results.

In this chapter, we describe the EASE configuration, a simple FEA system that has been useful to students of FEA and is available with this book. This configuration has all the essential features of a large-scale system except the ability to treat problems that involve many (more than 100) nodal variables. It has special features to minimize the time it takes for a student to involve the computer in finite element studies. It is designed to accommodate all the computer work problems that appear at the end of the chapters of this book.

We introduce EASE to assist in developing a skeleton of ideas on which to fasten all the capabilities of a general-purpose FEA computer configuration. We provide detailed information on EASE implementation capabilities, the structure and checking of problem descriptive data, and the highest level of computer configuration validation.

3.1 THE EASE COMPUTER CONFIGURATION

EASE implements static analysis of trusses, frames, membranes, and square cross-section torque tubes. It directs linear and stepwise linear analyses. It is limited to small sets of structural equations.

EASE includes logic for interactive input and output, for plotting the original and deformed geometry, and for assessing analysis accuracy and solution efficiency. It incorporates a simple database management system to facilitate changing just part of the data in updating an available data set to provide a new set.

EASE operates on an IBM PC that has at least two disk drives and 128K bytes of primary storage. In the two-disk drive configuration, the logic is contained on four $5\frac{1}{4}$ inch floppy disks. Problem data for computer work analyses in this book can be stored on one other floppy disk.

The steps to follow to install EASE under DOS are as follows:

1. Place the EASE "modeling disk" of interest in the default drive and a formatted disk (the "data disk") in the second drive. (Each disk must be label up and nearest to your right hand. Drive door must be closed.)
2. Turn on the scope, logic unit, and printer (if there is one).
3. Enter the data (in the form 11/24/89) and press the Return key.
4. When the menu appears, press the CapsLock key.
5. Respond to requests for input as they arise. Usually, one-keystroke entries are read directly. Multiple-stroke entries are read when the user presses the Return key.
6. To transfer a plot from the scope to the printer, hold the PrtSc key down and press down the up-arrow key.
7. To return to the DOS command level, press the Q key of the menu. To quit without returning to DOS, turn off the machine.

When installation is complete, all options of EASE are prompted on one of three levels—the modeling disk, operation, or parameter level.

The modeling disk level is the highest level. Here a menu facilitates electing one of the four model types, as follows:

1. LARS (Lattice Analysis and Redesign Subsystem)—models of trusses and frames
2. MELBA (Membrane Elements Behavior Analysis subsystem)—models of membranes
3. TESS (Torsion Elements Square Shaft subsystem)—models of torsion elements specialized for a shaft with a square cross section
4. LAIRD (Lattice Analysis for Incremental Reanalysis)—models for stepwise linear analysis of trusses and frames

The logic supporting each of the four types of models resides on a separate floppy disk.

The menu at the operation level prompts selection of the analysis step. Selection directs creating problem descriptive data, performing an analysis, or displaying analysis results using the data of a problem file.

Prompts at the parameter level elicit details of the problem description and user decisions on data disposition. The former are critical to problem solution interpretation. The latter facilitate future analyses.

3.2 PROBLEM-DESCRIPTIVE DATA

EASE, like most finite element codes, accepts problem-descriptive data with units undefined. The user can select any set of units to represent length, force, temperature, and time. In choosing numbers for problems in this text, inches, pounds, degrees Fahrenheit, and seconds were used. These units are not cited in problem descriptions or on computer displays of output.

The description of a structural problem must detail completely the geometry, material properties, and boundary conditions for the structure. Finite element codes require that the data be entered as connectivity data, element data, and nodal data.

Table 3.1 illustrates a two-bar structural problem and its rendition as finite element problem-descriptive data.

The connectivity data lists the node numbers of each element of the discretized model, thereby implying which element is connected to which. For EASE, the last line of connectivity data initiates with a negative sign to indicate the last connectivity data line. EASE logic checks that connectivity node numbers include no zero or fractional node numbers. It also checks that each node is connected to at least one other node.

The element data defines the stiffness parameters for each element. Since the only resistance offered by a truss element is due to axial stiffness, the rod stiffness is characterized by a single number that equals the product of the element cross-sectional area and Young's modulus. Because there are two lines of connectivity data in this sample problem, the logic recognizes that there will be just two lines of element data. Rigidity parameters, such as length, area, volume, Young's modulus, and shear modulus, are tested by the computer to ensure they are positive or zero.

Nodal data includes nodal coordinates, fixities, settlements, and loading components. The nodal coordinates refer to a common rectangular Cartesian coordinate system whose origin and orientation are selected by the analyst. Fixities define nodal variables that must remain zero as the structure deforms. Settlements define prescribed generalized displacements at the nodes under the loading.

For a two-dimensional truss, nodal variables are the displacements in the direction of each reference axis. These are assumed to be numbered 1 and 2, corresponding to displacements along the x and y axes. To indicate complete fixity at node 1, two lines of input are needed—one for the x component of displacement u and one for the y component v. Each line of fixity input data defines the node number and the number of the relevant nodal variable. The last line of fixity has a negative node number to signal that it is the last line of fixity data. The settlement data line, like a fixity line, includes the node and direction number but adds the

TABLE 3.1 PROBLEM-DESCRIPTIVE DATA FOR A TWO-ROD TRUSS

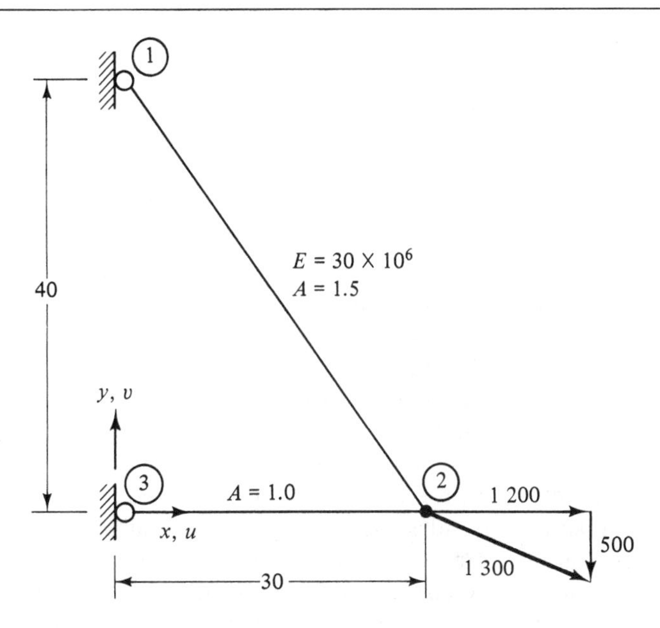

Connectivity Data		Element Data		Nodal Coordinates		
Element	Nodes	Element	Stiffness	Node	x	y
1	1, 2	1	45E6	1	0	40
2	−2, 3	2	30E6	2	30	0

Boundary Conditions	
Fixities	Point Loads
1, 1	2, 1, 1200
1, 2	−2, 2, 500
3, 1	
−3, 2	

value of the settlement. The last line of settlement data has a negative node number.

Fixity and settlement check logic makes sure that the node number is an integer that coincides with one of the connectivity nodes. The displacement component number must be an integer that corresponds to a nodal displacement variable. For the two-dimensional truss, this number must be 1 or 2.

A similar convention is used for describing nodal forces. Each component of force along a reference axis, corresponding to a displacement variable, is described separately. The node number of the last line is given a negative value.

Checks of loading node and component number are the same as these checks for fixity data.

3.3 VALIDATING THE COMPUTER CONFIGURATION

Validation of the computer configuration involves running test problems to demonstrate that implementation of each step and model of the analysis is adequate for all classes of problems of interest.

In a direct validation test, the values the computer configuration must generate are established independently of the configuration. These values may be calculated by an alternate computer configuration but are usually produced by a hand-held calculator. An advantage of direct checks is that a single exercising of the computer configuration can produce data for several direct checks. For example, data for most of the tests of Table 3.2 can be produced in a single run.

To illustrate direct tests for validating an FEA step, consider validation of the finite element process. Table 3.2 lists some tests of the process logic needed for validation. Besides checks of the process substeps, validation of error checks is necessary to prepare the analyst for interpreting aborted runs.

In addition to direct numerical tests of the process, validation requires checking data management logic. For example, at least two problems must be used to validate assembly. One problem should involve a stiffness matrix whose bandwidth is small compared with the matrix order and one for a fully populated matrix.

Validation is facilitated by configuration consistency tests. In these tests, computer logic integrity is assessed by comparing calculation results using data

TABLE 3.2 SOME DIRECT CHECKS OF FE PROCESS LOGIC

Substep	Feature to Test
Assembly	Bandwidth correctly calculated
	Zero coupling between nodes not directly connected
	All element stiffness coefficients added into the system matrix
	A diagnostic message notes when the number of equations is greater than the capability of the configuration
	A diagnostic message notes when the bandwidth is greater than the configuration capability
Force boundary conditions	Nodal force components are placed in the correct components of the vector and are the values assigned by the analyst
Displacement boundary conditions	Equations with prescribed displacements are removed from the set of structural equations
	Prescribed displacement appears in displacement vector of the solution
	Loading vector is added to by loading for prescribed displacements
	Prescription of force and displacement for the same nodal variable is rejected

from two analyses. The first analysis uses data for any problem the configuration can address. Part of the input data of the first analysis is changed to create input data for the second analysis. The change is selected so that the change in responses is predictable exactly.

Table 3.3 lists the basis and validation criteria for several consistency tests. For example, a scalability test could involve one run to establish displacements and internal forces for nominal boundary conditions. The element stiffnesses would be multiplied by a factor f to define element stiffnesses for the second analysis. Then the displacements of the second analysis must be $1/f$ times those of the first analysis, and the internal forces of both analyses must be the same if the input data is correct and the computer logic is valid.

As another example, a dimensional invariance test could involve changing the length unit of the first analysis—converting stiffness parameters (cross-sectional areas, Young's modulus, and nodal coordinates in a truss) and values of settlements by factors to modify the problem-descriptive data for the second analysis. Displacements calculated for the first and second analyses must match when results are scaled to the same length units.

One advantage of consistency testing is that all of the response values can be checked with few or no hand-held calculator evaluations. Also, since the choice of problem is not critical to the test, suitable problem input data can be created easily.

TABLE 3.3 BASES FOR SOME CONSISTENCY TESTS

Test Basis	Criterion for Validation
Scalability	A structural system is scalable in the sense that when all element stiffnesses or force and displacement boundary conditions are scaled by a common factor, all response displacements and, separately, internal forces are scaled by a common factor.
Invariance	The response of a structure must be independent of the dimensional units selected, the reference coordinates axes used, and the numbers assigned to nodes.
Linearity	The sum of the responses to two independent sets of boundary conditions must equal the responses from the sum of the boundary conditions.
Correspondence of boundary conditions	The internal forces due to a loading must equal the internal forces induced by the settlements caused by the loading.
Maxwell reciprocity	The inverse of the constrained stiffness matrix must be symmetric.
Symmetry of response	A symmetric structure with symmetric boundary conditions must respond in geometrically symmetric displacements, strains, internal forces, and stresses.
Determinancy	A determinate structure is a structure whose internal forces and stresses are unchanged by changes to element stiffnesses. An indeterminate structure is a structure in which there is at least one element whose stiffness change will change some internal forces and stresses in the system.

3.4 COMPUTERWORK†

***3-1.** EASE input for a truss.

Given: The truss of the figure.

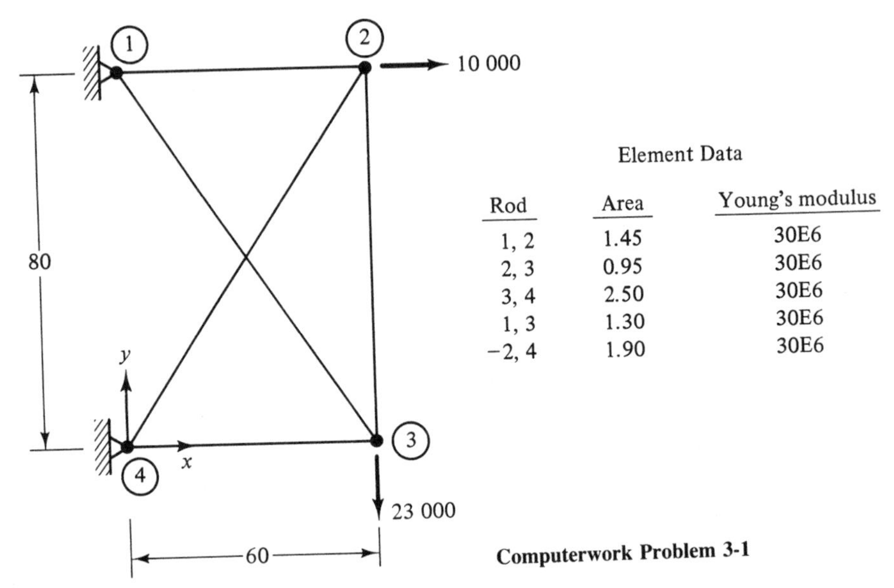

Element Data

Rod	Area	Young's modulus
1, 2	1.45	30E6
2, 3	0.95	30E6
3, 4	2.50	30E6
1, 3	1.30	30E6
−2, 4	1.90	30E6

Computerwork Problem 3-1

List: The EASE input for the truss.

3-2. EASE input for a frame.

Given: The frame defined here.

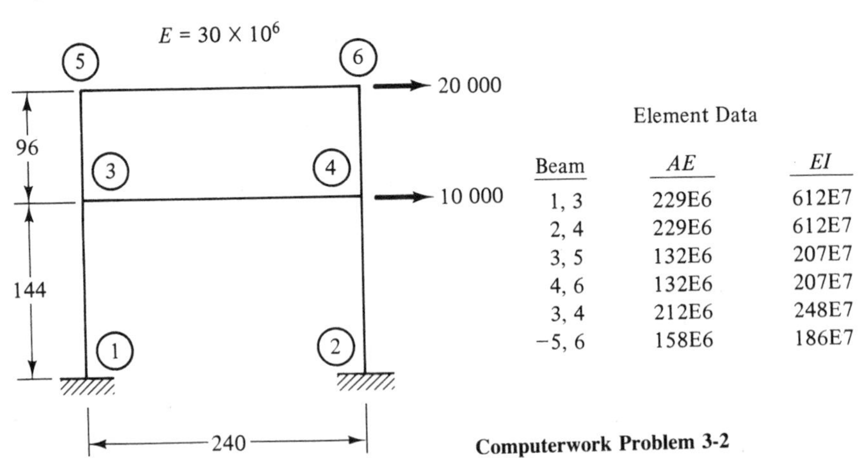

Element Data

Beam	AE	EI
1, 3	229E6	612E7
2, 4	229E6	612E7
3, 5	132E6	207E7
4, 6	132E6	207E7
3, 4	212E6	248E7
−5, 6	158E6	186E7

Computerwork Problem 3-2

List: The EASE input data for the frame.

†Computer work problems are labeled by chapter number and problem number separated by a hyphen. Therefore, 3-1 designates the first computerwork problem of Chapter 3.

3-3. Deducing problem geometry.

Given: The following EASE input data for a problem.

Element Data		Nodal Data		
Element	AE	Node	x	y
1, 2	30E6	1	0	0
2, 4	30E6	2	60	180
4, 5	30E6	3	120	0
5, 3	30E6	4	180	180
3, 1	30E6	5	240	0
3, 2	30E6			
−3, 4	30E6			

Fixities: 1, 2; 1, 2; 5, 1; −5, 2
Loads: 2, 2, −20 000; −4, 2, −20 000

3-4. EASE input error detection: Truss problem.

Given: The erroneous input data listed.

Element Data		Nodal Data		
Element	AE	Node	x	y
1, 2	26E6	1	240	−240
1, 3	−26E6	2	240	480
1, 3	21E6	3	480	480
3, 2	30E6	−4	480	240
−1, 3	14E6			
−3, 5	26E6			

Fixities: 1, 1; −1, 2
Loads: −2, 1, 0

Do the following:

(a) Identify four input errors. (There are five.)

(b) Enter the input into the computer and place a check beside each error found by the computer configuration.

***3-5.** EASE input error detection: Frame problem.

Given: The following erroneous input data.

Element Data			Nodal Data		
Element	AE	EI	Node	x	y
1, 2	270E6	5400E6	1	480	240
1, 4	270E6	5400E6	2	0	120
1, 3	390E6	9800E6	3	480	120
2, 3	240E6	6000E6	4	720	120
5, 2	390E6	9800E6	5	0	0
3, 4	150E6	2100E6	6	480	0
3, 6	390E6	9800E6	7	7200	0
4, 8	−390E6	9800E6	8	720	0

Fixities: 5, 1; 5, 3; 5, 2;
6, 3; 6, 2; 6, 1;
−7, 1; 7, 2; 7, 3

Settlements: −5, 1, 0

Loads: 1, 1, 30 000;
−1, 1, 15 000

Do the following:

(a) Identify four input errors.

(b) Enter the input into the computer and place a check beside each error also detected by the computer configuration.

***3-6.** Direct test of assembly.

Given: The input data for the frame shown.

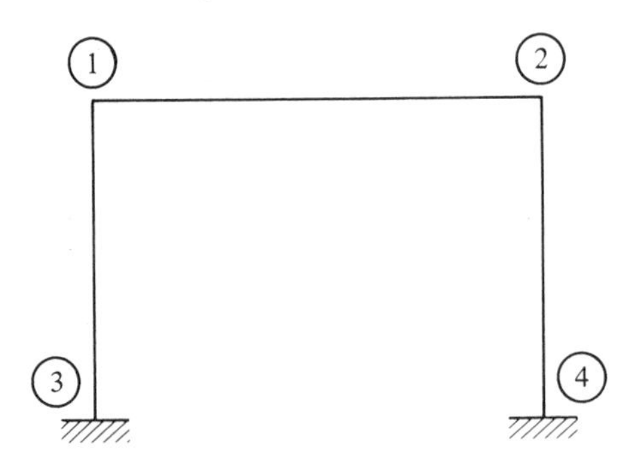

Element Data

Element	AE	EI
1, 2	15E6	23E7
1, 3	21E6	32E7
2, 3	23E6	38E7
−2, 4	23E6	38E7

Nodal Data

Node	x	y
1	0	168
2	240	168
3	0	0
4	480	0

Fixities: 3, 1; 3, 2; 3, 3;
4, 1; 4, 2; −4, 3

Loads: 1, 1, 1000; 1, 2, −6000;
2, 1, 8000; −2, 2, −6000

Computerwork Problem 3-6

Do the following

(a) Enter the input data into the computer, electing to print the element matrices and the intermediate results of assembly. (Choose the Debug printout level in EASE.)

(b) Assume that the element matrices are calculated correctly and evaluate

1. the moment at node 1 due to a unit rotation at node 1 and all other displacements and rotations zero
2. the force at node 1 due to a unit rotation at node 4 and all other displacements and rotations zero
3. the force at node 2 due to a unit vertical displacement at node 1

(c) Tabulate the values produced by the computer configuration for items 1–3 of task (b).

3-7. Combined scaling and dimensional units consistency test.

Given: The problem defined by the input data of Prob. 3.6.

Do the following:

(a) Use the given data as the basis for the first analysis.

(b) Change the force units of input data by a factor of 8 for the second analysis.

(c) Scale the input to reflect increasing Young's modulus by a factor of 2 for the second analysis.

(d) Predict what factors will apply to deflections and member moments to convert them to values expected in the second analysis.

(e) Perform the second analysis and assess the validity of the computer configuration.

3-8. Correspondence consistency test.

Given: The truss structure defined as follows.

Element Data			Nodal Data		
Element	AE	EI	Node	x	y
−1, 2	83.1E6	393E6	1	0	0
			2	120	50

Fixities: 1, 1; 1, 2; −1, 3
Loads: 2, 1; −5 000
−2, 2, −12 000

Do the following:

(a) Perform an analysis of the given structure as the first analysis.

(b) Eliminate the applied loading from the first analysis input and introduce the displacements of the first analysis as settlements in creating input data for the second analysis.

(c) Run the second analysis and determine if the internal forces of bars from the first analysis equal those of the second.

Chapter 4

Lattice Element Models

A lattice element is a two-noded finite element that, when used with like elements, does not evoke discretization error. Though all lattice elements have two nodes, not all two-noded elements are lattice elements.

A lattice structural model is an idealized structure all of whose discretized elements are lattice elements. From the system viewpoint, we require that only one lattice element connect any pair of nodes of the lattice structure. Furthermore, we require that each node in the set be connected to at least one lattice element.

Lattice elements comprise a very useful set of models for discretizing structures. Structures advantageously discretized by lattice elements include one-, two-, and three-dimensional trusses; frames; and axisymmetric membranes, plates, and shells.

This chapter presents samples of lattice element macro and micro models, classifies and characterizes them, defines algebraic and numerical tests for assessing whether a two-noded element is a lattice element, and examines means for developing the element macro equations.

4.1 TYPES OF LATTICE ELEMENT MODELS

Tables 4.1 and 4.2 present 10 samples of lattice element macro models.

Table 4.1 presents models that in the local coordinate axes of the figures involve only one displacement variable per node. Five of the seven models of this

TABLE 4.1 STIFFNESS EQUATIONS FOR LATTICE ELEMENTS THAT RESIST TENSION AND SHEAR

	Element Type and Equations	Equation Variables
Regular rod	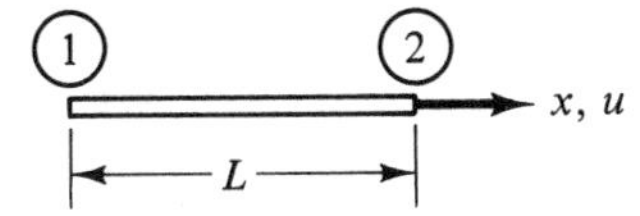 ① ② x, u; L $\frac{AE}{L}\,\mathbf{M}\,\mathbf{u} = \mathbf{p}$	A = cross-sectional area E = Young's modulus $\mathbf{u}$ = vector of node axial displacements $\mathbf{p}$ = vector of node axial forces
Tapered rod	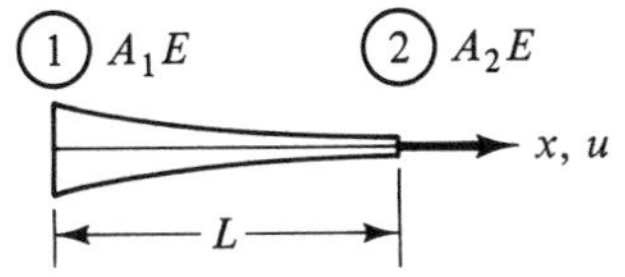 ① A_1E ② A_2E x, u; L $\frac{2}{L}\left(\frac{1}{A_1E} + \frac{1}{A_2E}\right)^{-1}\mathbf{M}\,\mathbf{u} = \mathbf{p}$	$\frac{1}{AE(x)} = \frac{1}{A_1E} + \left(\frac{1}{A_2E} - \frac{1}{A_1E}\right)\frac{x}{L}$ $\frac{L}{A_iE}$ = flexibility at node i
Pulley rope	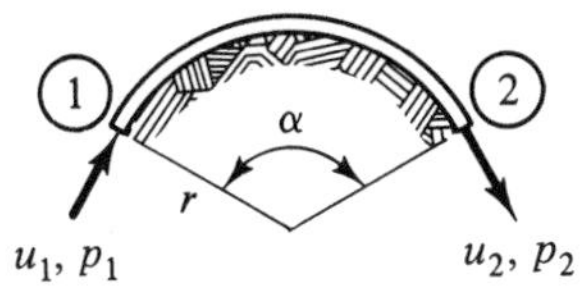 ① ② α, r, u_1, p_1, u_2, p_2 $\frac{AE}{r\alpha}\,\mathbf{M}\,\mathbf{u} = \mathbf{p}$	α = angle subtended (in radians) r = radius to rope midline $\mathbf{u}$ = tangential displacement vector $\mathbf{p}$ = tangential force vector
Tensioned string	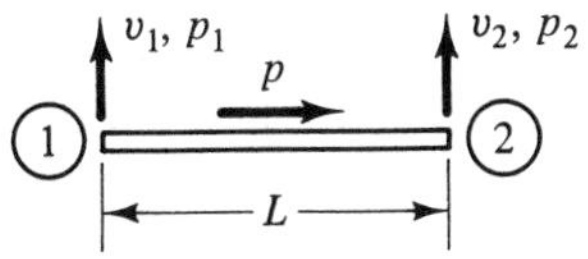 ① ② v_1, p_1, v_2, p_2, p; L $\frac{p}{L}\,\mathbf{M}\,\mathbf{v} = \mathbf{p}$	p = tension force in the string $\mathbf{v}$ = vector of normal displacements $\mathbf{p}$ = vector of normal forces
Bonded rod	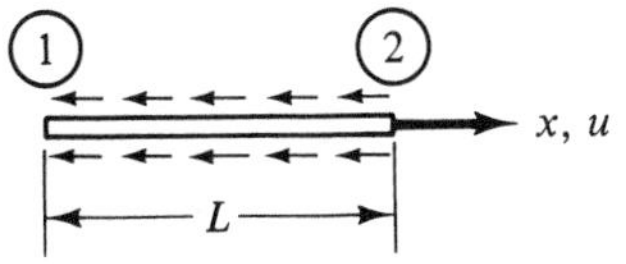 ① ② x, u; L $\frac{AE\gamma}{\sinh \gamma L}\begin{bmatrix}\cosh \gamma L & -1 \\ -1 & \cosh \gamma L\end{bmatrix}\begin{pmatrix}u_1 \\ u_2\end{pmatrix} = \begin{pmatrix}p_1 \\ p_2\end{pmatrix}$	γ = foundation stiffness, $\gamma = \sqrt{\frac{k}{AE}}$ k = bonding modulus force/length^2 u_i = axial displacement at node i p_i = axial force at node i

(Continued)

TABLE 4.1 (Continued)

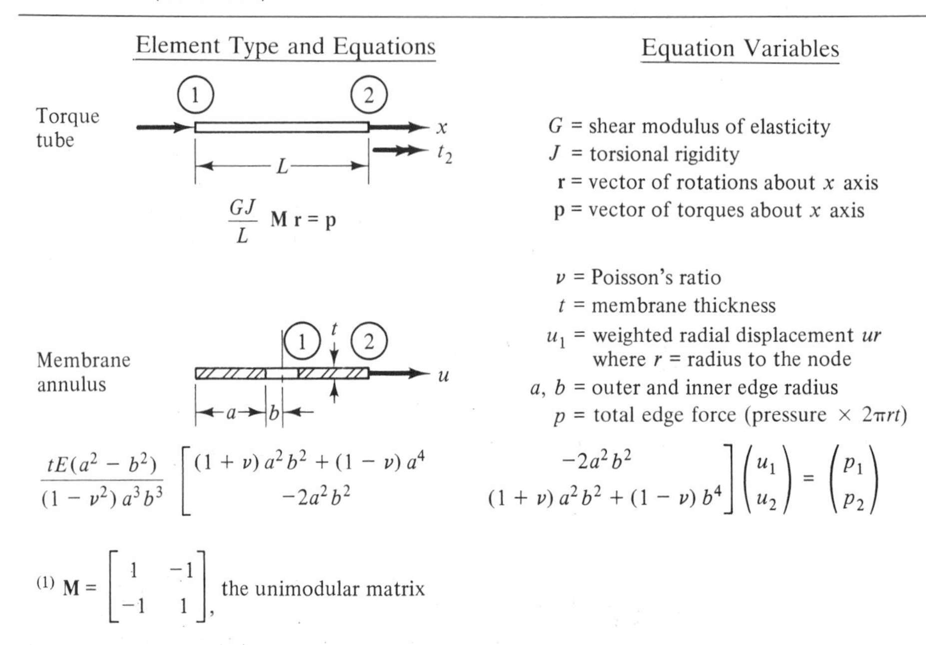

$$\frac{tE(a^2 - b^2)}{(1 - \nu^2)\,a^3 b^3}\begin{bmatrix}(1+\nu)\,a^2b^2 + (1-\nu)\,a^4 & -2a^2b^2 \\ -2a^2b^2 & (1+\nu)\,a^2b^2 + (1-\nu)\,b^4\end{bmatrix}\begin{pmatrix}u_1\\u_2\end{pmatrix} = \begin{pmatrix}p_1\\p_2\end{pmatrix}$$

(1) $\mathbf{M} = \begin{bmatrix} 1 & -1 \\ -1 & 1 \end{bmatrix}$, the unimodular matrix

table involve the unimodular matrix as the stiffness matrix. The regular rod element is the most commonly used element of Table 4.1. Its equations provide the basis for simulating truss members. The pulley rope element represents a string that slides, friction free, over a rigid cylindrical sheave. The tensioned string element represents a straight string that has axial stiffness only as a consequence of its prestress. This model implies that the change in tension in the string due to deformation is negligible compared with the pretensioning force. The bonded-rod model is based on the assumption that it is bonded all along its length to a medium that offers resistance proportional to the local relative movement between the point on the rod and the medium. This model is designed to represent the straight portion of concrete reinforcing bars or glass fibers embedded in a mastic—the fibers of a composite material. The annular axisymmetric flat membrane models a two-dimensional surface. Because this element implies symmetry, the loading at a node is the intensity of pressure applied around the periphery of the ring.

Of the element models of Table 4.1, the bonded-rod model is the only one that can deal with a structure idealized as semi-infinite. Setting the span infinite in this model and setting the displacement zero at infinity reduces this matrix to a one-by-one with stiffness $AE\gamma$. All the other elements have zero stiffness when the span is required to approach infinity.

The element macro equations of Table 4.2 are all beam element models. They differ because of changes in how the beam behavior is idealized. The Bernouilli-Euler model is the classical beam model. The Timoshenko model adds to the classical model consideration of shear flexibility.[1]

TABLE 4.2 STIFFNESS EQUATIONS FOR LATTICE ELEMENTS THAT RESIST BENDING

Element Type and Equations | Equation Variables

Bernouilli–Euler beam

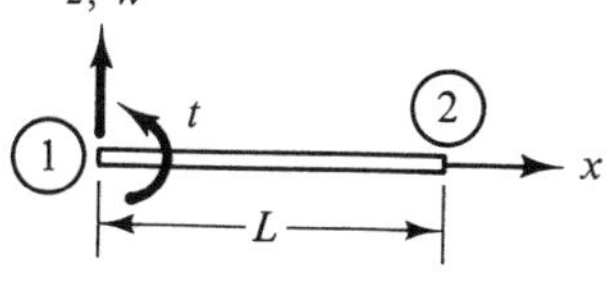

$$\frac{2EI}{L^3}\begin{bmatrix} 6 & & \text{Sym.} & \\ 3L & 2L^2 & & \\ -6 & -3L & 6 & \\ 3L & L^2 & -3L & 2L^2 \end{bmatrix}\begin{pmatrix} w_1 \\ t_1 \\ w_2 \\ t_2 \end{pmatrix} = \begin{pmatrix} W_1 \\ T_1 \\ W_2 \\ T_2 \end{pmatrix}$$

Tapered beam

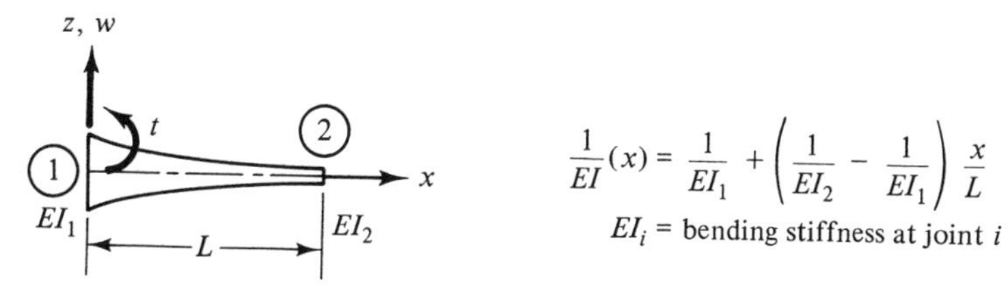

$$\frac{1}{EI}(x) = \frac{1}{EI_1} + \left(\frac{1}{EI_2} - \frac{1}{EI_1}\right)\frac{x}{L}$$

EI_i = bending stiffness at joint i

$$\frac{6E^2}{L^3\left(\frac{1}{I_1^2} + \frac{4}{I_1 I_2} + \frac{1}{I_2}\right)}\begin{bmatrix} \frac{6}{EI_1} + \frac{6}{EI_2} & & \text{Sym.} & \\ \frac{2L}{EI_1} + \frac{4L}{EI_2} & \frac{L^2}{EI_1} + \frac{3L^2}{EI_2} & & \\ \frac{-6}{EI_1} - \frac{6}{EI_2} & \frac{-2L}{EI_1} - \frac{4L}{EI_2} & \frac{6}{EI_1} + \frac{6}{EI_2} & \\ \frac{4L}{EI_1} + \frac{2L}{EI_2} & \frac{L^2}{EI_1} + \frac{L^2}{EI_2} & \frac{-4L}{EI_1} - \frac{2L}{EI_2} & \frac{3L^2}{EI_1} + \frac{L^2}{EI_2} \end{bmatrix}\begin{pmatrix} w_1 \\ t_1 \\ w_2 \\ t_2 \end{pmatrix} = \begin{pmatrix} W_1 \\ T_1 \\ W_2 \\ T_2 \end{pmatrix}$$

Braced Bernouilli–Euler beam

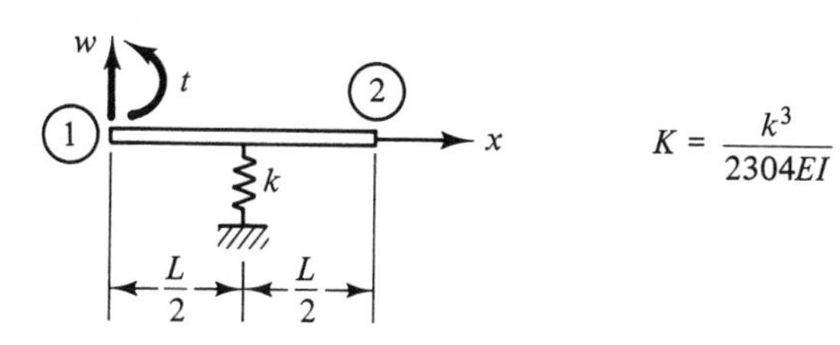

$$K = \frac{k^3}{2304EI}$$

$$\frac{2EI}{L^3(1+K)}\begin{bmatrix} 6+60K & & \text{Sym.} & \\ 3L+18KL & 2L^2+7KL^2 & & \\ -6+36K & -3L+6KL & 6+60K & \\ 3L-6KL & L^2-KL^2 & -3L-18KL & 2L^2+7KL^2 \end{bmatrix}\begin{pmatrix} w_1 \\ t_1 \\ w_2 \\ t_2 \end{pmatrix} = \begin{pmatrix} W_1 \\ T_1 \\ W_2 \\ T_2 \end{pmatrix}$$

(Continued)

TABLE 4.2 (Continued)

Timosherko beam

$\gamma = 3EI/(GA_v L^2)$

A_v = area resisting uniform shear

$$\frac{2EI}{L^3(1+4\gamma)}\begin{bmatrix} 6 & & \text{Sym.} & \\ 3L & 2L^2(1+\gamma) & & \\ -6 & -3L & 6 & \\ 3L & L^2(1-2\gamma) & -3L & 2L^2(1+\gamma) \end{bmatrix}\begin{pmatrix} w_1 \\ t_1 \\ w_2 \\ t_2 \end{pmatrix} = \begin{pmatrix} W_1 \\ T_1 \\ W_2 \\ T_2 \end{pmatrix}$$

Figure 4.1 is a diagram of the terms that may be used to characterize a lattice element model. The terms emphasize differences in geometries, structural behavior, and nodal variables.

Most of the elements for structural analysis involve nodal variables that describe the components of a displacement vector at a node. As the first model of Table 4.3 suggests, electrical circuits can be analyzed by lattice finite elements in which the nodal unknown is the magnitude of the potential. Similarly, thermal, and magnetic networks can be formulated as scalar variable problems.

The geometry that an element model can treat is a principal classification parameter. Classification details particularize the geometry, usually as some form of conic in two- or three-dimensional space.

The elements' behavioral characteristics are a much broader basis for classification. The major subclassification divides the elements into unbraced and braced types. In general, unbraced elements have singular macro stiffness matrices, while braced may be nonsingular. The bonded-rod and axisymmetric membrane models of Table 4.1 are braced elements. The elastically supported beam of Table 4.2 is braced. The remaining models of Tables 4.1 and 4.2 are unbraced.

The possible number of behavior states of a lattice model is infinite. For example, the beam cross section could be represented as deformed in any number of odd-ordered powers of the depth coordinate. Fortunately, the idealization that sections cut normal to the beam remain plane during infinitesimal deformation changes is usually acceptable for engineering predictions of behavior. Similarly, the models of material behavior are limited to a few that have served the profession well.

4.2 STIFFNESS MATRIX CHARACTERISTICS

The macro stiffness matrices are symmetric in conformity with the Maxwell-Betti theorem.

For linear elastic simulations, we require that the element stiffness matrix

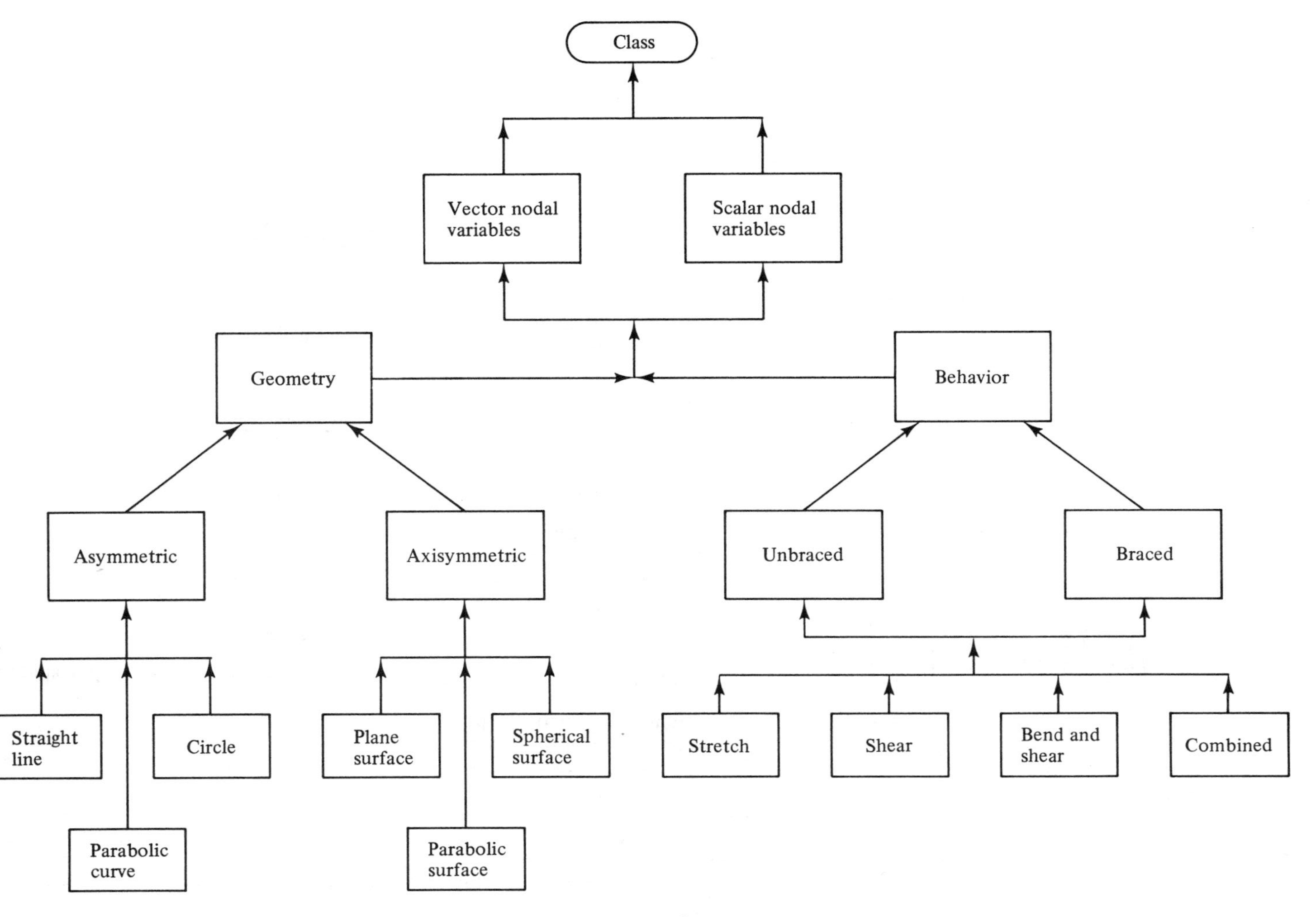

Figure 4.1 Classification of lattice models.

TABLE 4.3 EQUATIONS FOR SOME SCALAR-VARIABLE LATTICE ELEMENTS

Application	Constitutive Law	Equation Variables	Branch Equations
Electrical circuits	Ohm's law $\Delta e = i\, rLA$	e = electrical potential i = electrical circuit rAl = branch resistance A = area; L = length	$\frac{1}{rLA}\begin{bmatrix} 1 & -1 \\ -1 & 1 \end{bmatrix}\begin{pmatrix} e_i \\ e_j \end{pmatrix} = \begin{pmatrix} i_i \\ i_j \end{pmatrix}$
Capillary flow	Darcy's law $\frac{pA}{L}\,\Delta h = f$	h = hydrostatic head f = fluid flux $\frac{l}{pA}$ = flow resistance A = area; L = length	$\frac{pA}{L}\begin{bmatrix} 1 & -1 \\ -1 & 1 \end{bmatrix}\begin{pmatrix} h_i \\ h_j \end{pmatrix} = \begin{pmatrix} f_i \\ f_j \end{pmatrix}$
Heat pipe flow	Fourier's law $\frac{kA}{L}\,\Delta T = A\,\frac{dT}{dx}$ 1 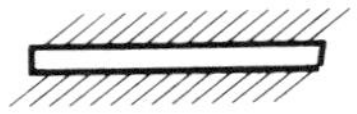 2	T = temperature $A\,\frac{dT}{dx}$ = heat flux $\frac{kA}{L}$ = conductivity A = area; L = length	Insulated $\frac{k}{L}\begin{bmatrix} 1 & -1 \\ -1 & 1 \end{bmatrix}\begin{pmatrix} T_i \\ T \end{pmatrix} = \begin{pmatrix} \left.\frac{dT}{dx}\right\|_i \\ \left.\frac{dT}{dx}\right\|_j \end{pmatrix}$
	$\frac{dT}{dy} = h(T - T_\infty)$ 1 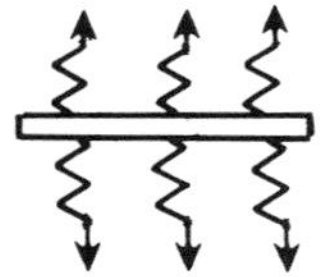 2 T_∞	h = heat transfer coefficient T_∞ = ambient temperature $\alpha = 2\pi rhL$	Convective $\begin{bmatrix} \frac{kA}{L} - \frac{\alpha}{3} & -\frac{kA}{L} - \frac{\alpha}{6} \\ -\frac{kA}{L} - \frac{\alpha}{6} & \frac{kA}{L} - \frac{\alpha}{3} \end{bmatrix}\begin{pmatrix} T_i \\ T_j \end{pmatrix} = \begin{pmatrix} A\frac{dT}{dx_i} \\ A\frac{dT}{dx_j} \end{pmatrix} \frac{\alpha T}{2}\begin{pmatrix} 1 \\ 1 \end{pmatrix}$

associate with positive strain energy whenever the element undergoes strain. Thus we require that

$$SE = 0.5\, \mathbf{g}_e^T \mathbf{K}_e \mathbf{g}_e \geq 0 \tag{4.1}$$

for all possible choices of a non-null displacement vector **g**. Thereby we require that each element stiffness matrix be "positive semidefinite."

Since setting the displacement vector to an influence vector $\mathbf{g}_e^T = (0, 0, \ldots, 0, 1, 0, \ldots, 0)$ implies some strain, the requirement of positive semidefiniteness means, in part, that the diagonal coefficients of the macro stiffness matrix must be zero or positive.

The requirement that element stiffness matrices be positive semidefinite restricts the class of structures that we can analyze. It limits the material models, the prestress states, and the geometry changes.

From the viewpoint of the lattice element models of Tables 4.1 and 4.2, we exclude materials with negative Young's modulus, E, or shear modulus, G. For virtually all structural materials—structural metals, concrete, glass, stone, refractories, and most plastics—these moduli are positive while the strains are less than the proportional limit strain.

The macro equations of the tensioned string of Table 4.1 suggest the prestress constraint. If the prestress is compressive, this stiffness matrix is not positive semidefinite. This is symptomatic of the more general prestress limitation. The definiteness requirement for lattice models limits us to ignoring prestresses altogether or ignoring them when they reduce stiffness.

Stiffness is also reduced when loading induces compressive stresses with significant changes to the original structural geometry. To ensure positive semidefiniteness, we require that the geometry changes be so small that we can relate equilibrium equations to the undisplaced geometry. We thereby restrict ourselves from predicting buckling loads, for the present.

In summary, the restriction to positive semidefinite element stiffness matrices limits modeling to conventional structural materials undergoing strains much less than one, to prestress states sufficiently small that they do not reduce stiffness, and to infinitesimal geometry changes.

Limiting models to positive semidefinite element matrices ensures that the system stiffness matrix will be positive semidefinite. Interpreting the assembly as adding element energies is the basis for proving this statement.

4.3 TESTING ELEMENTS FOR THE LATTICE PROPERTY

The most important characteristic of lattice elements is that they evoke no discretization error. If we are analyzing a lattice structure, the values of behavior predictions will be unaffected by the number and disposition of nodes and elements over the structure.

The lattice property element test is an algebraic test that proves that a given two-noded element stiffness matrix is a lattice element model.

This test is illustrated, for the rod element of Table 4.1, in Table 4.4. The steps of the test are as follows:

1. Write the element equation for a structure discretized by a single two-nodal element of the model under scrutiny. Writing these equations provides Eqs. (a) of Table 4.4. We omit external loads applied between the nodes. The lattice property test addresses the element stiffness matrix and end loads only.

2. Write the system stiffness equations for the structure discretized by two elements. This action results in Eqs. (b) of Table 4.4. The interior node has been located at an arbitrary point that bisects the structure of Eqs. (a) into elements of lengths a and b.

3. Perform row operations on the system stiffness equations to decouple the equations relating to nodal variables of the end nodes from the variables of the intermediate node. (The row operations are represented by the matrix **C**, Eqs. (c) of Table 4.4. Multiplying Eqs. (b) by matrix **C** transforms Eqs. (b) to Eqs. (d).)

4. Simplify the equations seeking to match the equations of step 1. The simplification includes extracting the end node equations from the set and relating the bisected geometry parameters to the undivided element parameters. (In the example of Table 4.4., this step simplifies Eqs. (d) to Eqs. (e).)

5. Compare the simplified equations of the bisected element with the equations of the undivided element. (Compare Eqs. (e) and (a) of the example.) If all the coefficients match, coefficient by coefficient, the element stiffness matrix represents a lattice element.

Because the bisection point is arbitrary in the test, the test ensures that the representation of the element to the rest of the structure is unchanged by bisection. Since we can represent any subdivision as a sequence of arbitrary bisections, if the test is satisfied, the element involves no discretization error regardless of how the structure is discretized.

All the elements of Tables 4.1 through 4.3 satisfy the lattice property test, except the braced Bernouilli-Euler Beam. This braced beam fails the test because subdivision implies addition of new discrete supports.

Although analytical proof that a given element model is a lattice model is important, as structural analysts we are more concerned with whether a particular element model of the code we are using is a lattice model. A numerical lattice property test provides a means of addressing this concern.

As an illustration of a numerical test, consider the following activities:

1. Choose a structural problem that can be discretized by two elements of the type under consideration. The end nodes of this structure will be loaded and have determinate boundary conditions.

2. Solve the structural problem, using the production code, evaluating the unknown displacements of the end nodes for an arbitrarily bisected structure.

TABLE 4.4 LATTICE TEST OF ROD ELEMENT

Undivided element

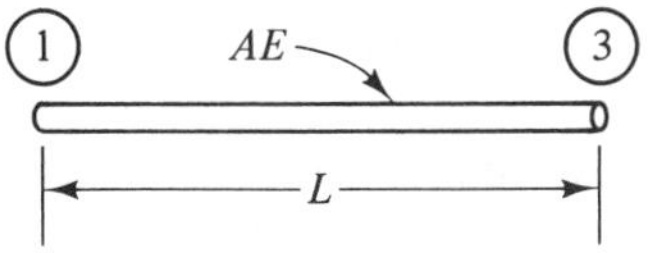

$$\frac{AE}{L}\begin{bmatrix} 1 & -1 \\ -1 & 1 \end{bmatrix}\begin{pmatrix} g_1 \\ g_3 \end{pmatrix} = \begin{pmatrix} p_1 \\ p_3 \end{pmatrix} \qquad \text{(a)}$$

Subdivided element

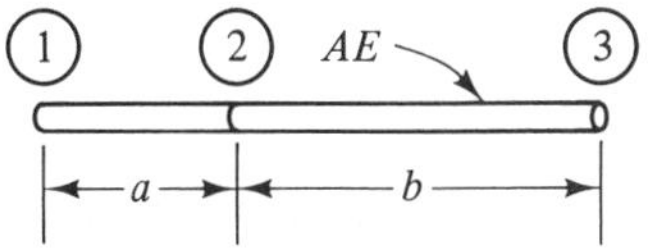

$$AE\begin{bmatrix} \frac{1}{a} & -\frac{1}{a} & 0 \\ -\frac{1}{a} & \frac{1}{a}+\frac{1}{b} & -\frac{1}{b} \\ 0 & -\frac{1}{b} & \frac{1}{b} \end{bmatrix}\begin{pmatrix} g_1 \\ g_2 \\ g_3 \end{pmatrix} = \begin{pmatrix} p_1 \\ 0 \\ p_2 \end{pmatrix} \qquad \text{(b)}$$

Premultiplying Eq. (b) by

$$\begin{bmatrix} 1 & \frac{b}{a+b} & 0 \\ 0 & 1 & 0 \\ 0 & \frac{a}{a+b} & 1 \end{bmatrix} = \mathbf{S} \qquad \text{(c)}$$

yields

$$AE\begin{bmatrix} \frac{1}{a+b} & 0 & -\frac{1}{a+b} \\ -\frac{1}{a} & \frac{1}{a}+\frac{1}{b} & -\frac{1}{b} \\ -\frac{1}{a+b} & 0 & \frac{1}{a+b} \end{bmatrix}\begin{pmatrix} g_1 \\ g_2 \\ g_3 \end{pmatrix} = \begin{pmatrix} p_1 \\ 0 \\ p_3 \end{pmatrix} \qquad \text{(d)}$$

Noting that $L = a + b$ and that the first and third equations of Eqs. (c) do not involve g_2, we can write Eqs. (d) as

$$\frac{AE}{L}\begin{bmatrix} 1 & -1 \\ -1 & 1 \end{bmatrix}\begin{pmatrix} g_1 \\ g_3 \end{pmatrix} = \begin{pmatrix} p_1 \\ p_3 \end{pmatrix} \qquad \text{(e)}$$

We see that Eq. (e) matches Eq. (a). Q.E.D.

3. Solve the structural problem, using the production code, evaluating end node displacements for a different arbitrarily bisected structure than used in activity 2.

4. Compare the end node displacements of activities 2 and 3. If displacements match, we have reason to believe that the element stiffness being tested is a lattice model.

The circumspect analyst may wish to strengthen the numerical test by considering all possible linearly independent end loadings and support conditions. For the rod case, this consideration multiplies the number of cases by a factor of 8.

The results of numerical experiments cannot constitute a mathematical proof that a model has the lattice property. Since the test involves minuscule computer resources, however, the analyst can usually make as many arbitrary subdivisions as required to become convinced that the element stiffness model satisfies the requirements to be a lattice model.

This test has the desirable feature that it is a self-consistency test, which, from the viewpoint of the theory, emphasizes the fact that the lattice property test does not concern idealization error.

4.4 FINDING EQUIVALENT LOADS

Given a lattice element stiffness matrix, we can extend the element equations to evaluate the equivalent loading vector for any loads applied between the end nodes.

The equations of Table 4.5 illustrate the process using the rod element case. The steps of this process are as follows:

1. Bisect the element, placing a new node at the point where the interior load is to be applied. (This step yields Eqs. (b) of Table 4.5.)

TABLE 4.5 ROD ELEMENT EQUIVALENT LOADING

No interior load

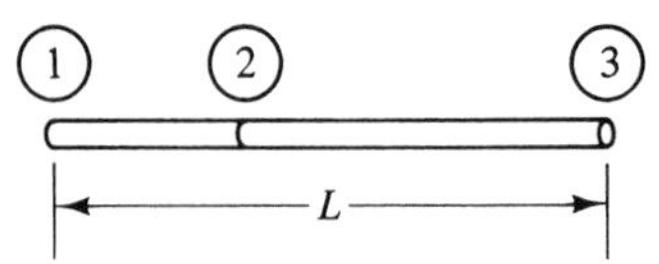

$$\mathbf{p} = \begin{pmatrix} p_1 \\ 0 \\ p_3 \end{pmatrix} \tag{a}$$

Interior load, magnitude p

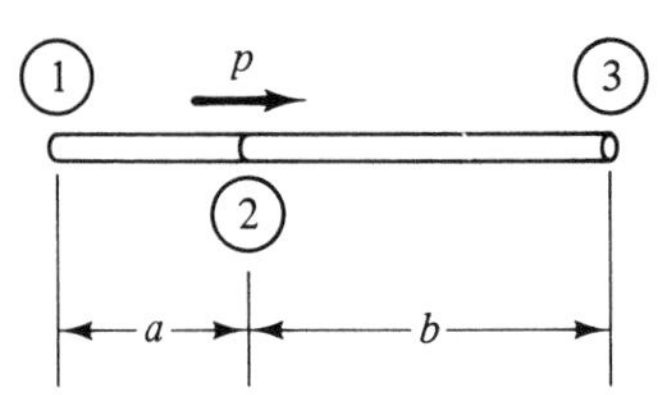

$$\mathbf{p} = \begin{pmatrix} p_1 \\ p \\ p_3 \end{pmatrix} \tag{b}$$

Multiplying the loading of Eq. (b) by **S** of Table 4.4, Eq. (c) gives

$$\mathbf{p}' = \begin{pmatrix} p_1 + \dfrac{b}{a+b} p_2 \\ p_2 \\ p_3 + \dfrac{a}{a+b} p_2 \end{pmatrix} \tag{c}$$

By comparing Eqs. (a) and (c) and setting aside the interior node equations, we see that the equivalent loading is

$$\mathbf{q} = \begin{pmatrix} q_1 \\ q_3 \end{pmatrix} = \begin{pmatrix} -\dfrac{b}{a+b} p \\ -\dfrac{a}{a+b} p \end{pmatrix} \tag{d}$$

TABLE 4.6 SOME EQUIVALENT LOAD FORMULAS FOR THE UNIMODULAR STIFFNESS MATRIX

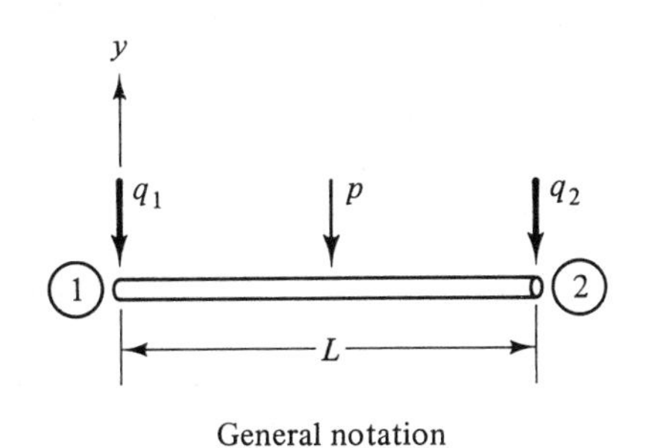

General notation

Case	q_1	q_2
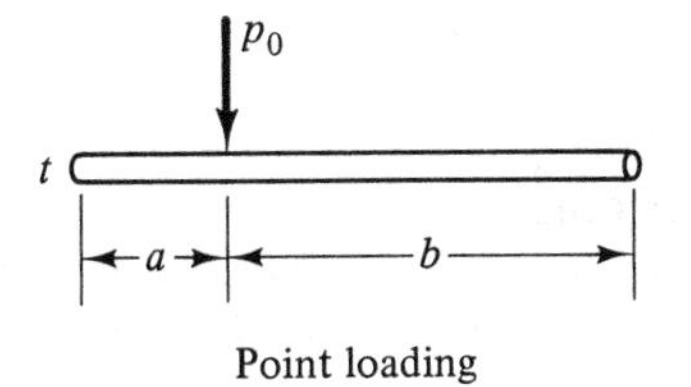 Point loading	$-\dfrac{bp_0}{L}$	$-\dfrac{ap_0}{L}$
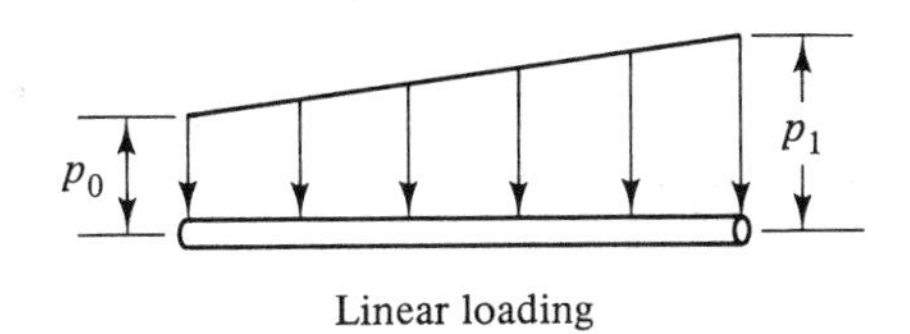 Linear loading	$-\dfrac{L}{6}(2p_0 + p_1)$	$-\dfrac{L}{6}(p_0 + 2p_1)$
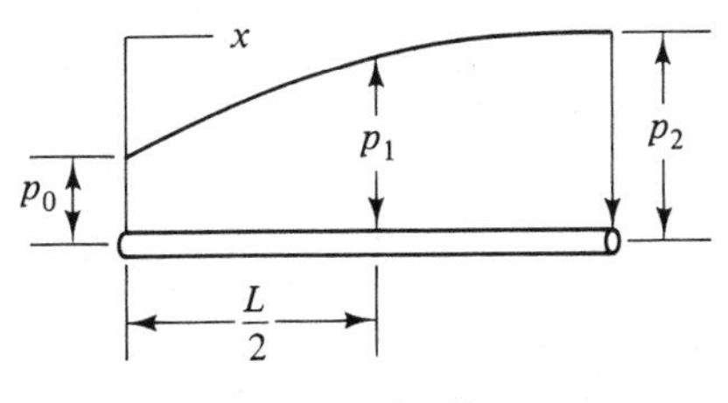 Quadratic loading	$L^2 p = 2\left(x - \dfrac{L}{2}\right)(x - L)p_0 - 4x(x - L)p_1 + 2x\left(x - \dfrac{L}{2}\right)p_2$ $-\dfrac{L}{6}(p_0 + 2p_1)$	 $-\dfrac{L}{6}(2p_1 + p_2)$
Sinusoidal loading	$p = p_0 \sin \dfrac{n\pi x}{L}$, n-an integer $-\dfrac{Lp_0}{n\pi}$	 $-\dfrac{Lp_0}{n\pi} \cos n\pi$

2. Apply the decoupling transformation to the loading. (In the example, the **C** matrix of Table 4.4 transforms Eqs. (b) of Table 4.5 to Eqs. (c) of Table 4.5.)

3. Subtract the transformed loading vector of the bisected element, setting aside the components at the interior node, from the loading of the undivided element. (This step defines the equivalent loading vector, Eqs. (d) of Table 4.5.)

By this process we discover the equivalent loading that is consistent with a given lattice stiffness matrix. This loading is uniquely defined. Use of this loading to represent intermediate point loading means that we do not need to place a node at every point loading site to avoid discretization error.

Given the equivalent load formulas for a concentrated loading applied at a point between the end nodes, we can create the loading equivalent to any distributed loading by integration. The integration exploits the fact that the equivalent loads of two or more interior loads can be superimposed.

Table 4.6 lists results of integrating the formulas of Eqs. (d) for a variety of load distributions. Since these formulas are independent of the stiffness matrix scalar multiplier, they apply to all the element matrices of Table 4.1 that involve the unimodular stiffness matrix. For clarity, the loading intensity of Table 4.6 has been plotted as if the loading were acting normal to the element axis. This interpretation is direct only for the tensioned string. In other cases, the interpretation changes with the element under consideration.

We conclude that equivalent loads for lattice finite elements are uniquely defined by the element stiffness matrix. Equivalent loads of lattice models can be interpreted to be the negatively signed reactions at the end nodes from an FEA of a subdivided element that has interior loading. This interpretation facilitates calculation of equivalent loads without recourse to matrix operations.

4.5 MICROSCOPIC MODELS

Table 4.7 presents lattice element microscopic models for some of the elements of Tables 4.1 and 4.2.[2] These models provide a measure of stress in the cross section as a function of element section geometry parameters and the assumptions of the stress model. More than one model may exist for a given section, as suggested by the two bending stress models.

These formulas define stresses in terms of the end loads on a finite element. To account for an external loading at intermediate cross sections, formulas are developed considering subdivision. Assessment of internal forces at intermediate cross sections is a by-product of the process for evaluating equivalent loads. For the rod example, taking the left-hand side of the second equation of Eqs. (d) of Table 4.4 and equating it to the right-hand side of Eqs. (c) of Table 4.5 provides the equation to find the displacement of the intermediate node. Given the displacement, the element macro equations can be used to calculate $p_{2'}$ and $p_{2''}$.

Considering subdivisions we can develop influence equations that express the

TABLE 4.7 SOME LATTICE ELEMENT STRESS FORMULAS

Stress State	Cross Section	Stress Formula
Stretch		$\sigma_{xx} = \dfrac{p_i}{A}$
Torsion (solid cylinder)		$\sigma_{yz} = \dfrac{2p_i}{\pi r^3}$ Max. stress
Torsion (thin-walled closed tube)		$\sigma_{yz} = \dfrac{p_i}{2tA} \quad t \ll \sqrt{A}$ Mean stress
Torsion (thin-walled open tube)		Max. stress $\sigma_{xy} = \dfrac{D}{1 + \dfrac{\pi^2 D^4}{16A^2}} \left[1 + 0.15\left(\dfrac{\pi^2 D^4}{16A^2} - \dfrac{D}{2r}\right)\right]$ D = diameter of largest inscribed circle r = radius at boundary point
Bending (Bernouilli–Euler)		$\sigma_{xx} = \dfrac{M_i c}{c_i}$ $\sigma_{yz} = \dfrac{W_i a \bar{y}_i}{I_i t}$
Bending (Timoshento)		$\sigma_{xx} = \dfrac{M_i c}{I_i}$ $\sigma_{yz} = \dfrac{W_i}{A_{vi}}$

TABLE 4.8 INTERNAL LOADS FOR INTERMEDIATE LOADINGS

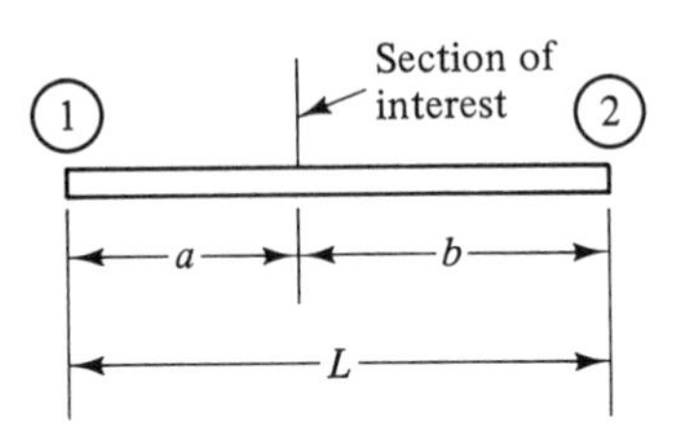

Case	Force on Cross Section

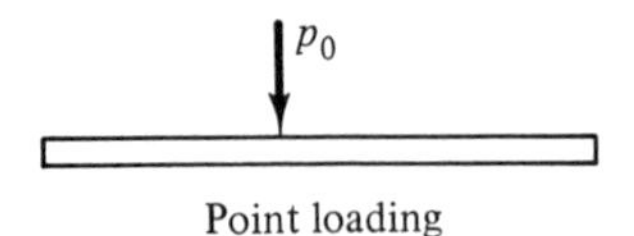

Point loading

$$p_x = \frac{AE}{L}[1 \quad -1]\begin{pmatrix} u_1 \\ u_2 \end{pmatrix} + \frac{bp_o}{L} \qquad x < a$$

$$p_x = \frac{AG}{L}[1 \quad -1]\begin{pmatrix} u_1 \\ u_2 \end{pmatrix} - \frac{ap_o}{L} \qquad x > a$$

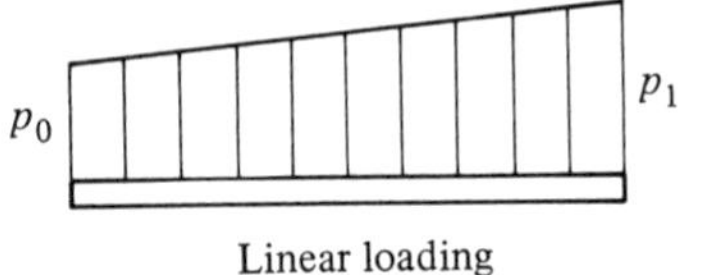

Linear loading

$$p_x = \frac{AE}{L}[1 \quad -1]\begin{pmatrix} u_1 \\ u_2 \end{pmatrix} + p_o\left(\frac{L}{6} - b + \frac{b^2}{L}\right) + p_1\left(\frac{L}{3} - \frac{b^2}{L}\right)$$

p_0 p_1 p_2

$L/2$

Quadratic loading

$$p_x = \frac{AE}{L}[1 \quad -1]\begin{pmatrix} u_1 \\ u_2 \end{pmatrix} + \frac{\left(\frac{L}{2} - \bar{x}\right)}{L}\int_{-L/2}^{L/2} p(x)\,dx - \int_{-L/2}^{L/2+\bar{x}} p(x)\,dx$$

where

$$\int_{-L/2}^{L/2} p(x)\,dx = \frac{L}{6}(p_o + 4p_1 + p_2)$$

$$\int_{-L/2}^{L/2} p(x)x\,dx = \frac{L^2}{12}(-p_o + p_2)$$

$$\int_{-L/2}^{L/2+\bar{x}} p(x)\,dx = \frac{2}{L^2}\left\{\frac{\left(\frac{L}{2} + \bar{x}\right)^3}{3}p_o - 2\left[\frac{\left(\frac{L}{2} + \bar{x}\right)^3}{3} - \frac{L^2\left(\frac{L}{2} + \bar{x}\right)}{4}\right]p_1 + \frac{\left(\frac{L}{2} + \bar{x}\right)^3}{3}p_2\right\}$$

and

$$\bar{x} = \int_{-L/2}^{L/2} p(x)x\,dx \Bigg/ \int_{-L/2}^{L/2} p(x)\,dx$$

value of loads at any cross section in terms of the element and node displacements, the position of the cross section, and the equivalent load parameters. Table 4.8 lists these formulas for the loadings of Table 4.6.

4.6 UNBRACED AND BRACED ELEMENTS

For lattice elements, the degeneracies of the stiffness matrix engender only rigid body displacements. In rigid motion, the distance between any two points on the element does not change. Confining displacements to quantities so small that sines of angular changes equal the angle and cosines equal 1 means that at most deflections can vary linearly along the span and rotations must be constant. Thus if both ends of the element displace the same amount, all points along the length move that amount. If a straight element rotates, it must remain straight.

An element may have at most six linearly independent rigid motions. Six may occur in three-dimensional systems, three representing linear combinations of translations in the x, y, and z directions, three representing linear combinations of rigid rotations about the axes. Some of these are eliminated when displacements are limited to the plane, still more when the displacements are uniaxial. Because of these eliminations, it is convenient to establish the rank of element stiffness matrices in the space of smallest dimension, that is, the space used in defining the element matrices of Tables 4.1 through 4.3.

As a consequence of the fact that only zero or rigid motions associate with lattice stiffness matrix degeneracies, determination of the rank of the matrix is a simple task because the candidate $\mathbf{r}^T$ vectors are easily defined. The variables of the rod element limit the candidate rigid modes to translation along the rod axis. The tensioned rod can only involve degeneracy for translation normal to the rod; the torque tube, for rigid rotation about the axis. For the beam, translation normal to the axes and rotation in the plane are the candidate rigid modes for degeneracies of each of these matrices.

In general, the unbraced elements have macro stiffness matrix rank equal to the number of displacement variables at a node. Thus all but the last two models of Table 4.1 and the last model of Table 4.2 are unbraced elements.

The properties of unbraced element matrices lead to memorable interpretations. The element macro equations define the coefficients of the stiffness matrices as forces. A column of this matrix evaluates the forces when one of the nodal variables has a value of 1 and all other nodal variables are zero. Premultiplication of the column by a row vector corresponding to the displacements of rigid translation sums the forces in one direction. Similarly, premultiplication by a displacement vector corresponding to rigid rotation is consistent with taking moments of the nodal forces about some point. Noting that an element matrix has a rigid motion degeneracy corresponds to the observation that the element matrix coefficients ensure that the end loads will be in equilibrium regardless of the values assigned to the nodal displacements.

Table 4.9 illustrates these characteristics for the Bernouilli-Euler beam element. A similar exercise shows that each of the unbraced elements of Tables 4.1 and 4.2 implies equilibrium of the forces and moments in any column of its stiffness matrix; the matrix columns satisfy macroscopic equilibrium.

Continuing the force interpretation, we observe that the assembly process adds forces without changing their point of action. Therefore, the structural system stiffness matrix, when composed only of unbraced lattice elements, echoes the satisfaction of macroscopic force and moment equilibrium equations for the system.

This argument extends to equivalent loads. As negative reactions to intermediate loadings, equivalent loads of the unbraced beam element balance the total

TABLE 4.9 RIGID MOTIONS OF THE UNBRACED BEAM ELEMENT MODEL

Given: The Bernouilli-Euler beam element.

Show: That the model nodal forces satisfy equilibrium for rigid motions in two-dimensional space.

1. The element equations are

$$\frac{2EI}{L^3}\begin{bmatrix} 6 & & & \text{Sym.} \\ 3L & 2L^2 & & \\ -6 & -3L & 6 & \\ 3L & L^2 & -3L & 2L^2 \end{bmatrix}\begin{pmatrix} v_1 \\ t_1 \\ v_2 \\ t_2 \end{pmatrix} = \begin{pmatrix} V_1 \\ T_1 \\ V_2 \\ T_2 \end{pmatrix} \tag{a}$$

2. Consider a rigid translation mode,

$$[v_1\ t_1\ v_2\ v_2]^T = [\alpha\ 0\ \alpha\ 0]^T \tag{b}$$

where α is a nonzero constant. Multiplying Eqs. (a) by the right-hand vector of Eq. (b) gives

$$0 = \alpha(\mathrm{V}_1 - \mathrm{V}_2) \tag{c}$$

or, since $\alpha \neq 0$, $\mathrm{V}_1 = \mathrm{V}_2$; that is, the coefficients of the stiffness matrix imply satisfaction of the lateral force equilibrium equation for every column of the stiffness matrix.

3. Consider a mode involving rigid translation and rotation:

$$[v_1\ t_1\ v_2\ t_2]^T = [0\ \alpha\ L\alpha\ -\ \alpha]^T \tag{d}$$

Multiplying Eqs. (a) by the right-hand vector of Eq. (d) gives

$$0 = \alpha\,(T_1 + V_2\,L - T_2) \tag{e}$$

since $\alpha \neq 0_0$. Eq. (e) says that the coefficients of the stiffness matrix imply satisfaction of the moment equilibrium equation.

force and moment of the intermediate loading because zero strain energy affiliates with rigid translation and rotation of the element.

The rank of the element stiffness matrix of a braced element is usually greater than the number of displacement variables at a node. As a consequence, at least one of the rigid motions of the corresponding unbraced element does not imply macroscopic equilibrium. For braced elements there are still six rigid motions that are candidates for stiffness matrix degeneracies. However, due to an implied support between the element nodes, one or more of these motions is excluded.

For the bonded rod of Table 4.1, unit displacement of each node does not correspond to rigid translation because the continuous support along the member induces strains. The ring membrane centerline is not permitted to migrate from its original position, and thus this element model has rank 2. The intermediate support of the Bernouilli-Euler beam braced element retains the degeneracy associated with rigid rotation but not that of rigid translation normal to the members' axis.

Though at least one equilibrium condition is not an explicit part of the braced element macro equations, these models are in all respects valid finite elements. They have symmetric and positive semidefinite (sometimes positive definite) element stiffness matrices. They imply an energy process and are therefore constrained by the contragradient transformation. They are suitable for use with the finite element process.

4.7 TRANSFORMING ELEMENT MODELS FOR GENERAL USE

The element macro models of Tables 4.1 through 4.3 are in their simplest form. They are expressed with respect to element coordinate axes: axes embedded in the element. To make these models useful for other than one-dimensional structures, we need to expand the nodal variables to reflect more general displacement variables, and we need to relate the equations to axes of arbitrary orientation.

The diagram of Fig. 4.2 indicates the element and more general nodal variables of interest in lattice analysis. This diagram defines three classes of analysis nodal variables: in-plane, out-of-plane, and three-dimensional. In structural engineering, in-plane analyses address simulation or planar trusses and frames that remain in the plane under loading. Out-of-plane nets and grillages deflect only normal to their original plane. Points on a three-dimensional structure require three coordinates to describe the initial or final geometry.

There are two types of lattice displacement variables at a node: displacements only or displacements and angular rotations. Displacements are sufficient to describe behavior of trusses and nets. Rotations must be added to describe structural elements that can bend or twist.

The equations of Table 4.10 illustrate the use of the contragradient transfor-

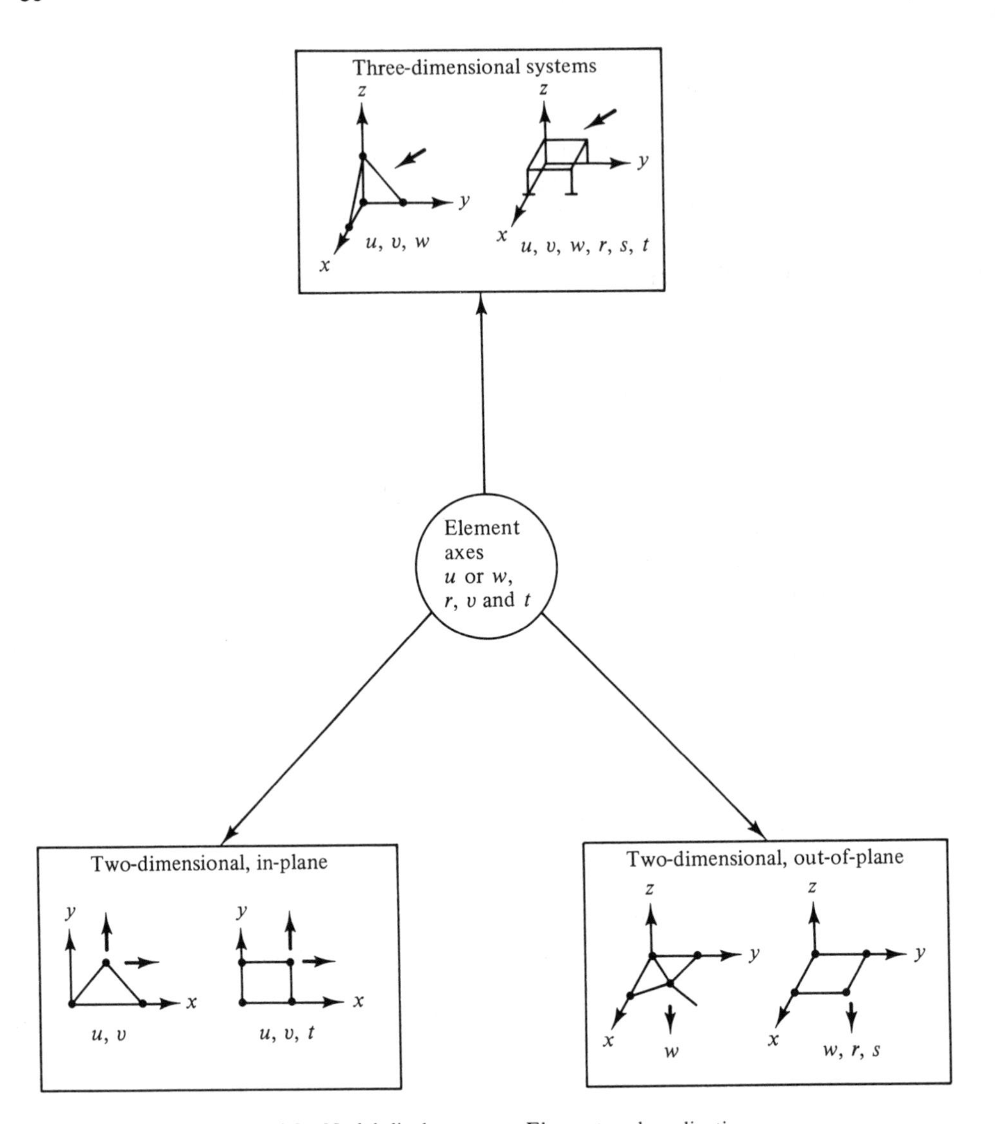

Figure 4.2 Nodal displacements: Element and application axes.

mation to increase the number of nodal variables for the rod of a planar truss. The result of this transformation is the addition of null rows and columns to the stiffness matrix. The rank of the stiffness matrix is preserved. The degeneracy increases by the excess of nodal variables on the right-hand side of the constraint transformation matrix compared with the variables on the left.

Table 4.11 illustrates the transformation to change Eqs. (c) of Table 4.10 so that they relate to force and displacement components along the rotated axes. We

TABLE 4.10 EXPANSION OF THE ROD ELEMENT EQUATIONS FOR IN-PLANE USE

1. The equations referred to an axis along the centerline of the rod	$\frac{AE}{L}\begin{bmatrix} 1 & -1 \\ -1 & 1 \end{bmatrix}\begin{pmatrix} u_1 \\ u_2 \end{pmatrix} = \begin{pmatrix} p_1' \\ p_2' \end{pmatrix}$	(a)
2. The element in-plane constraint equations	$\begin{pmatrix} u_1 \\ u_2 \end{pmatrix} = \begin{bmatrix} 1 & 0 & 0 & 0 \\ 0 & 0 & 1 & 0 \end{bmatrix}\begin{pmatrix} u_1 \\ v_1 \\ u_2 \\ v_2 \end{pmatrix}$ where u_1 and u_2 are global x' displacements, v_1 and v_2 are global y' displacements	(b)
3. The contragradient transformation of Eqs. (a) by Eqs. (b)	$\frac{AE}{L}\begin{bmatrix} 1 & 0 & -1 & 0 \\ 0 & 0 & 0 & 0 \\ -1 & 0 & 1 & 0 \\ 0 & 0 & 0 & 0 \end{bmatrix}\begin{pmatrix} u_1 \\ v_1 \\ u_2 \\ v_2 \end{pmatrix} = \begin{pmatrix} p_{u1} \\ p_{v1} \\ p_{u2} \\ p_{v2} \end{pmatrix}$	(c)

observe that the transformation of forces and displacement components, Eqs. (a) and (b), involve the same transformation matrix. Furthermore, the rotation matrix $\mathbf{R}$ of Eqs. (d) transforms variables at each node independently. The stiffness matrix of the transformed Eqs. (f) is symmetric and fully populated for arbitrary rotation, as compared with the sparsely populated element axis stiffness matrix Eqs. (c) of Table 4.10.

There are only five particular rotation matrices of interest for lattice analysis. These are listed explicitly in Fig. 4.3 for the in-plane and out-of-plane rotations and the three-dimensional rotation matrix $\mathbf{R}_3$. The data of Fig. 4.3 indicates that the rotation and moment vectors transform like the displacement and force vectors.

The fact that the rotation matrix treats each node independently (Table 4.11, Eqs. (d) and (e)) means that we can select a different orientation of the reference axes at each node. This facility is exploited when nodal external forces or displacement constraints have special directional characteristics. Some of these cases are suggested by the sketches in Fig. 4.4

In most finite element computer codes, however, a single reference axis is used for all nodes. Given input describing directional characteristics at a node, the boundary conditions are transformed to the common axes using rotation matrices. The codes require special input data if response generalized displacements and forces are to reference local axes.

In any case, all elements sharing a node must reference the same axes so that the addition of forces in the assembly process implies the addition of parallel force and moment vector components.

TABLE 4.11 ROTATION OF AXES: IN-PLANE

Given: The rod element stiffness matrix referred to element axes.

Find: The matrix coefficients when referenced to differently rotated coordinate axes at each node in a plane.

1. The forces at a node referred to rotated axes are given by

$$\begin{pmatrix} p_x \\ p_y \end{pmatrix} = \begin{bmatrix} \cos\gamma & \sin\gamma \\ -\sin\gamma & \cos\gamma \end{bmatrix} \begin{pmatrix} p_{xe} \\ p_{ye} \end{pmatrix} \qquad \text{(a)}$$

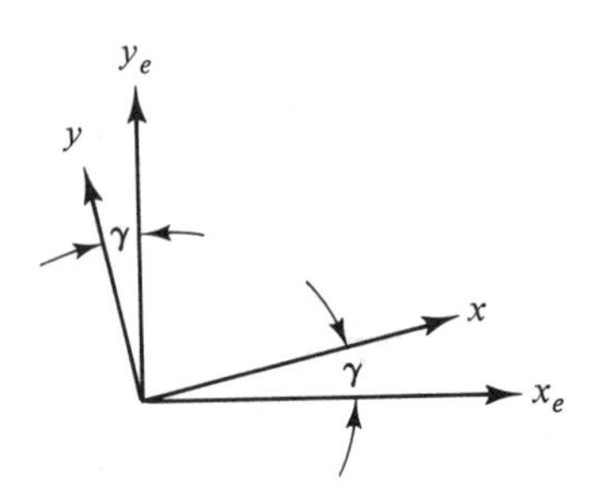

2. The displacement components referred to the rotated axes are

$$\begin{pmatrix} u \\ v \end{pmatrix} = \begin{bmatrix} \cos\gamma & \sin\gamma \\ -\sin\gamma & \cos\gamma \end{bmatrix} \begin{pmatrix} u_e \\ v_e \end{pmatrix}, \quad \text{or} \quad \begin{pmatrix} u_e \\ v_e \end{pmatrix} = \begin{bmatrix} \cos\gamma & -\sin\gamma \\ \sin\gamma & \cos\gamma \end{bmatrix} \begin{pmatrix} u \\ v \end{pmatrix} \qquad \text{(b)}$$

3. Since the forces at each node can be related to differently rotated axes, we can write the second of Eqs. (b) as

$$\begin{pmatrix} v_{e1} \\ v_{e1} \\ u_{e2} \\ v_{e2} \end{pmatrix} = \begin{bmatrix} \cos\gamma & -\sin\gamma & 0 & 0 \\ \sin\gamma & \cos\gamma & 0 & 0 \\ 0 & 0 & \cos\beta & -\sin\beta \\ 0 & 0 & \sin\beta & \cos\beta \end{bmatrix} \begin{pmatrix} u_1 \\ v_1 \\ u_2 \\ v_2 \end{pmatrix} = \mathbf{R}^{-1} \begin{pmatrix} u_1 \\ v_1 \\ u_2 \\ v_2 \end{pmatrix} \qquad \text{(c)}$$

4. Multiplying Eqs. (c) of Table 4.10 by **R** and substituting in Eq. (c) gives

$$\frac{AE}{L} \begin{bmatrix} \cos^2\gamma & & \text{Sym.} & \\ -\sin\gamma\cos\gamma & \sin^2\gamma & & \\ -\cos\gamma\cos\beta & \sin\gamma\cos\beta & \cos^2\beta & \\ \sin\gamma\cos\beta & -\sin\gamma\sin\beta & -\sin\beta\cos\beta & \sin^2\beta \end{bmatrix} \begin{pmatrix} u_1 \\ v_1 \\ u_2 \\ v_2 \end{pmatrix} = \begin{pmatrix} p'_{x1} \\ p'_{y1} \\ p_{x1} \\ p_{y1} \end{pmatrix} \qquad \text{(d)}$$

4.8 COGRADIENT TRANSFORMATION OF FEA

Because the force and displacement and moment and angular displacement vectors transform under the same linear transformation, they are cogradient. Thus, as illustrated by Table 4.11, the components of the nodal variables are related to the components of the variables referenced to a rotated axis by

$$\mathbf{g}_e = \mathbf{R}_e \mathbf{g}'_e \qquad (4.2)$$

Figure 4.3 Special Forms of the Rotation Matrix.

Space	Coordinates	Form of the Rotation Matrix
In-plane	u, v	$\begin{bmatrix} \cos\gamma & \sin\gamma \\ -\sin\gamma & \cos\gamma \end{bmatrix}$
	u, v, t	$\begin{bmatrix} \cos\gamma & \sin\gamma & 0 \\ -\sin\gamma & \cos\gamma & 0 \\ 0 & 0 & 1 \end{bmatrix}$ 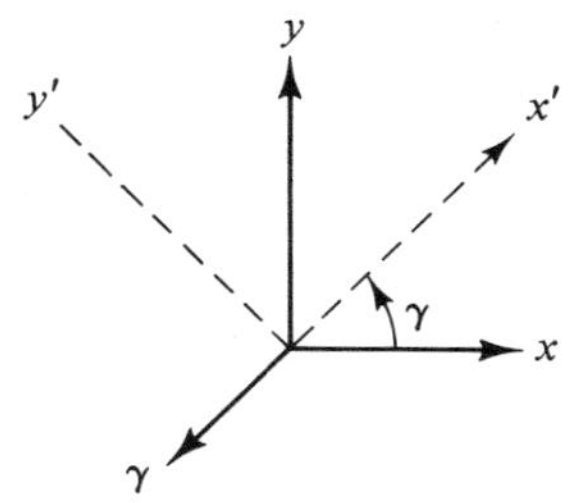
Out-of-plane	w	$[1]$
	w, r, s	$\begin{bmatrix} 1 & 0 & 0 \\ 0 & \cos\gamma & \sin\gamma \\ 0 & -\sin\gamma & \cos\gamma \end{bmatrix}$ 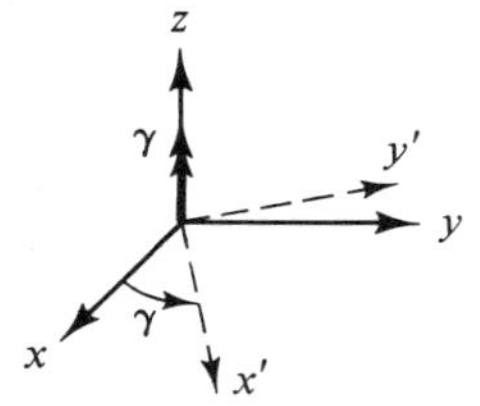
Three-dimensional	u, v, w	$\mathbf{R}_3 = \begin{bmatrix} 1 & 0 & 0 \\ 0 & \cos\gamma & \sin\gamma \\ 0 & -\sin\gamma & \cos\gamma \end{bmatrix} \begin{bmatrix} \cos\beta & 0 & \sin\beta \\ 0 & 1 & 0 \\ -\sin\beta & 0 & \cos\beta \end{bmatrix} \begin{bmatrix} \cos\alpha & \sin\alpha & 0 \\ -\sin\alpha & \cos\alpha & 0 \\ 0 & 0 & 1 \end{bmatrix}$
	u, v, w, r, s, t	$\begin{bmatrix} R_3 & 0 \\ 0 & R_3 \end{bmatrix}$ 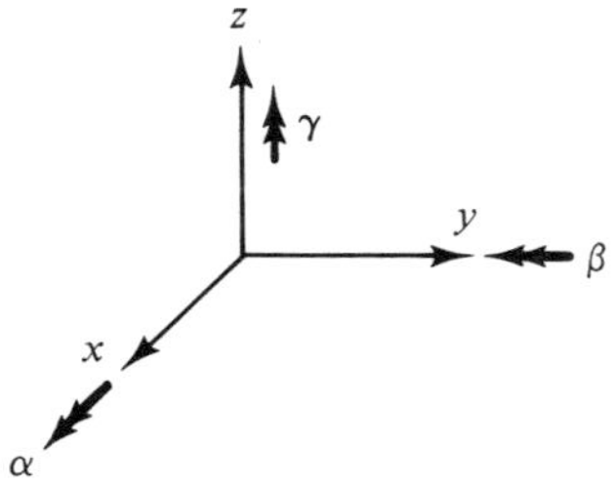

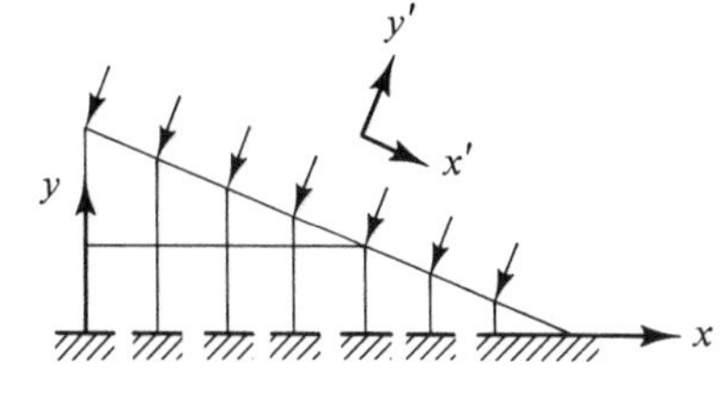

Reduces description of force vectors to definition of magnitude and reference axes

(A) To simplify defining force boundary conditions

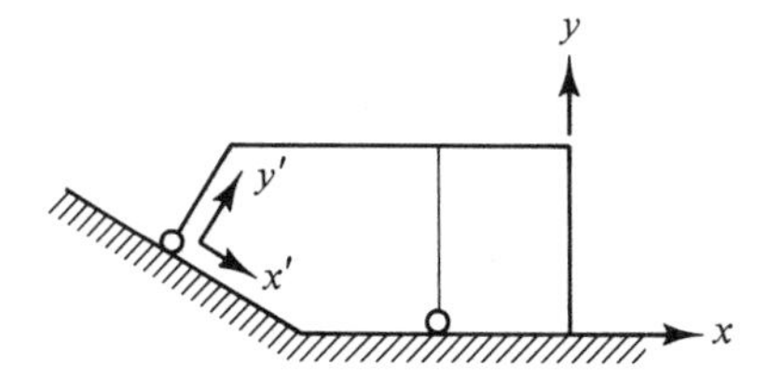

Permits describing displacement constraints as fixities rather than general linear constraints

(B) To simplify defining displacement boundary conditions

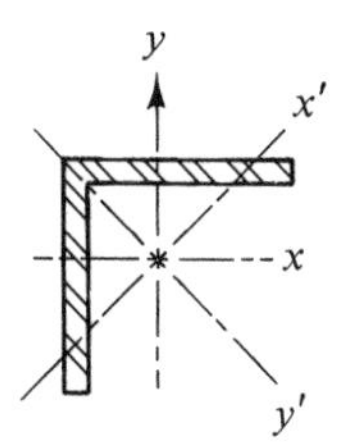

Permits describing properties with respect to principal axes

(C) To simplify defining nonsymmetric cross sections

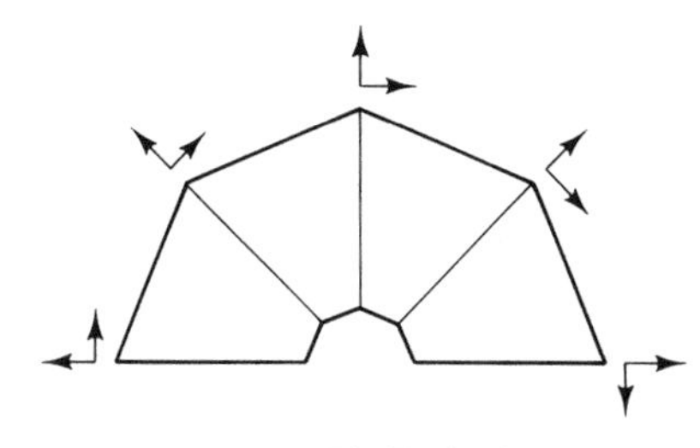

Permits developing the stiffness matrix for a substructure and using this directly for other substructures

(D) To facilitate exploitation of repetitive geometry

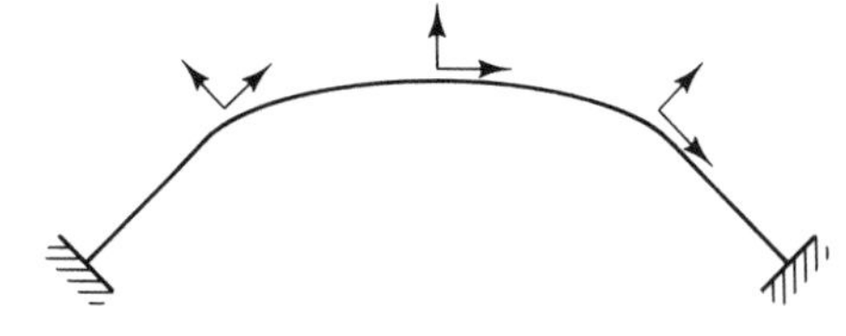

Permits interpreting displacements as those normal and tangential to the surface of the structure

(E) To facilitate interpretation of analysis results

Figure 4.4 Uses for local reference axes.

$$\mathbf{p}_e = \mathbf{R}_e\mathbf{p}'_e \qquad \text{and} \qquad \mathbf{q}_e = \mathbf{R}_e\mathbf{q}'_e \tag{4.3}$$

where

the prime designates the vectors referenced to the rotated axes, and
$\mathbf{R}_e$ is the rotation matrix of element e.

Substituting Eqs. (4.2) in Eqs. (1.1) and premultiplying by $\mathbf{R}^{-1}$ gives the equations that are the cogradient transformation of Eqs. (1.1):

$$\mathbf{R}_e^{-1}\,\mathbf{K}_e\mathbf{R}_e\mathbf{g}'_e = \mathbf{R}_e^{-1}\mathbf{p}_e + \mathbf{R}_e^{-1}\mathbf{q}_e \tag{4.4}$$

or

$$\mathbf{K}'_e\mathbf{g}'_e = \mathbf{p}'_e \tag{4.5}$$

where

$\mathbf{K}'_e = \mathbf{R}_e^{-1}\mathbf{K}_e\mathbf{R}_e$ is the similarity transform of $\mathbf{K}_e$,
$\mathbf{p}'_e = \mathbf{R}_e^{-1}\mathbf{p}_e$ is the vector transformation of $\mathbf{p}_e$, and
$\mathbf{q}'_e = \mathbf{R}_e^{-1}\mathbf{q}_e$ is the transformed $\mathbf{q}_e$.

In the special case where the axes are rectangular Cartesian, the rotation matrix $\mathbf{R}$ is orthogonal; that is, if

$$\mathbf{R}_e^T = \mathbf{R}_e^{-1} \tag{4.6}$$

then $\mathbf{K}'_e$ must be symmetric because $\mathbf{K}_e$ is symmetric. To ensure this symmetry and its consequent benefits to data processing efficiency, we will insist on the general use of rectangular Cartesian axes.

Because changing the axes to which equations are referenced cannot change the intrinsic properties of the equations, it has been proved that the coordinate transformation preserves the order, rank, and degeneracy of $\mathbf{K}_e$ in $\mathbf{K}'_e$. It preserves the definiteness (and eigenvalues) of $\mathbf{K}_e$ in $\mathbf{K}'_e$. The length and spatial direction of the $\mathbf{p}_e$ vector is preserved in $\mathbf{p}'_e$, and $\mathbf{q}_e$ in $\mathbf{q}'_e$.

4.9 DERIVING MACRO MODELS FROM TEST DATA

If a structure is idealizable as a lattice, we can deduce the coefficients of the element macro equations from physical measurements. If unbraced elements are appropriate, nondestructive tests will furnish the necessary data. When braced elements are necessary, measurement of internal forces is required.

Table 4.12 illustrates the procedure for deducing the element stiffness matrix from tests on a single unbraced element. We suppose that we measure the influ-

TABLE 4.12 DEDUCTION OF UNBRACED ELEMENT STIFFNESS COEFFICIENTS

Given: The influence matrix for a cantilevered beam, as shown.
Find: The coefficients of the element stiffness matrix.

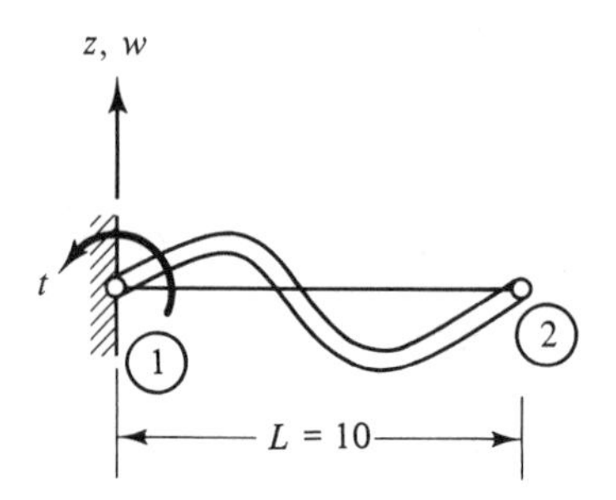

$$\begin{pmatrix} w_2 \\ t_2 \end{pmatrix} \begin{bmatrix} 2.0 & 0.3 \\ 0.3 & 0.06 \end{bmatrix} \begin{pmatrix} W_2 \\ T_2 \end{pmatrix} \tag{a}$$

Consider the partitioned stiffness matrix

$$\begin{bmatrix} \mathbf{K}_{11} & \mathbf{K}_{12} \\ \mathbf{K}_{21} & \mathbf{K}_{22} \end{bmatrix} \begin{pmatrix} w_1 \\ t_1 \\ w_2 \\ t_2 \end{pmatrix} = \begin{pmatrix} W_1 \\ T_1 \\ W_2 \\ T_2 \end{pmatrix} \tag{b}$$

1. We find $\mathbf{K}_{22}$, by inverting the relationship of Eqs. (a):

$$\begin{bmatrix} 2 & -10 \\ -10 & \frac{200}{3} \end{bmatrix} \begin{pmatrix} w_2 \\ t_2 \end{pmatrix} = \mathbf{K}_{22} \begin{pmatrix} w_2 \\ t_2 \end{pmatrix} = \begin{pmatrix} W_2 \\ T_2 \end{pmatrix} \tag{c}$$

2. We find the coefficients of $\mathbf{K}_{12}$ so that the forces and moments in the last two column vectors are in equilibrium, thus

$$\mathbf{K}_{12} = \begin{bmatrix} -2 & 10 \\ -10 & \frac{100}{3} \end{bmatrix} \tag{d}$$

3. We find $\mathbf{K}_{21}$ by requiring symmetry of the stiffness matrix; then

$$\mathbf{K}_{21} = \mathbf{K}_{12}^T = \begin{bmatrix} -2 & -10 \\ 10 & \frac{100}{3} \end{bmatrix} \tag{e}$$

4. We find $\mathbf{K}_{11}$ coefficients, so each of the first two column vectors satisfy force and moment equilibrium. Therefore,

$$\mathbf{K}_{11} = \begin{bmatrix} 2 & 10 \\ 10 & \frac{200}{3} \end{bmatrix} \tag{f}$$

ence coefficients for the physical system, thereby particularizing the matrix on the right-hand side of Eq. (a).

Consider the beam matrix partitioned by node:

$$\begin{bmatrix} \mathbf{K}_{11} & \mathbf{K}_{12} \\ \mathbf{K}_{21} & \mathbf{K}_{22} \end{bmatrix} \begin{bmatrix} \mathbf{g}_1 \\ \mathbf{g}_2 \end{bmatrix} = \begin{bmatrix} \mathbf{p}_1 \\ \mathbf{p}_2 \end{bmatrix} \tag{4.7}$$

where

$\mathbf{K}_{11}$, $\mathbf{K}_{12}$, $\mathbf{K}_{21}$, and $\mathbf{K}_{22}$ are two-by-two matrices, and

$\mathbf{g}_1$, $\mathbf{g}_2$, $\mathbf{p}_1$, and $\mathbf{p}_2$ are two-by-one vectors affiliated with nodes 1 and 2.

Then the tests, with node 1 fixed, establish the coefficients of $\mathbf{A}$ of

$$\mathbf{K}_{22}\mathbf{g}_1 = \mathbf{A}_{22}\mathbf{p}_1 \tag{4.8}$$

where $\mathbf{A}$ is the influence matrix.

So $\mathbf{K}_{22} = \mathbf{A}_{22}^{-1}$. Given $\mathbf{K}_{22}$, $\mathbf{K}_{12}$ is constructed so that columns of the stiffness matrix relating to node 2 satisfy equilibrium, that is,

$$\mathbf{K}_{12} = \mathbf{C}\,\mathbf{K}_{22} \tag{4.9}$$

where the $\mathbf{C}$ coefficients ensure equilibrium of nodal forces and moments. Thus the $\mathbf{C}$ coefficients depend only on the distance between nodes 1 and 2.

Continuing the development, we find $\mathbf{K}_{21}$ by symmetry, that is, by

$$\mathbf{K}_{21} = \mathbf{K}_{12}^T \tag{4.10}$$

and find $\mathbf{K}_{11}$ by the equilibrium transformation,

$$\mathbf{K}_{11} = \mathbf{C}\,\mathbf{K}_{21} = \mathbf{C}\,\mathbf{K}_{22}\mathbf{C}^T \tag{4.11}$$

Eqs. (4.11) imply few analysis assumptions. The process can be applied without change to elements with curved axes, to nonprismatic members, and to elements formed of inhomogeneous materials.

This process enables deduction of unbraced element stiffness coefficients and equivalent loading for any structure that behaves like a lattice. Assume that influence coefficients for the structure have been measured for the nodes of interest. Inverting this matrix produces the constrained system stiffness matrix. If the lattice idealization is appropriate, the inverse will be a sparsely populated matrix. Each of the off-diagonal partitions of this matrix can be used as values of an element's $\mathbf{K}_{12}$ coefficients to deduce element stiffness matrices for most of the structure's elements. The $\mathbf{K}_{11}$ or $\mathbf{K}_{22}$ partition coefficients of any element not yet identified can usually be established by subtracting diagonal partitions of elements of known coefficients. If this is impossible, adding nodes and measuring additional influence coefficients can always make it possible.

The process extends easily to encompass deduction of equivalent loads. Then measurement of influence coefficients for interior point loads provides the basic data for deducing equivalent loading values.

We observe that coefficients in corresponding rows and columns of the $\mathbf{K}_{11}$ and $\mathbf{K}_{22}$ partitions are equal in magnitude. This is true regardless of whether the element is prismatic or irregular, has a straight or curved axis, or is formed of inhomogeneous materials. Hence it is true for every element of Tables 4.1 and 4.2. Since proof of this property depends on the uniqueness of solutions in the linear structural theory, we term this property "uniqueness symmetry."

Figure 4.5(A) illustrates the geometry of a beam that associates with the first column of a beam element's stiffness matrix. The deformed geometry is that of a beam cantilevered at node 2 and loaded with a force and couple at node 1. The load components are such that there is no rotation at node 1, only a unit lateral displacement.

Figure 4.5(B) shows the geometry of the beam associated with the third column of the beam element's stiffness matrix when the node 2 lateral deflection is -1. In this case, the beam is cantilevered at node 1 and loaded at node 2 so no rotation and unit lateral displacement exists at node 1.

But the beams of Fig. 4.5 must have a common deformed shape. Both are fixed at an end and have a relative lateral displacement of 1 between the ends. Therefore, if a linear differential equation defines the deformed geometries, the solution is unique and the displacement shapes must match all along the span.

Since uniqueness symmetry depends only on linearity of the differential equation, it applies to braced as well as unbraced models. For braced models, its use reduces the number of measurements needed to deduce element stiffness coefficients or assists in validating element models.

Assume that the beam element of Table 4.12 is braced by supports lying between the ends. We require measurements to deduce coefficients of the $\mathbf{K}_{11}$

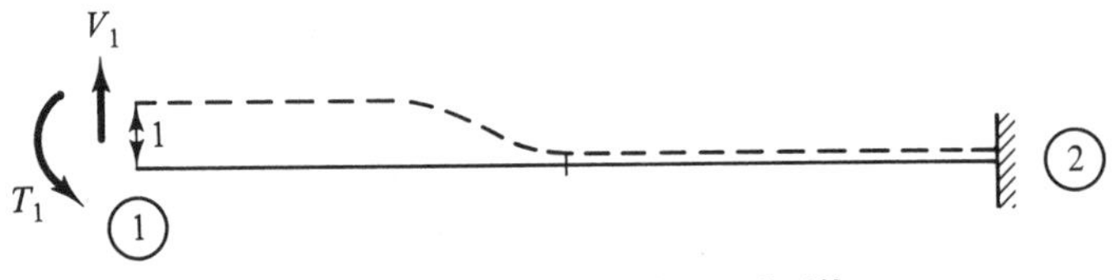

(A) Deflected shape, first column of stiffnesses

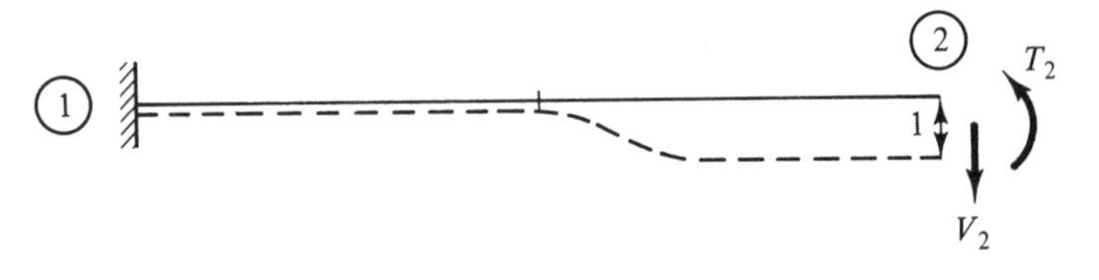

(B) Deflected shape, third column of stiffnesses

Figure 4.5 Deflected beam shapes.

and $\mathbf{K}_{12}$ or $\mathbf{K}_{21}$ and $\mathbf{K}_{22}$ partitions to establish the stiffness model. Given either of these data sets, matrix and uniqueness symmetry will serve to fill in missing coefficients. We need to perform $n_n(n_n + 1)/2 + n_n^2$ measurements to identify the stiffnesses, where n_n is the number of displacement variables at a node.

For a single braced element, we can measure enough coefficients without invading the structure only if we can make measurements for more than one set of boundary conditions. If we are limited to one set, we must measure reactions or internal forces in the element in addition to influence coefficients.

Similarly, if we have a structure suitable for idealization as a collection of braced lattice elements, displacement influence data will not be sufficient to deduce element stiffnesses.

4.10 FORMULATING LATTICE ELEMENT MODELS

The source of all the lattice element models of Tables 4.1 and 4.2 is the structural component theory created in the late 1800s and early 1900s. We can regard this theory as derived from the theory of elasticity by making additional assumptions and approximations, though historically much of it preceded elasticity. These assumptions and approximations reduce the equations of elasticity to uniaxial linear differential equations relating generalized displacements to applied loads. Using exact solutions of these differential equations invariably leads to element macro models with the lattice property.

Accordingly, the means for formulating lattice models can be simply by superimposing solutions of the structural component differential equations consistent with the interpretation of coefficients in a column of the element stiffness matrix. For unbraced elements, use of the equilibrium conditions reduces the algebra. In the case of braced element models, formulation is simplified by exploiting uniqueness symmetry as well as matrix symmetry.

Table 4.13 presents essential data for developing lattice models for a circular arch segment in bending and a beam on an elastic foundation. The stiffness models of Tables 4.1 through 4.3 are given in explicit form. By contrast, Table 4.13 presents data that, when supported by the procedures of Sec. 4.9, provides the basis for computer development of additional lattice models. Though development of models on the computer takes more calculations than evaluation of formulas for the coefficients, the procedural form facilitates a far richer lattice library than the explicit form.

The approximations used to devolve the lattice differential equations are venerable. They have been well tested by applications to real structures and are well accepted by the profession. Thus the analyst can expect additions to the lattice element library primarily for the purpose of dealing with variable geometry and material properties within the context of the classical structural component theory. Be extremely wary of elements based on new component theory.

TABLE 4.13 PROCEDURAL BASIS FOR SOME LATTICE ELEMENTS

Element

Circular beam

$$\begin{pmatrix} u_1 \\ v_1 \\ s_1 \end{pmatrix} = \frac{1}{EI}\begin{bmatrix} r^2(\frac{1}{2} - \cos\theta - \sin^2\theta + \frac{1}{2}\cos^2\theta + \theta\sin\theta\cos\theta) & & \text{Sym.} \\ r^2(\frac{\pi}{2} - 2\sin\theta + \frac{\theta}{2} + \theta\sin^2\theta + \frac{3}{2}\sin\theta\cos\theta) & r^2(\frac{1}{2} - \cos\theta - \sin^2\theta + \frac{1}{2}\cos^2\theta - \theta\sin\theta\cos\theta) & \\ r^2(1 - \theta\sin\theta - \cos\theta) & r^2(\sin\theta - \theta\cos\theta) & r\theta \end{bmatrix}\begin{pmatrix} U_1 \\ V_1 \\ S_1 \end{pmatrix}$$

Beam on elastic foundation

1. Solve for a, b, c, d from

$$\begin{bmatrix} 1 & 0 & 1 & 0 \\ 1 & 1 & -1 & 1 \\ -2\beta^2 e^{\beta L}\sin\beta L & 2\beta^2 e^{\beta L}\cos\beta L & 2\beta^2 e^{-\beta L}\sin\beta L & -2\beta^2 e^{-\beta L}\cos\beta L \\ -2\beta^3 e^{\beta L}(\sin\beta L + \cos\beta L) & 2\beta^3 e^{\beta L}(\cos\beta L - \sin\beta L) & -2\beta^3 e^{-\beta L}(\sin\beta L + \cos\beta L) & 2\beta^3 e^{\beta L}(\sin\beta L - \cos\beta L) \end{bmatrix}\begin{pmatrix} a \\ b \\ c \\ d \end{pmatrix} = \begin{pmatrix} 0 \\ 0 \\ S_2 \\ V_2 \end{pmatrix}$$

$\beta = \sqrt{k/(4EI)}$
k = foundation modulus, force/length

2. Solve for generalized displacements from

$$v_1 = e^{\beta L}(a\cos\beta L + b\sin\beta L) + e^{-\beta L}(c\cos\beta L + d\sin\beta L)$$
$$s_1 = \beta e^{\beta L}[a(\cos\beta L - \sin\beta L) + b(\sin\beta L + \cos\beta L)] + \beta e^{-\beta L}[-c(\cos\beta L + \sin\beta L) - d(\sin\beta L - \cos\beta L)]$$

3. Solve for support reactions from

$$S_1 = 2\beta^2(b - d)$$
$$V_1 = 2\beta^3(b - a) + 2\beta^3(c + d)$$

4. Deduce $\mathbf{K}e$ using symmetry.

4.11 DISCRETIZATION OF LATTICE STRUCTURES

In theory, any lattice structure can be modeled without discretization error by a single lattice element. In practice, a given computer program includes logic for a limited number of lattice models with limited generality. These limitations force discretization errors upon the practitioner.

For example, suppose that we wish to predict the deflection at the center of a tapered concrete parabolic arch when loaded by its own weight. Assume that the computer code supports all the lattice element models of this chapter. Since none of the models represents a parabolically curved beam or deals with this taper, we could not model the structure as a lattice. Since the necessary lattice macro model would not be available, we would have to approximate the equivalent loading as well. We might approximate the structure using a stepped arch model with straight element segments, but the estimate of behavior would vary with the number and location of nodes of the discretized model.

If we are satisfied that the structural component theory is adequate for the members of a structure, analysis reliability, accuracy, and economy advocate lattice analysis. A computer configuration with the widest variety of lattice element models will support the most discretization error–free analyses and is therefore an excellent investment.

4.12 HOMEWORK

***4:1.** Static condensation.

Given: The following stiffness equations for the braced lattice structure of the figure shown.

① ② ③

$$\gamma\begin{bmatrix} 12 & 5 & 3 \\ 5 & 8 & 6 \\ 3 & 6 & 8 \end{bmatrix}\begin{pmatrix} u_1 \\ u_2 \\ u_3 \end{pmatrix} = \begin{pmatrix} p_1 \\ p_2 \\ p_3 \end{pmatrix}$$

Find: The exact representation of stiffness expressed only in terms of the u displacements of the exterior nodes.

Method: Selective Gauss elimination.

4:2. Matrix interpretation of static condensation.

Given: Problem 4.1 and its solution.

Find: The transformation matrix $\mathbf{C}$ such that

$$\lfloor p_1\ p_2\ p_3 \rfloor\,\mathbf{C} = \lfloor p_{1'}\ p_{3'} \rfloor$$

and the contragradient constraint of the equations given for Prob. 4:1 produces the reduced equations of Prob. 4:1.

4:3. Lattice element subdivisibility.

Given: The following element macro equations for a lattice element.

$$L^n \begin{bmatrix} 1 & -1 \\ -1 & 1 \end{bmatrix} \begin{pmatrix} g_1 \\ g_2 \end{pmatrix} = \begin{pmatrix} p_1 \\ p_2 \end{pmatrix}$$

Do the following:

(a) Find a value of n that associates with the lattice property.

(b) Prove that your solution is unique.

***4:4.** Equivalent loads.

Given: Formulas for the beam reactions of the figure.

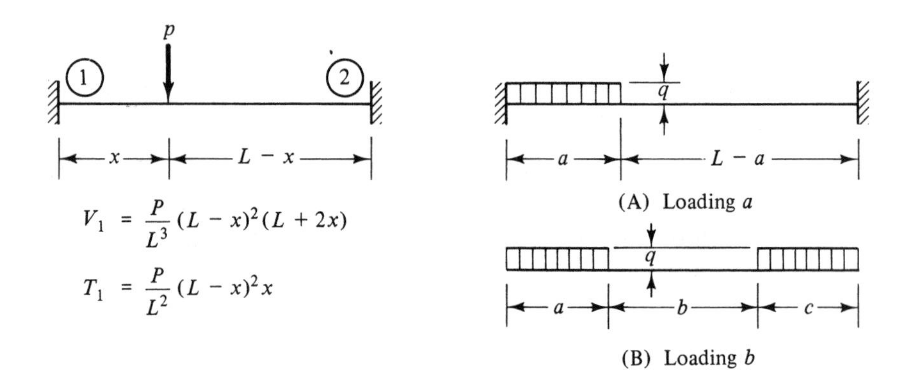

Homework Problem 4:4

Do the following:

(a) Find the equivalent loads for node 1 of the beam loading of part (B) of the figure.

(b) Find the equivalent vertical force for node 1 of the beam loading of part (B) of the figure.

***4:5.** Rank of truss stiffness matrices.

Given: The trusses as sketched here.

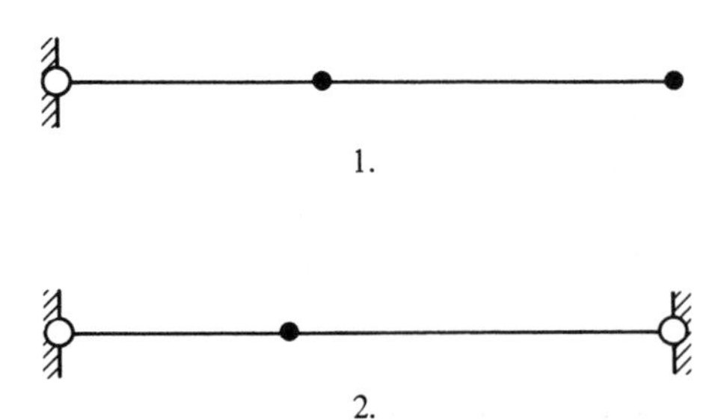

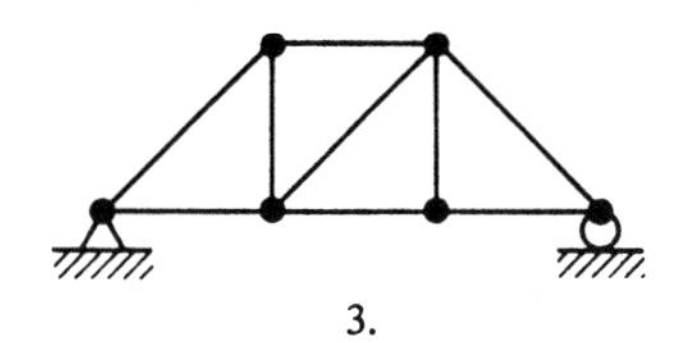

3.

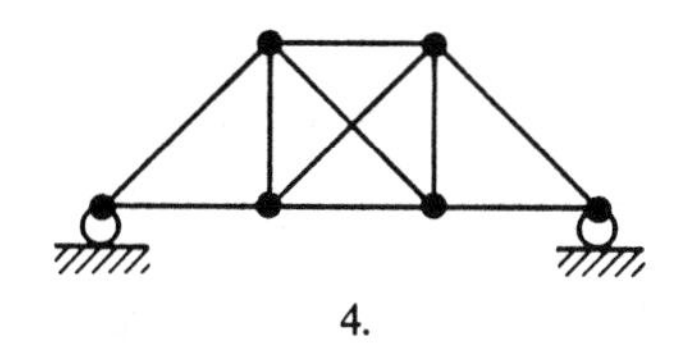

4.

Homework Problem 4:5

Determine for each case:

(a) The order of the constrained system stiffness matrix $\mathbf{K}_c$.

(b) The maximum rank of $\mathbf{K}_c$, that is, the rank of the element stiffness matrix times the number of elements.

(c) The actual rank of $\mathbf{K}_c$.

(d) The source of degeneracy, u and/or v.

4:6. Rank of frame stiffness matrices.

Given: The frame element systems (beam + rod stiffnesses) of the figure.

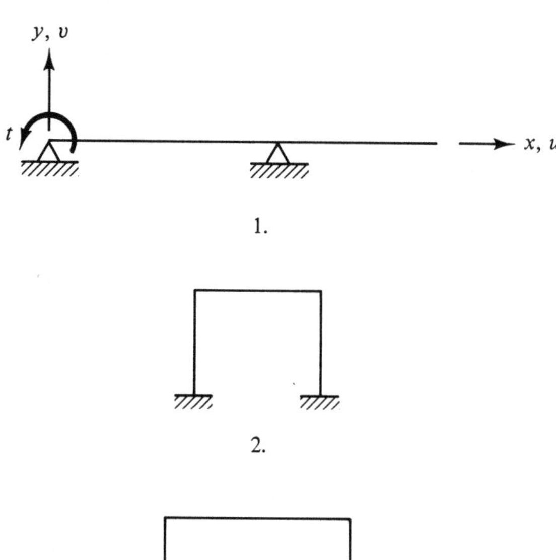

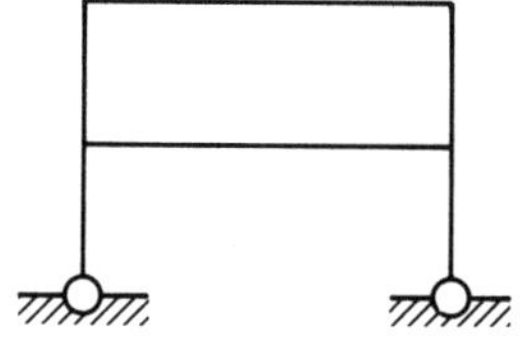

3.

(Continued)

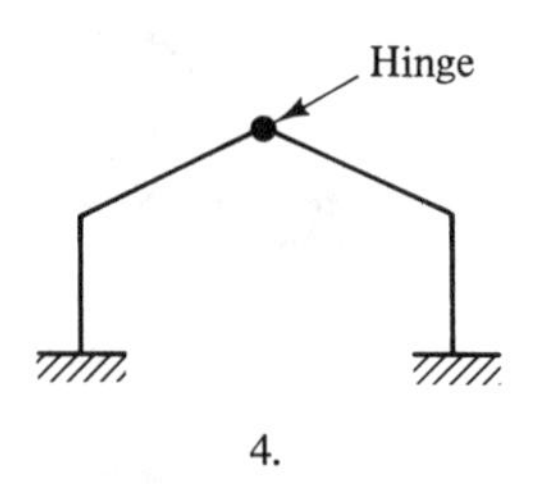

4.

Homework Problem 4:6

Determine for each case:

(a) The order of $\mathbf{K}_c$.

(b) The maximum rank of $\mathbf{K}_c$, that is, the rank of the element stiffness matrix times the number of elements.

(c) The actual rank of $\mathbf{K}_c$.

(d) The source of degeneracy, u, v, and/or t.

***4:7.** Beam stiffness matrix expansion.

Given: The following beam element stiffness equations.

$$\frac{2EI}{L^3}\begin{bmatrix} 6 & & \text{Sym.} & \\ 3L & 2L^2 & & \\ -6 & -3L & 6 & \\ 3L & L^2 & -3L & 2L^2 \end{bmatrix}\begin{pmatrix} v_1 \\ t_1 \\ v_2 \\ t_2 \end{pmatrix} = \begin{pmatrix} pv_1 \\ pt_1 \\ pv_2 \\ pm_2 \end{pmatrix}$$

Do the following:

(a) Particularize the coefficients of the constraint matrix that defines v_1, t_1, v_2, and t_2 in terms of u_1, v_1, t_1, u_2, v_2, and t_2.

(b) Perform the congruent transformation of the beam stiffness matrix to find the expanded stiffness matrix.

4:8. Expanded stiffness matrices.

Given: The rod and beam stiffness matrices of Tables 4.1 and 4.2.

Do the following:

(a) Find the bonded-rod stiffness matrix when expanded to three-dimensional space.

(b) Find the Timoshenko beam stiffness matrix when expanded to three-dimensional space.

4:9. Orthogonality of rotation matrix.

Given: The planar rotation matrix $\mathbf{R}$ of Eqs. (a) of Table 4.11.

Prove: $\mathbf{R}$ is orthogonal regardless of the value of the rotation angle.

4:10. Similarity transformation of rotation.

Given: The rod element stiffness matrix of the rod sketched here.

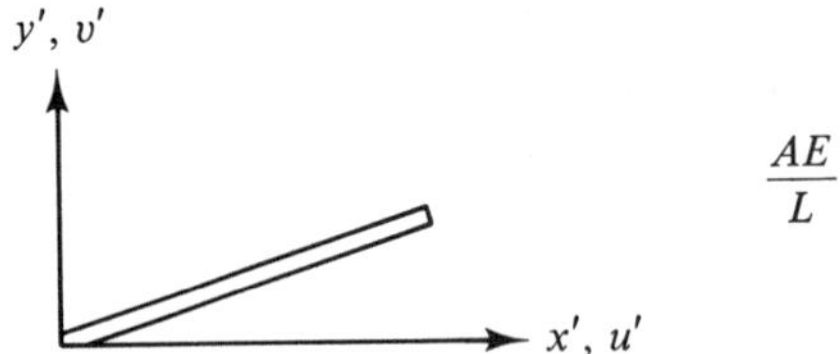

$$\frac{AE}{L}\begin{bmatrix} 0.36 & 0.48 & -0.36 & -0.48 \\ 0.48 & 0.64 & -0.48 & -0.64 \\ -0.36 & -0.48 & 0.36 & 0.48 \\ -0.48 & -0.64 & 0.48 & 0.64 \end{bmatrix}\begin{pmatrix} u_{1'} \\ v_{1'} \\ u_{2'} \\ v_{2'} \end{pmatrix} = \mathbf{P}$$

Homework Problem 4:10

Find: The coefficients of the stiffness matrix when the reference axes are rotated such that the cosine of the angle of rotation is 0.8 and the sine is 0.6.

***4:11.** Cogradient transformation.

Given: The torque tube stiffness equations of Table 4.1.

Find: The stiffness equations for the tube expanded to two-dimensional space when the x axis is aligned with the tube axis at node 1, but the x axis is rotated counterclockwise 30 degrees from the tube axis at node 2, as suggested by the following sketch.

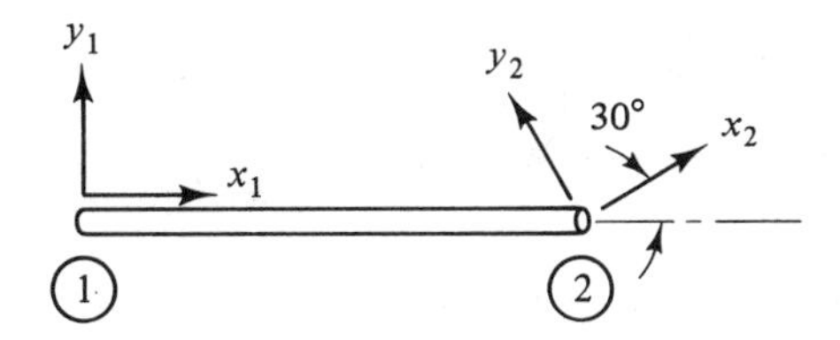

Homework Problem 4:11

4:12. Development of $\mathbf{K}_e$ for an unbraced beam

Given: The deflection influence coefficients measured for the cranked beam of the figure.

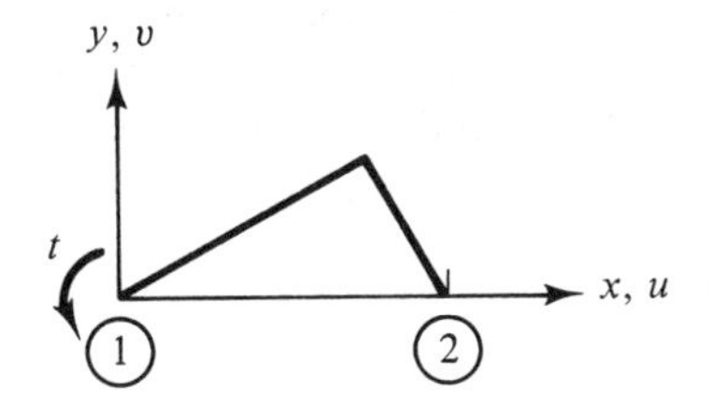

$$\begin{pmatrix} v_1 \\ t_1 \end{pmatrix} = \begin{bmatrix} 3 & 3 \\ 3 & 6 \end{bmatrix} \begin{pmatrix} V_1 \\ T_1 \end{pmatrix}$$

Homework Problem 4:12

Find: The beam's stiffness matrix.

4:13. Formulation of a lattice element stiffness matrix.

Given: The tapered rod of Table 4.1.

Show: That the stiffness matrix of the table is correct.

Method: Superposition of exact solutions of the rod complying with direct interpretation of each stiffness coefficient.

4.13 COMPUTERWORK

4-1. Direct validation of a program for truss analysis.
Given: The truss of the figure.

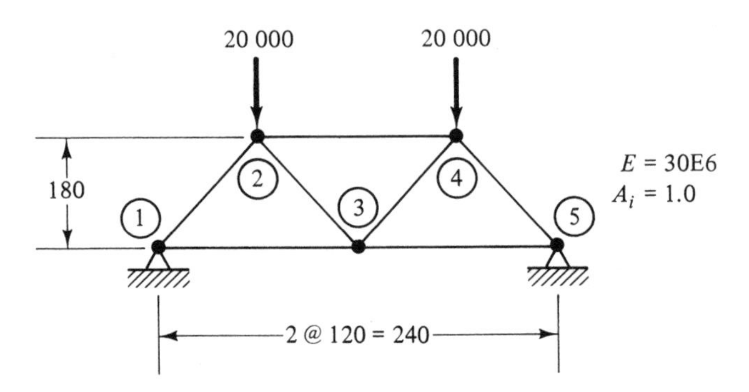

Computerwork Problem 4-1

Do the following:
(a) Determine the forces in the rods without using your computer program.
(b) As a partial validation, determine the forces in the rods using your program and compare with your independently calculated ones.

4-2. Direct validation of a program for frame analysis.
Given: The bent of the figure.

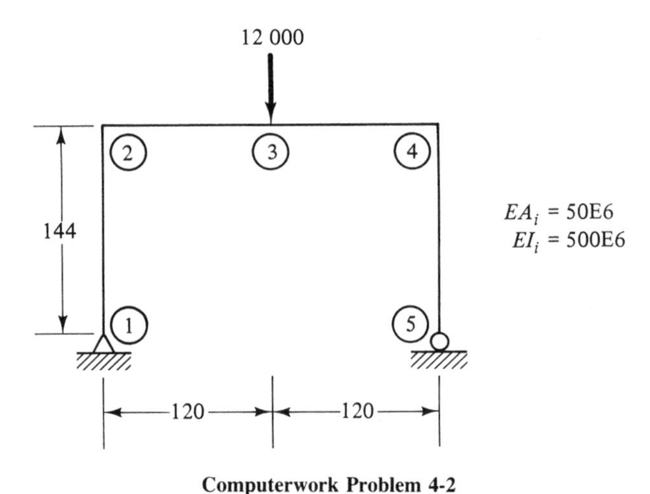

Computerwork Problem 4-2

Do the following:
(a) Find the generalized internal forces (axial force, bending moment, and shear force) at the higher-node-number end of each beam without using the computer code.

(b) Solve the problem using the code and compare generalized forces with your independently calculated ones.

4-3. Beam element numerical subdivisibility test.

Given: The beam problem of part (A) of the figure.

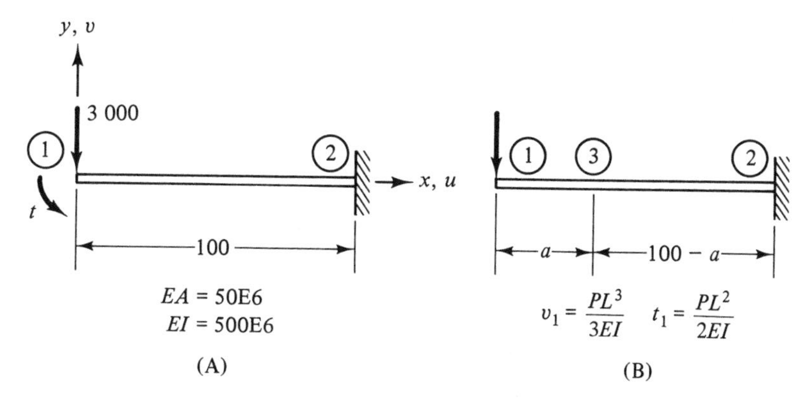

Computerwork Problem 4-3

Do the following:

(a) Choose the location of node 3 somewhere between nodes 1 and 2 and run the problem with the grid of part (B) of the figure.

(b) Choose a different location of node 3 and rerun the problem of part (B).

(c) Compare the results of parts (A) and (B) to complete a self-consistency test of the beam model of the computer code.

(d) Compare tip deflection and rotation of part (A) with their values using the formulas, to perform a direct validation test of the code.

***4-4.** Beam element rank test.

Given: The beam element of the figure.

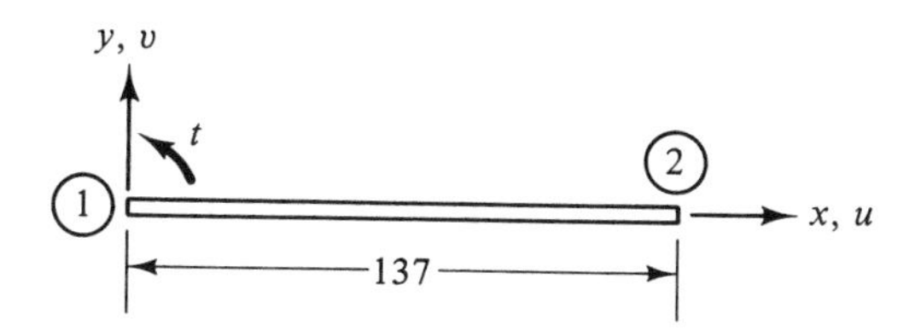

Computerwork Problem 4-4

Find: The rank of the stiffness matrix by using the code for the following kinematic boundary condition cases: $u_1 = 0$; $u_1 = t_1 = 0$; $u_1 = t_1 = t_2 = 0$; $u_1 = u_2 = v_1 = 0$; $u_1 = v_1 = v_2 = 0$.

4-5. Beam equivalent loads test.

Given: The equivalent load formulas of part (A) of the figure.

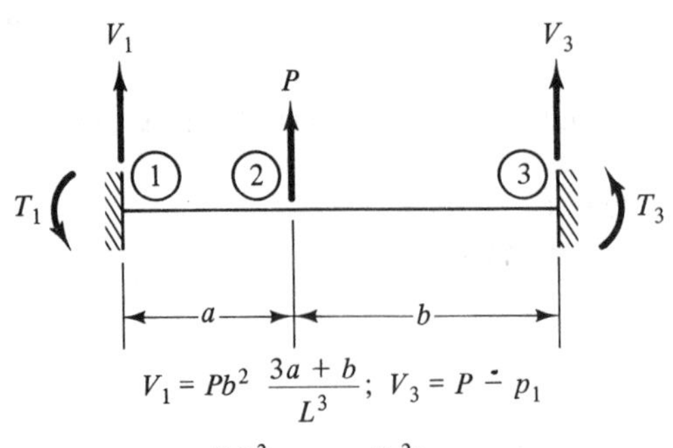

$$V_1 = Pb^2 \frac{3a + b}{L^3}; \; V_3 = P - p_1$$

$$T_1 = \frac{Pab^2}{L^2}; \; T_3 = \frac{Pa^2 b}{L^2}$$

(A) Loads formulas

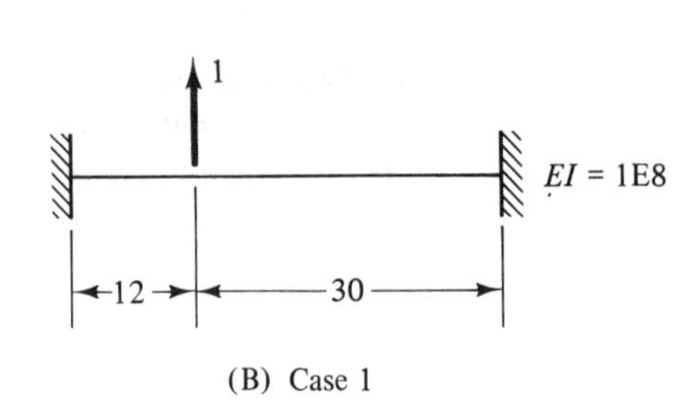

(B) Case 1

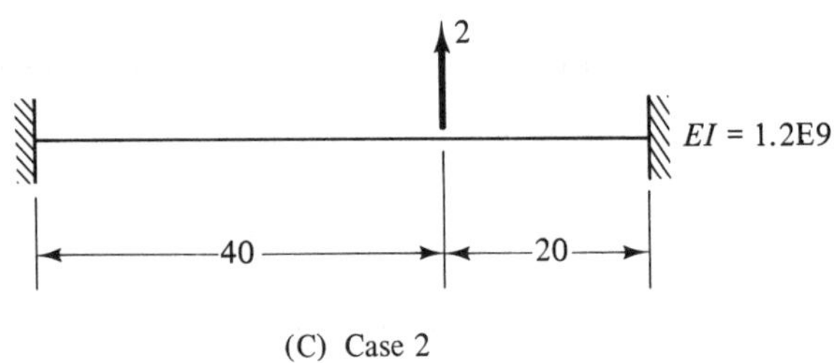

(C) Case 2

Computerwork Problem 4-5

Do the following:

(a) Run the problems of parts (B) and (C) of the figure.

(b) Show that the reactions of parts (B) and (C) runs agree with those you calculate using the formulas of part (A).

4-6. Loads equivalent to multiple loads.

Given: The real loading on the beam of part (A) of the figure.

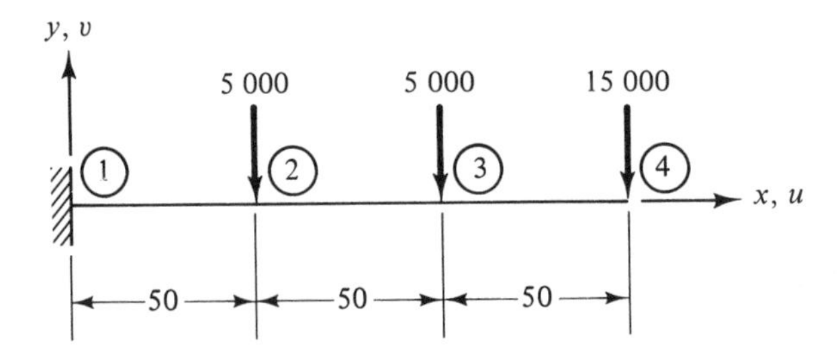

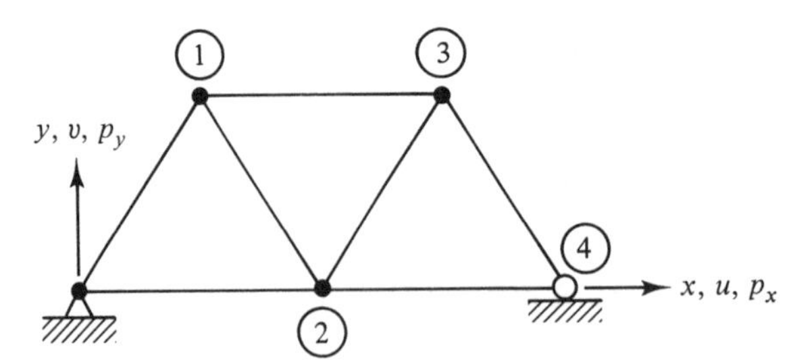

Computerwork Problem 4-6

Find:

(a) The tip displacements and the support reactions for the part (A) beam.

(b) The equivalent loading at the tip of the cantilever, using the computer code.

(c) The tip displacements and support reactions using the equivalent loading of activity (b).

4-7. Secondary behavior of a truss.

Given: The measured influence coefficients for the described structure, relating u and v displacements of successive nodes to U and V forces at the nodes. That is, the flexibility matrix $\mathbf{A} = \mathbf{K}_c^{-1}$ is given for the equations $\mathbf{g} = \mathbf{Ap}$.

	p_{x1}	p_{y1}	p_{x2}	p_{y2}	p_{x3}	p_{y3}	p_{x4}
u_1	22.7254	−4.3750	7.5000	−5.0000	17.7254	−3.1250	10.0000
v_1	−4.3750	12.6691	−3.7500	9.4877	−1.8750	5.0564	−5.0000
u_2	7.5000	−3.7500	10.0000	−2.5000	7.5000	−1.2500	10.0000
v_2	−5.0000	9.4877	−2.5000	17.7254	0.0000	9.4877	−5.0000
u_3	17.7254	−1.8750	7.5000	0.0000	22.7254	−0.6250	10.0000
v_3	−3.1250	5.0564	−1.2500	9.4877	−0.6250	12.6691	−5.0000
u_4	10.0000	−5.0000	10.0000	−5.0000	10.0000	−5.0000	20.0000

Determine: If the structure is satisfactorily modeled as pinned at the joints, that is, if joint moments are negligible.

Method: Find the stiffness matrix, and interpret the sparsity pattern. Given a finite element analysis code, the solution can be found by creating a finite element model whose stiffness matrix will be the influence matrix given. Then, using a set of influence loading vectors, the code will produce the inverse of the **A** matrix. The EASE input below defines the structural model whose system stiffness matrix is the given **A** matrix.

Input Data For Equivalent Model

Element Data		Nodal Data		
Element	AE	Node	x	y
1, 2	−4.3750	1	0	0
1, 3	15.0000	2	1.0	0
1, 4	−15.0000	3	2.0	0
1, 5	70.9016	4	3.0	0
1, 6	−16.6250	5	4.0	0
1, 7	60.0000	6	5.0	0
2, 3	−3.7500	7	6.0	0
2, 4	18.9754	8	1.0	0
2, 5	−5.6250	9	2.0	0
2, 6	20.2256	10	3.0	0
2, 7	−25.0000	11	4.0	0
3, 4	−2.5000	12	5.0	0

(Continued)

Input Data For Equivalent Model

Element Data		Nodal Data		
Element	AE	Node	x	y
3, 5	15.0000	13	6.0	0
3, 6	−3.7500	14	7.0	0
3, 7	40.0000			
4, 6	18.9754	Fixities: 1, 2; 2, 2; 3, 2; 4, 2; 5, 2; 6, 2; 7, 2; 8, 1; 8, 2; 9, 1; 9, 2; 10, 1; 10, 2; 11, 1; 11, 2; 12, 1; 12, 2; 13, 1; 13, 2; 14, 1; −14, 2		
4, 7	−15.0000			
5, 6	−0.6250			
5, 7	20.0000			
6, 7	−5.0000			
2, 9	13.1250			
3, 10	−7.5000			
4, 11	11.2500	Loads: Use influence loadings of the form (0, 0, . . . , 0, 1, 0, . . . , 0). Thus the first loading would be (1, 0, 0, . . . , 0). This loading would be defined for the computer by the input −1, 1, 0.		
5, 12	−10.0000			
6, 13	8.1250			
−7, 14	5.0000			

4.14 REFERENCES

1. Timoshenko, S., *Strength of Materials*, New York: Van Nostrand, 1941, pp. 209–211.
2. Roark, R. J.; Young, W. C., *Formulas for Stress and Strain*, 5th ed., New York: McGraw-Hill, 1975.

Chapter 5

Solving System Equations

Equation solving is an essential step that consumes more computer resources than any other step of FEA. The choice of the equation-solving process is the key to establishing data processing efficiency because the step takes more calculations and data transfers than element model generation, the finite element process, or stress evaluation. The role of equation solving becomes increasingly more dominant in implementation, as the number of simultaneous structural equations increases, until it becomes the only important concern in analysis efficiency and computer cost.

The primary mathematical objective in linear static analysis is to solve the linear simultaneous equations

$$\mathbf{K}_c \mathbf{g}_c = \mathbf{p}_c \tag{5.1}$$

where

$\mathbf{K}_c$ is the symmetric positive definite constrained system stiffness matrix and

$\mathbf{p}_c$, the loading vector, includes the equivalent loading.

The loading vector $\mathbf{p}_c$ may be a single vector, but usually the integrity of the structure must be assayed for many vectors. Building codes require considering 5 to 12 independent loadings; Bridge, ship, and railroad structures, 8 to 16 loadings; and highway vehicles and manned space vehicles, 12 to 22. Aircraft integrity analyses may require 40 to 80 loadings depending on the flight missions.

The secondary mathematical objective in linear static analysis is to assess the definiteness of the stiffness matrix and characterize nonpositiveness. This assessment is necessary to ensure that the problem is well posed. If the stiffness matrix is indefinite, the engineer needs additional information to suggest the input or computer error. If the matrix is singular, extra information will assist in establishing appropriate corrective action to eliminate the singularity.

The computer science problem of FEA is the problem of effecting equation solution using a minimum of computer resources. This problem requires exploiting the symmetry and sparseness of the stiffness matrix to reduce calculations, data transfers, and storage needs.

The engineer's responsibilities in equation solving include validating each equation solver to ensure that it produces accurate solutions over the spectrum of problems to be addressed. In solving a particular problem, the engineer's goal is to select the computer configuration and solution process, within the program, that is most appropriate for the analysis purpose and to verify equation-solving results.

The mathematical algorithm for solving Eqs. (5.1) may be direct, iterative, or N-step iterative. For an algorithm of the direct class, the number of calculations to evaluate the solution is determinable before calculations are initiated. For an iterative algorithm, the number of calculations depends on the values of the numbers of the problem and the required solution accuracy. The number of calculations cannot be determined *ab initio*. For an N-step iterative process, where N is the number of equations, it is guaranteed that no more than N iteration cycles will be required to obtain the solution.

Our examination of equation solving will be from the perspective of the engineer who directs the finite element analysis. It will focus on solution efficiency and validation of the solver.

5.1 TRIANGULAR FACTORIZATION: A DIRECT METHOD

Triangular factorization is the most popular choice of solution algorithm among computer program developers. It is supported by all FEA production computer codes. It requires the least number of calculations of all direct methods.[1] It attains the secondary mathematical objective with a negligible increase in calculations over the number to develop the equation solution.

The solution of the equations requires implementing three sequential steps: factorization, forward substitution, and back substitution. In the first step, the stiffness matrix is resolved into three factors:

$$\mathbf{L}\;\mathbf{D}\;\mathbf{L}^T = \mathbf{K}_c \tag{5.2}$$

where

L is a matrix that may have nonzero terms only on or below the diagonal, a lower triangular matrix, and has 1.0 for each diagonal coefficient,

$\mathbf{D}$ is a diagonal matrix of D_{ii} coefficients, and
$\mathbf{L}^T$ is the transpose of $\mathbf{L}$.

Forward substitution involves solving the equations

$$\mathbf{L}\,\mathbf{d}_c = \mathbf{p}_c \tag{5.3}$$

for $\mathbf{d}_c$. This task is "forward substitution" because the unknown $\mathbf{d}_c$ components are evaluated in cardinal order, starting from the first equation. Back substitution involves solving for $\mathbf{g}_c$ by solving the equations

$$\mathbf{g}_c = \mathbf{D}\,\mathbf{L}^T\,\mathbf{d}_c \tag{5.4}$$

This is "back substitution" because least calculation evaluation of the $\mathbf{g}_c$ coefficients starts by solving for the last $\mathbf{g}_c$ component and proceeding in reverse cardinal order.

Once the triangular factors are available, only the calculations of forward and back substitution are necessary to find the solution for new loading vectors.

Gauss elimination provides a general approach to triangularization, using successive row operations on $\mathbf{K}_c$ to transform it to $\mathbf{D}\,\mathbf{L}^T$. Each $\mathbf{D}$ is then simply the coefficient on the diagonal of $\mathbf{D}\,\mathbf{L}^T$.

Table 5.1 illustrates the stages of triangularization of an eighth-order stiffness matrix with a semibandwidth of 2. Matrix 1 is the original constrained stiffness matrix. Matrices 2 through 8 are the results of row operations that lead to $\mathbf{D}\,\mathbf{L}^T$. Matrix 9 represents $\mathbf{L}^T$.

The row operations that transform matrix 1 to matrix 2 are chosen so that each coefficient below the first diagonal coefficient is replaced by a zero, thereby "decoupling" the first variable from all but the first equation. In the example, this is achieved by multiplying the first row by 1 and adding it to the second row and multiplying the first row by -2 and adding it to the third row to create matrix 2.

Similarly, we multiply the second row of matrix 2 by factors and add that row to successive rows to decouple the second variable from successive equations and create matrix 3.

This process is repeated to decouple each variable from successive equations leading to matrix 8. In matrix 4, the fourth variable becomes decoupled; in matrix 5, the fifth variable, and so on until the matrix becomes the upper triangular matrix in matrix 8.

The cumulative effect of the row operations is to multiply the stiffness matrix by $\mathbf{L}^{-1}$. Multiplying each row of the $\mathbf{D}\,\mathbf{L}^T$ representation, matrix 8, by the reciprocal of the diagonal coefficient exposes the matrix $\mathbf{L}^T$, except for null rows. When the row is null, the row of $\mathbf{L}^T$ has 1 as the diagonal and zeros elsewhere, as matrix 9 shows.

Many other sequences of coefficient replacement result in the same $\mathbf{L}^T$. Of these, it has been proved that none can require fewer calculations than the Gauss data processing sequence when the band is populated only with nonzero coefficients.[1] Thus triangularization with Gauss processing is especially attractive.

TABLE 5.1 SAMPLE FACTORIZATION

Matrix 1, **K**

$$\begin{bmatrix} 1 & -1 & 2 & 0 & 0 & 0 & 0 & 0 \\ -1 & 3 & 2 & -2 & 0 & 0 & 0 & 0 \\ 2 & 2 & 12 & -4 & 0 & 0 & 0 & 0 \\ 0 & -2 & -4 & 3 & -2 & 1 & 0 & 0 \\ 0 & 0 & 0 & -2 & 7 & -11 & 3 & 0 \\ 0 & 0 & 0 & 1 & -11 & 30 & -5 & 6 \\ 0 & 0 & 0 & 0 & 3 & -5 & 12 & 10 \\ 0 & 0 & 0 & 0 & 0 & 6 & 10 & 28 \end{bmatrix}$$

Matrix 2

$$\begin{bmatrix} 1 & -1 & 2 & 0 & 0 & 0 & 0 & 0 \\ 0 & 2 & 4 & -2 & 0 & 0 & 0 & 0 \\ 0 & 4 & 8 & -4 & 0 & 0 & 0 & 0 \\ 0 & -2 & -4 & 3 & -2 & 1 & 0 & 0 \\ 0 & 0 & 0 & -2 & 7 & -11 & 3 & 0 \\ 0 & 0 & 0 & 1 & -11 & 30 & -5 & 6 \\ 0 & 0 & 0 & 0 & 3 & -5 & 12 & 10 \\ 0 & 0 & 0 & 0 & 0 & 6 & 10 & 28 \end{bmatrix}$$

Matrix 3

$$\begin{bmatrix} 1 & -1 & 2 & 0 & 0 & 0 & 0 & 0 \\ 0 & 2 & 4 & -2 & 0 & 0 & 0 & 0 \\ 0 & 0 & 0 & 0 & 0 & 0 & 0 & 0 \\ 0 & 0 & 0 & 1 & -2 & 1 & 0 & 0 \\ 0 & 0 & 0 & -2 & 7 & -11 & 3 & 0 \\ 0 & 0 & 0 & 1 & -11 & 30 & -5 & 6 \\ 0 & 0 & 0 & 0 & 3 & -5 & 12 & 10 \\ 0 & 0 & 0 & 0 & 0 & 6 & 10 & 28 \end{bmatrix}$$

Matrix 4

$$\begin{bmatrix} 1 & -2 & 1 & 0 & 0 & 0 & 0 & 0 \\ 0 & 2 & 4 & -2 & 0 & 0 & 0 & 0 \\ 0 & 0 & 0 & 0 & 0 & 0 & 0 & 0 \\ 0 & 0 & 0 & 1 & -2 & 1 & 0 & 0 \\ 0 & 0 & 0 & -2 & 7 & -11 & 3 & 0 \\ 0 & 0 & 0 & 1 & -11 & 30 & -5 & 6 \\ 0 & 0 & 0 & 0 & 3 & -5 & 12 & 10 \\ 0 & 0 & 0 & 0 & 0 & 6 & 10 & 28 \end{bmatrix}$$

Matrix 5

$$\begin{bmatrix} 1 & -2 & 1 & 0 & 0 & 0 & 0 & 0 \\ 0 & 2 & 4 & -2 & 0 & 0 & 0 & 0 \\ 0 & 0 & 0 & 0 & 0 & 0 & 0 & 0 \\ 0 & 0 & 0 & 1 & -2 & 1 & 0 & 0 \\ 0 & 0 & 0 & 0 & 3 & -9 & 3 & 0 \\ 0 & 0 & 0 & 0 & -9 & 29 & -5 & 6 \\ 0 & 0 & 0 & 0 & 3 & -5 & 12 & 10 \\ 0 & 0 & 0 & 0 & 0 & 6 & 10 & 28 \end{bmatrix}$$

Matrix 6

$$\begin{bmatrix} 1 & -2 & 1 & 0 & 0 & 0 & 0 & 0 \\ 0 & 2 & 4 & -2 & 0 & 0 & 0 & 0 \\ 0 & 0 & 0 & 0 & 0 & 0 & 0 & 0 \\ 0 & 0 & 0 & 1 & -2 & 1 & 0 & 0 \\ 0 & 0 & 0 & 0 & 3 & -9 & 3 & 0 \\ 0 & 0 & 0 & 0 & 0 & 2 & 4 & 6 \\ 0 & 0 & 0 & 0 & 0 & 4 & 9 & 10 \\ 0 & 0 & 0 & 0 & 0 & 6 & 10 & 28 \end{bmatrix}$$

Matrix 7

$$\begin{bmatrix} 1 & -2 & 1 & 0 & 0 & 0 & 0 & 0 \\ 0 & 2 & 4 & -2 & 0 & 0 & 0 & 0 \\ 0 & 0 & 0 & 0 & 0 & 0 & 0 & 0 \\ 0 & 0 & 0 & 1 & -2 & 1 & 0 & 0 \\ 0 & 0 & 0 & 0 & 3 & -9 & 3 & 0 \\ 0 & 0 & 0 & 0 & 0 & 2 & 4 & 6 \\ 0 & 0 & 0 & 0 & 0 & 0 & 1 & -2 \\ 0 & 0 & 0 & 0 & 0 & 0 & -2 & 10 \end{bmatrix}$$

Matrix 8, $\mathbf{LDL}^T$

$$\begin{bmatrix} 1 & -2 & 1 & 0 & 0 & 0 & 0 & 0 \\ 0 & 2 & 4 & -2 & 0 & 0 & 0 & 0 \\ 0 & 0 & 0 & 0 & 0 & 0 & 0 & 0 \\ 0 & 0 & 0 & 1 & -2 & 1 & 0 & 0 \\ 0 & 0 & 0 & 0 & 3 & -9 & 3 & 0 \\ 0 & 0 & 0 & 0 & 0 & 2 & 4 & 6 \\ 0 & 0 & 0 & 0 & 0 & 0 & 1 & -2 \\ 0 & 0 & 0 & 0 & 0 & 0 & 0 & 6 \end{bmatrix}$$

Matrix 9, $\mathbf{L}^T$

$$\begin{bmatrix} 1 & -2 & 1 & 0 & 0 & 0 & 0 & 0 \\ 0 & 1 & 2 & -1 & 0 & 0 & 0 & 0 \\ 0 & 0 & 1 & 0 & 0 & 0 & 0 & 0 \\ 0 & 0 & 0 & 1 & -2 & 1 & 0 & 0 \\ 0 & 0 & 0 & 0 & 1 & -3 & 1 & 0 \\ 0 & 0 & 0 & 0 & 0 & 1 & 2 & 3 \\ 0 & 0 & 0 & 0 & 0 & 0 & 1 & -2 \\ 0 & 0 & 0 & 0 & 0 & 0 & 0 & 1 \end{bmatrix}$$

$D_{ij} = (1, 2, 0, 1, 3, 2, 1, 6)$

The D_{ii} coefficients have a useful structural interpretation based on the fact that they are the diagonals of the upper triangular matrix $\mathbf{D}\,\mathbf{L}^T$. D_{ii} is the stiffness for unit changes of the displacement variable i when all variables less that i not prescribed by displacement boundary conditions are free to change and all variables greater that i are zero.

By inspecting the diagonal coefficients of $\mathbf{D}$, we determine the definiteness of $\mathbf{K}$ according to the following rules:

Rules of Definiteness

FORM OF D_{ii}	DEFINITENESS OF MATRIX
$D_{ii} > 0$, all i	Positive definite
Some $D_{ii} > 0$, some $= 0$	Positive semidefinite
Some $D_{ii} > 0$, some < 0	Indefinite
Some $D_{ii} < 0$, some $= 0$	Negative semidefinite
$D_{ii} < 0$, all i	Negative definite

Negative definite and negative semidefinite matrices can be changed to positive definite and positive semidefinite, respectively, by multiplying the matrix by -1. Nonpositive matrices can arise in linear static structural analysis when we subtract one stiffness matrix from another. This subtraction may also produce an indefinite matrix.

Since the $\mathbf{L}$ matrix is lower triangular, its determinant is simply the product of its diagonal terms. Since its diagonals are ones, its determinant is 1. Since it has a nonzero determinant, $\mathbf{L}^{-1}$ exists, and we can rewrite Eq. (5.2) as

$$\mathbf{D} = \mathbf{L}^{-1}\,\mathbf{K}_c\,\mathbf{L}^{-1T} \tag{5.5}$$

Eq. (5.5) emphasizes that factorization implies a similarity transformation: a change in displacement variables. The change casts the strain energy into the uncoupled form

$$SE = 0.5\,\mathbf{d}^T\,\mathbf{D}^{-1}\,\mathbf{d} = \frac{0.5\,\Sigma\,d_{ii}^2}{D_{ii}} \tag{5.6}$$

where summation is over the order of $\mathbf{D}$. In this form, the rules of definiteness can be applied directly for determining definiteness.

The stiffness matrix is singular if at least one D_{ii} is zero. Since the determinant of the product of square matrices is the product of the determinants, and the determinant of $\mathbf{L}$ and $\mathbf{L}^T = 1$, singularity of $\mathbf{K}$ for a zero D_{ii} follows from Eq. (5.2).

Most production codes implementing triangular factorization produce a diagnostic message identifying the first D_{ii} that is zero. The message is usually sufficient for the engineer to interpret the source of the zero as a deformation-free mode of structural behavior.

To illustrate the interpretation, consider the planar frame and truss structures of Fig. 5.1. In this figure, each of the subfigures has two sketches: The first illustrates the structure; the second, the rigidly displaced geometry (in dashed lines) in relation to the undisplaced.

The supports of the frame structure of Fig. 5.1(A) tolerate rigid translation in the horizontal direction. Thus the mode of displacement is 1 for all u variables and zero for all other variables. This displacement vector is a "mode" in the sense that displacements are characterized by the relative values of the vector components rather then their actual values. Therefore, multiplying the vector by any particular nonzero scalar does not change the mode.

While the supports of the truss of Fig. 5.1(B) are indeterminate, thereby precluding rigid body modes of the structure as a whole, the hinge at node 2 tolerates folding of the structure at that point. For infinitesimal deflections, the structural model includes no resistance to deflections normal to a line between nodes 1 and 3. Therefore, a local rigid mode exists in the model and will be reflected by a zero D_{ii} coefficient.

Similarly, since the model of the wind-bracing truss of Fig. 5.1(C) offers no resistance to differential horizontal displacements at the roof and second level, it will evoke a zero D_{ii}. In fact, as Fig. 5.1 suggests, there are two independent zero strain modes, so this model will result in 2 zero D_{ii}.

Mathematically speaking, a single zero D_{ii} means that displacements are a function of one undefined parameter. This parameter is the multiplier of the rigid mode vector. The value of this parameter must be defined by the analyst in order to proceed with the problem.

For lattice structures, it is often so easy for the engineer to predict the degeneracy of the stiffness matrix that it should be done to avoid the relatively high cost of computer calculation of **D**. The degeneracy of the constrained stiffness matrix can usually be determined by visualizing the number of strain-free modes of the structure. A check is available by calculating the minimum degeneracy: the matrix order minus the sum of the ranks of the element matrices of the system.

To illustrate this analysis, consider the structure in Fig. 5.1(C). To describe nodal displacements of this planar truss, there will be two displacement variables at each node. Therefore, the order of the system stiffness matrix of the two-story frame will be 18. Since the vertical and horizontal displacements are prevented at nodes 7, 8, and 9, the order of the constrained stiffness matrix will be 12. The rank of each rod model is 1, the number of displacement variables at the lattice element node when defined with respect to its natural axis. Thus the maximum rank of $\mathbf{K}_c$ is 10, the number of elements, and the degeneracy must be at least the matrix order minus the maximum rank; $12 - 10 = 2$.

The type of constraint to introduce to eliminate the degeneracy will depend on whether the singularity associates with an inconsistent or a consistent equation. When inconsistent, the model is self-contradictory, and the engineer must remove the contradiction. When consistent, the analysis can proceed if the compute

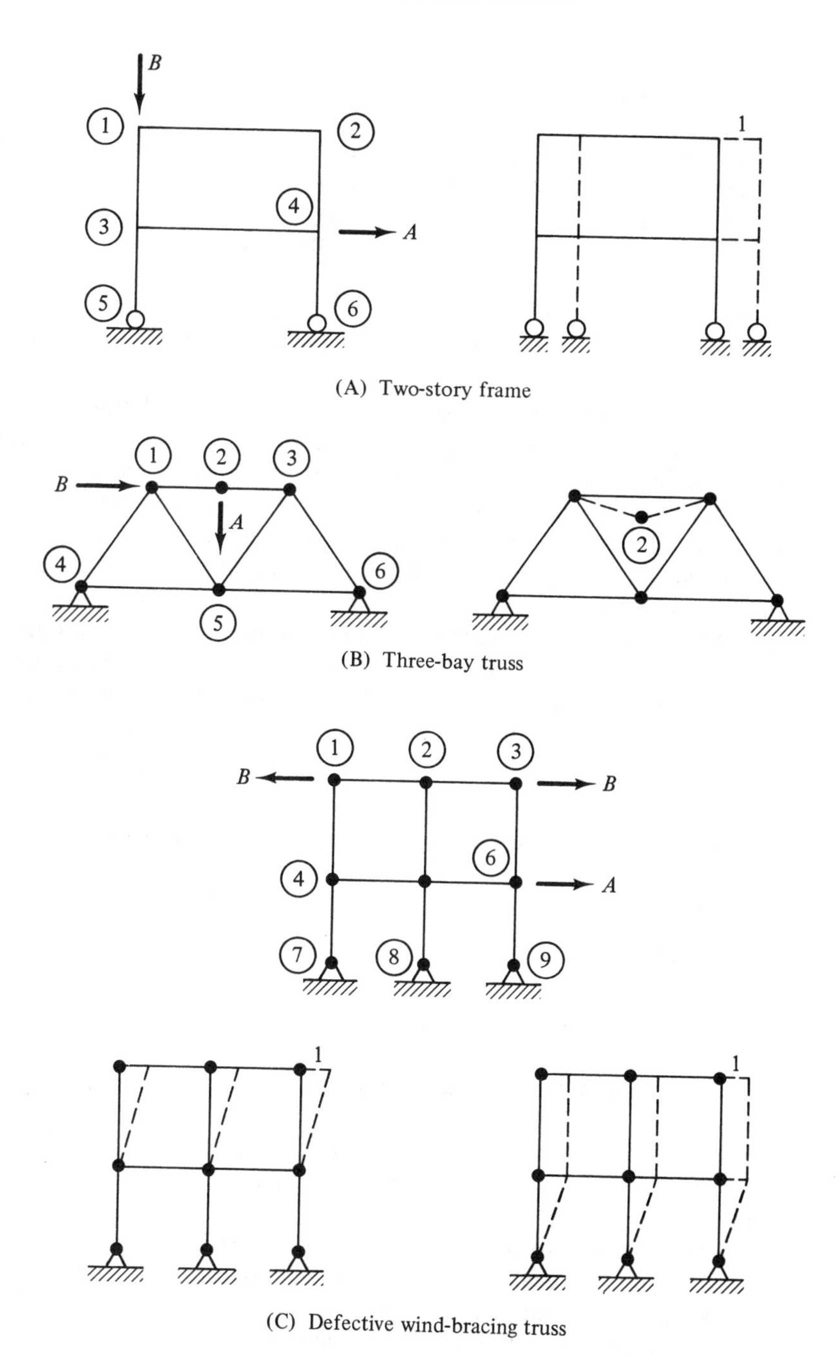

Figure 5.1 Sample lattices and unconstrained rigid modes.

program or analyst defines a datum to which displacements can be related. The usual datum adopted is that the variable associated with the zero diagonal is zero.

The equations are inconsistent when, after completing the forward substitution, we discover that the equation with $D_{ii} = 0$ has a nonzero p_i. This equation implies the nonsense that zero is not zero.

The structural interpretation of this mathematical nonsense is that the model reflects an active kinematic instability. This instability will occur for each of the Fig. 5.1 structural models under the A loading. Under the A loading, the mechanism modes on the right half of each subfigure are excited with undefined amplitude. To stabilize the model, either the structural model or the loading must be changed.

The equations are "consistent" when, after completing the forward substitution, we discover that an equation with $D_{ii} = 0$ has $p_i = 0$. Because p_i equals zero, the instability is not excited. The solution can proceed with g_i set to an arbitrary constant that defines the amplitude of the deformation-free mode. The solution will yield the correct values for internal forces. The displacements found will include the arbitrary magnitude deformation–free mode. The equations of the structures of Fig. 5.1 are consistent for the B loadings.

Occasionally, the engineer will need more information to define appropriate changes to the analysis model or loading. In particular, examination of the deformation-free mode may be necessary. Removing the applied loading and imposing a unit settlement in the ith displacement will produce the modal vector.

The calculations to effect a triangular factorization solution are as follows:

Calculations of a Triangular Factorization Solution

OPERATION	CALCULATIONS
Triangularization	Nb^2
Substitutions	$4Nbn_p$
Assessment of definiteness	$2N$

where n_p is the number of independent loading vectors. Here we assume $n >> b >> 1$; the large problem class of analyses. We define a calculation as a single arithmetic operation: add, subtract, multiply, or divide. Since the basic arithmetic operation is a dot product, about half the calculations are additions and subtractions and half multiplications and divisions.

Triangular factorization with Gauss processing is accepted as the most efficient direct method. Because nonzero coefficients can exist only within the band at each stage of triangularization, calculations and storage space are allocated only for coefficients within the band. Because in every stage of factorization, as illustrated in Table 5.1, the matrix has either zeros or symmetric coefficients below the diagonal, calculation and storage savings can be achieved. Only substitution calculations must be added to deal with more than one loading vector. The number of calculations to assess definiteness is negligible compared with all other calculations.

In practice, the efficiency of triangular factorization is enhanced by other processing logic. Wavefront, frontal, and skyline logic improve efficiency by exploiting the existence of zero coefficients within the band of the stiffness matrix.[2-4] Nodal resequencing algorithms change the size of the band or pattern of sparseness so that the matrix requires less computer resources than otherwise.[5,6] Use of any of these algorithms does not change the interpretation of the results of equation solving.

While most of the advantages of triangular factorization apply to other direct methods, alternatives usually require more calculations. Crout's method and Choleski's method are available in production codes. Examination of wavefront transformation, cyclic reduction, generalized cyclic reduction, the WF method, and transfer-reduction suggest that methods other than triangular factorization with Gauss processing may offer improved efficiency in parallel-processing computers.[7,8]

5.2 GAUSS-SEIDEL ITERATION

The computer resources needed for an iterative method depend on the efficiency of data processing per cycle of iteration and the number of cycles needed to produce a solution of acceptable accuracy. Processing in each cycle can take full advantage of sparseness and symmetry of the stiffness matrix to conserve calculations, data storage, and data transfers. Algorithms offer considerable opportunity to exploit facilities for parallel processing[9] and problem adaptation.[10] Since the number of cycles depends on the structural model, there exist special problem classes that justify the use of an iterative solution process, though they are rare.

Among itertive algorithms, accelerated Gauss-Seidel is the most popular. We will examine it in some detail to establish the concepts that associate with iteration methods.

Assume that the stiffness matrix is synthesized of three matrices

$$\mathbf{K}_c = \mathbf{L}_I + \mathbf{D}_I + \mathbf{L}_I^T \tag{5.7}$$

where

$\mathbf{L}_I$ is a lower triangular matrix with zeros on the diagonal,

$\mathbf{D}_I$ is a diagonal matrix with the value of $\mathbf{K}_{cii}$ on each diagonal, and

the subscript I distinguishes the fact that the matrices associate with iteration methods.

Accelerated Gauss-Seidel iteration is defined by

$$\mathbf{g}_{j+1} = \mathbf{g}_j + \alpha\, \mathbf{D}_I^{-1} \{\mathbf{p} - [(\mathbf{L}_I + \mathbf{D}_I)\mathbf{g}_j + \mathbf{L}_I^T\, \mathbf{g}_{j+1}]\} \tag{5.8}$$

where

α is the acceleration factor, and

j, as a subscript, notes the iteration cycle number.

We imply that immediately after we calculate the i component of $\mathbf{g}_{j+1}$, we update $\mathbf{g}$ for subsequent components. Equation (5.8) thereby defines the calculations and calculation sequence for each Gauss-Seidel cycle.

Additional cycles of calculations are performed until the error in $\mathbf{g}_{j+1}$ is sufficiently small. This error can be measured by taking the absolute sum of the values of the components of the vector in brackets in Eq. (5.8)—the absolute sum of the equation residuals. This measure is a "direct error measure" because it provides a necessary and sufficient measure of the ability of the $\mathbf{g}_i$ to satisfy the stiffness equations. The absolute sum of the differences of successive $\mathbf{g}_j$ may also be used. This is an "indirect error measure" because it provides only a necessary check of solution accuracy.

When $\alpha = 1$, Eq. (5.8) defines the Gauss-Seidel algorithm. When $\alpha > 1$, the algorithm is accelerated Gauss-Seidel. Convergence of the sequence $\mathbf{g}_j, \mathbf{g}_{j+1}$, . . . is guaranteed as long as $\mathbf{K}_c$ is positive definite and $\alpha = 1$. Below some value of $\alpha > 1$, convergence will occur; above this value convergence cannot be guaranteed. In general, convergence will not occur if $\alpha > 2$.

Rewriting Eq. (5.8) for the $j + 2$ solution, subtracting Eq. (5.8) from the result, and reorganizing terms yields

$$(I + \alpha \mathbf{D}_I^{-1}\mathbf{L}_I^T)\,(\mathbf{g}_{j+2} - \mathbf{g}_{j+1}) = [(1 - \alpha)I - \alpha \mathbf{D}_I^{-1}\mathbf{L}_I]\,(\mathbf{g}_{j+1} - \mathbf{g}_j) \qquad (5.9)$$

This equation shows that the rate of convergence is independent of the loading vector and a nonlinear function of the acceleration factor and the values of coefficients of the stiffness matrix.

Table 5.2 defines the results for each step of accelerated Gauss-Seidel equation solving for a four-element uniaxial-rod problem. The first part of the table defines the problem, and the second part gives the results of each calculation. For clarity, the acceleration factor of these calculations is 1.5, and the initial guess of u_i has a value of 1.0 for all i.

Consider the treatment of equation 2 in cycle 1. To find the residual, we multiply the second row of the stiffness matrix by the current estimate of $\mathbf{u}$ and subtract the result from the equation 2 loading component. To calculate the change in the second $\mathbf{u}$ component, we divide this residual by the second diagonal coefficient and multiply the result by the acceleration factor. The new value of the second $\mathbf{u}$ component is reset to its current value plus its change.

The data of Table 5.2 illustrates the evaluation of both the residual and displacement change scalar error measures. These measures provide independent evaluations of the error of the solution. To save processing time for the residual force calculation, the absolute sum of the residuals of each equation is calculated during the treatment of each equation.

TABLE 5.2 SAMPLE ACCELERATED GAUSS-SEIDEL SOLUTION

3 ← (1) (2) (3) (4) (5)

4 @ 120 = 480

$AE_i = 120$

$$\begin{bmatrix} 1 & -1 & 0 & 0 \\ -1 & 2 & -1 & 0 \\ 0 & -1 & 2 & -1 \\ 0 & 0 & -1 & 2 \end{bmatrix} \begin{pmatrix} u_1 \\ u_2 \\ u_3 \\ u_4 \end{pmatrix} = \begin{pmatrix} 3 \\ 0 \\ 0 \\ 0 \end{pmatrix}$$

Cycle	Eq.	$\mathbf{u}_1$*	$\mathbf{u}_2$*	$\mathbf{u}_3$*	$\mathbf{u}_4$*	$\mathbf{r}_i$†	$\mathbf{u}_i$‡
1	1	1.0000	1.0000	1.0000	1.0000	3.3000	5.5001
	2	5.5000	1.0000	1.0000	1.0000	4.5000	4.3750
	3	5.5000	4.3750	1.0000	1.0000	3.3750	3.5312
	4	5.5000	4.3750	3.5312	1.0000	1.5312	2.1484
						12.4062	

$$\text{Relative force error} = \frac{\Sigma|r_i| - \Sigma|p_i|}{\Sigma|p_i|} = \frac{12.4062 - 3}{3} = 3.1354$$

$$\text{Relative displacement change} = \frac{\Sigma u_{k+1} - \Sigma u_k}{\Sigma u_k} = \frac{14.4062 - 4}{4} = 2.6016$$

Cycle	Eq.	$\mathbf{u}_1$*	$\mathbf{u}_2$*	$\mathbf{u}_3$*	$\mathbf{u}_4$*	$\mathbf{r}_i$†	$\mathbf{u}_i$‡
2	1	5.5000	4.3750	3.5312	2.1484	2.9750	9.9625
	2	9.9625	4.3750	3.5312	2.1484	4.7437	7.9278
	3	9.9625	2.9278	3.5312	2.1484	. . .	
	⋮						

Cycle		
3	Relative force error = −0.1853	Relative displacement change = 0.6995
4	Relative force error = +0.1464	Relative displacement change = 1.1008
5	Relative force error = +0.5985	Relative displacement change = 1.5779
⋮		

*Displacement variable values input to equation.

†r_i = i^{th} equation residual, $\mathbf{r}_j = \mathbf{p}_j - [(\mathbf{L}_I + \mathbf{D}_I)\mathbf{g}_j + \mathbf{L}_I^T\mathbf{g}_{j+1}]$ where $\mathbf{g}$ is the vector u.

‡Updated estimate of displacement variable for $a = 1.5$.

Figure 5.2 shows how the displacement change error measure changes for the sample problem as a function of the number of iterative cycles and the acceleration factor. The data indicates that, on the average, convergence is logarithmic. The number of cycles required is proportional to the number of digits of accuracy required in the solution. Experience with a variety of structural problems indicates that this conclusion is valid in general. Accordingly, an initial guess reduces calculations by the number of accurate digits in the guess compared with the number desired.

Figure 5.3 shows how the number of cycles to obtain results of six-digit

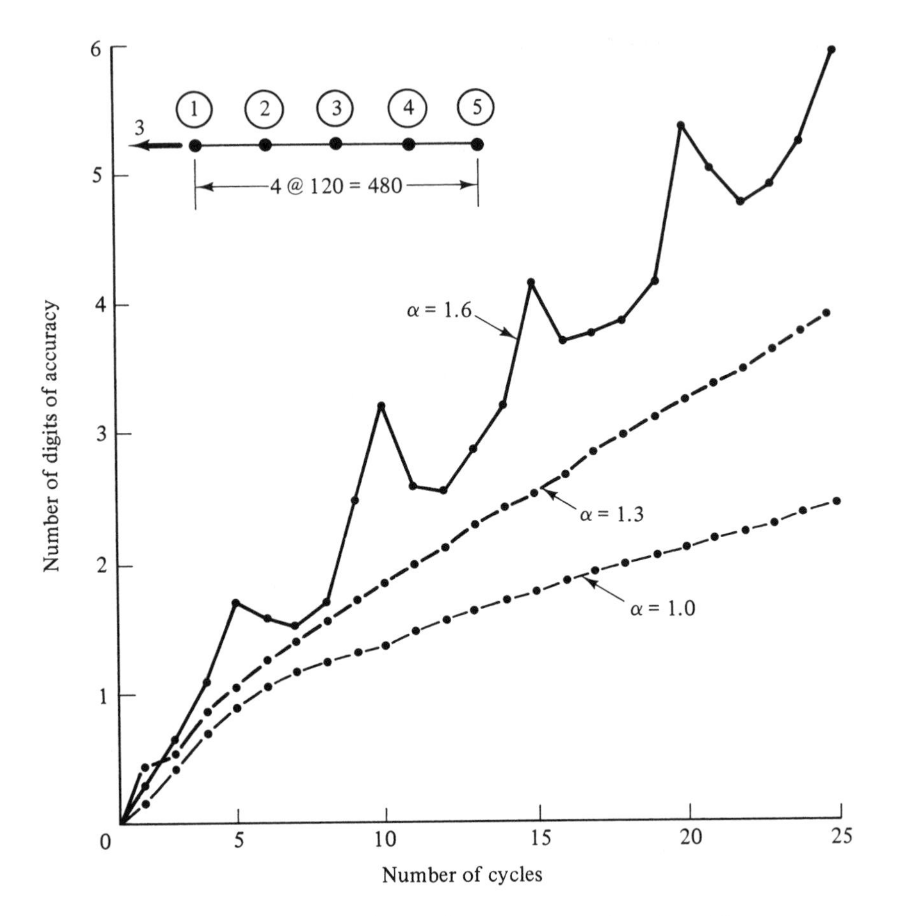

Figure 5.2 Gauss-Seidel solution convergence: Sample problem.

accuracy varies with the acceleration factor for the sample problem. This curve confirms that there is an optimum acceleration factor. Furthermore, the number of cycles is dramatically less at the optimum acceleration than with $\alpha = 1$. Experience indicates that both these generalizations are valid for structural analyses.

Making minor changes to the algorithm provides for sensing consistency and inconsistency when singularity occurs. If the system of equations is consistent the residual force test will report convergence, while the displacement change test will not. If the equations are singular and inconsistent, neither the residual force nor the displacement change measure will reflect convergence. These characteristics also associate with excessive acceleration factors compared with the optimum.

A test of positive definiteness can be integrated into each cycle of the iteration based on the requirement that the strain energy associated with $\mathbf{K}_c$ must be greater than zero.

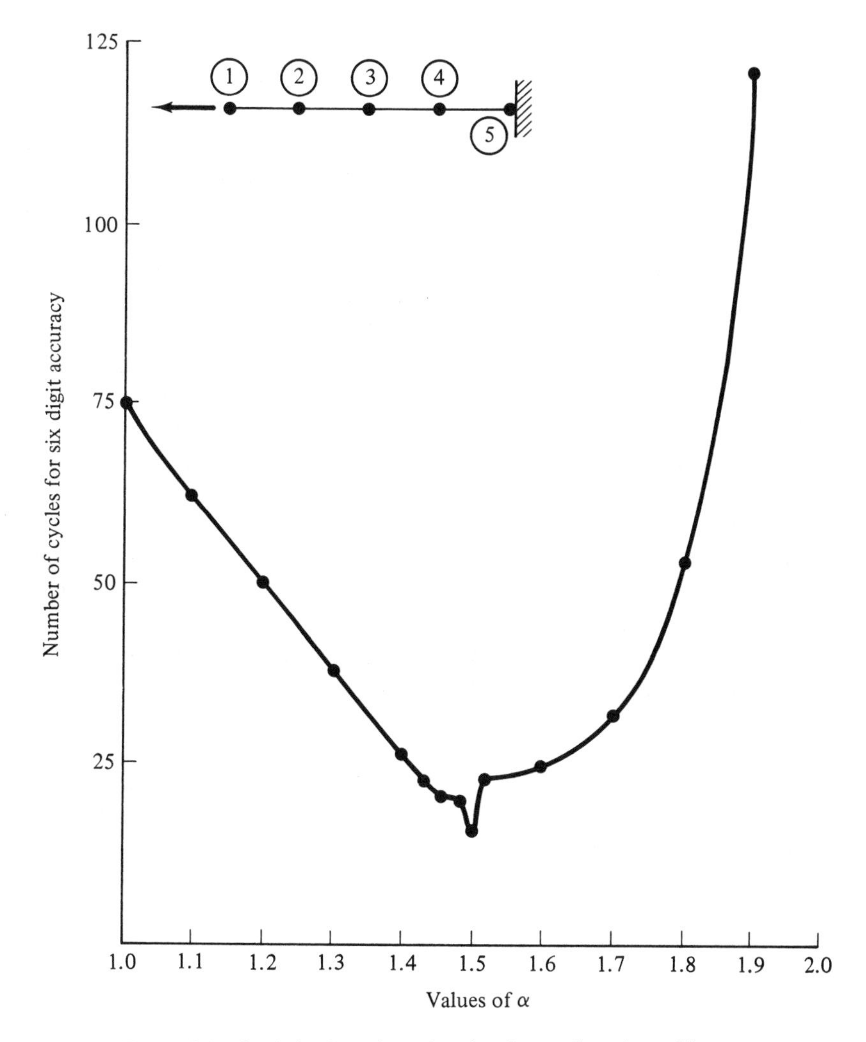

Figure 5.3 Optimization of acceleration factor: Sample problem.

The calculations to evolve an accelerated Gauss-Seidel solution for the large problem class are as follows.

Calculations of Gauss-Seidel Iteration

OPERATION	CALCULATIONS
Solution cycle calculations	$4Nbn_pn_c$
Evaluation of errors	$2Nn_pn_c$
Assessment of definiteness	$2Nn_pn_c$

where n_c is the number of iteration cycles. The arithmetic mix is half additions or subtractions and half multiplications or divisions. The n_p factor reflects the fact that in iterative methods, the number of calculations is directly proportional to the number of loading vectors. The number of calculations does not reflect the fact that since the optimum acceleration factor is a function of only the stiffness matrix, knowledge developed on the value of the optimum factor is relevant in reducing n_c when multiple loading vectors exist.

When $b >> 1$, almost all the calculations are involved in evaluating the dot product of a row of the stiffness matrix and the column of current estimates of the unknowns. Most of the advantages of Gauss-Seidel iteration stem from this fact. The arithmetic of not only zero stiffness coefficients outside the band, but also zeros within the band can be avoided. High-speed storage needs can readily be limited to just two vectors, if desired, without extensive logic for data management. In Gauss-Seidel iteration, any random errors introduced in one cycle tend to be reduced in subsequent cycles.

The principal characteristics of Gauss-Seidel iteration are shared by all iterative methods. The principal resource-consuming calculation is the calculation of the equation residuals. The modular unit of calculations is evaluation of the dot product of a pair of vectors.

Generally, the formulas for evaluating improved estimates and the sequence in which equations are treated are the principal changes among iteration algorithms. Some other methods include Jacobi's method, the power method, the Monte Carlo method, relaxation, and the method of alternating directions.

Since the data processing features of iteration algorithms are common to virtually all the iterative methods, the critical distinction between methods is in their rates of convergence. On a given problem, the rate of convergence will vary from method to method. Unfortunately, selection of the best iterative method, the one with the smallest n_c for the problem at hand, is unlikely *ab initio*, because the number is unknown until after the method has been applied.

5.3 *N*-STEP PROCESSES

There are a handful of iterative solution algorithms for which the maximum number of cycles to obtain the solution will not exceed N. These include the methods of Hestenes-Stiefel, Fletcher-Powell, Fletcher-Reeves, and Jacobson-Ocksman.[11–13] These methods differ from iteration methods by calculations to condition the choice of the vector of changes of the unknowns, in each iteration cycle, to guarantee the solution in N cycles. If accurate solution estimates are generated in less than N cycles, the process is stopped. Experiences with these methods for structural analyses indicate that it is unusual for these algorithms to obtain satisfactorily accurate solution estimates without approximately N iterations.

A check on the positive definiteness of $\mathbf{K}_c$ can be based on the fact that each of the change vectors must imply positive strain energy. If the strain energy test is satisfied for N successive vectors linearly independent, the definiteness of $\mathbf{K}_c$ is guaranteed.

The dominant number of calculations in each cycle in each of these N-step algorithms is the multiplication of the coefficient matrix times a vector. Accordingly, the arithmetic mix is about half additions or subtractions and half multiplications or divisions. The calculations for $N >> b >> 1$ are as follows.

Calculations for *N*-Step Iteration

OPERATION	CALCULATIONS PER CYCLE
Solution cycle calculations	$4N^2b$
Evaluation of error	$3N^2$
Assessment of definiteness	N^2
Additional loading vectors	$3N^2(n_p - 1)$

Unlike conventional iteration methods, N-step methods can reuse the vectors developed in each cycle to construct a solution for alternative loading vectors if all N cycles are performed. The table lists the additive calculations in the last line.

5.4 SELECTING THE SOLVER

Because all methods are capable of obtaining the solution of the equations and assessing definiteness, the selection of the equation solver reduces to comparison of the relative amount of computer resources needed for equation solution. We measure resources by the number of calculations for the large problem class, $N >> b >> 1$.

We conclude that triangular factorization is much better than an N-step method. The relative number of calculations of factorization is about b/n that of an N-step process for both the first loading vector and subsequent vectors.

We conclude that the factorization method is better than an iteration method as long as the number of iterations is greater than $b/4$ in the case of a single loading vector and greater than 1 for subsequent vectors. This means that factorization is the choice when multiple loadings are involved and for structural problems with more than 100 equations.[9]

For parallel processing computer configurations, an iterative algorithm may result in reduced wall clock time compared with use of a direct equation-solving algorithm. While the parallel processor will require more calculations than the uniprocessor, speedup will be effected by the use of concurrent processing.[9]

Comparisons based on the number of calculations imply the that the computer programmer's skill and data management strategy do not affect the computer resources needed. This assumption is unrealistic. Accordingly, measurements of resources needed for particular computer configurations should provide the final basis for selecting the equation solver.

More than 1000 publications describe advances in equation solution implementation for the finite element method.[14] The focus of this research is to reduce the computer resources needed for solution. The 29-year cumulative effect of this research has been computer code that, by using the best mathematical algorithm and supporting it with the best data management procedures, requires less than 1/100 the calculations needed with the 1960 implementations. Most of the improvements associate with avoiding calculations that would result in zero-valued intermediate results. Although the use of the latest equation-solving technology is not necessary to achieve accurate results, comparisons of equation-solving costs are useful to the analyst in engineering a cost-efficient finite element analysis.

5.5 VALIDATING THE SOLVER

General validation of the equation solver should include checks to ensure the following for each solver of the computer configuration:

1. When all the generalized displacements are prescribed, the solution vector contains the correct values.
2. A single equation is solved correctly.
3. The correct solution is obtained for the case of a fully populated stiffness matrix when the number of equations is greater than 1.
4. The correct solution is obtained for a sparsely populated stiffness matrix when the number of equations is greater than 1.
5. An equation set that has too many equations or too large a bandwidth for the computer configuration results in a pertinent error message.
6. A nonpositive stiffness matrix results in an error message.
7. A singular stiffness matrix results in an error message.
8. A singular stiffness matrix leads to an analysis and report of the consistency and inconsistency of the set of equations.
9. Any error measure used to assess the accuracy of the solution indicates large errors when the solution is inaccurate.

As the computerwork problems at the end of the chapter suggest, both direct and program consistency checks can be used for validation testing.

If any of these tests indicates an error, the engineer may wish to isolate the logic error to a subset of the equation-solving logic. In the case of triangular factorization, the error may be associated with the factorization or the substitutions. In the case of iteration, the error may be partitioned to the calculation of the error measures or calculation of the improved g_i of an equation.

5.6 HOMEWORK

5:1. Equation set consistency.

Given: The following equations.

$$\begin{bmatrix} 2 & -2 & -4 \\ -2 & 6 & 6 \\ -4 & 6 & 9 \end{bmatrix} \begin{pmatrix} g_1 \\ g_2 \\ g_3 \end{pmatrix} = \begin{pmatrix} 3 \\ 7 \\ -8 \end{pmatrix}$$

Determine: If the coefficient matrix is singular and, if so, whether the set of equations is consistent or inconsistent.

Method: Triangular factorization.

5:2. Definiteness.

Given: The following equations.

$$\begin{bmatrix} 2 & -2 & -4 \\ -2 & 6 & 6 \\ -4 & 6 & 9 \end{bmatrix} \begin{pmatrix} g_1 \\ g_2 \\ g_3 \end{pmatrix} = \begin{pmatrix} 3 \\ -7 \\ -8 \end{pmatrix}$$

Determine: The definiteness of the matrix.

Method: Triangular factorization.

5:3. Equation set consistency.

Given: The following equations.

$$\begin{bmatrix} 2 & -2 & -4 \\ -2 & 6 & 6 \\ -4 & 6 & 9 \end{bmatrix} \begin{pmatrix} g_1 \\ g_2 \\ g_3 \end{pmatrix} = \begin{pmatrix} 3 \\ 7 \\ -8 \end{pmatrix}$$

Determine: If the coefficient matrix is singular and, if so, whether the set of equations is consistent or inconsistent.

Method: Triangular factorization.

***5:4.** Definiteness.

Given: The following equations.

$$\begin{bmatrix} 3 & 6 & 6 \\ 6 & 11 & 9 \\ 6 & 9 & 11 \end{bmatrix} \begin{pmatrix} g_1 \\ g_2 \\ g_3 \end{pmatrix} = \begin{pmatrix} 3 \\ 0 \\ -2 \end{pmatrix}$$

Determine: The definiteness of the matrix of coefficients.

Method: Triangular factorization.

5:5. Definiteness.

Given: The following equations.

$$\begin{bmatrix} 1 & -2 & -1 \\ -2 & 6 & -4 \\ 1 & -4 & 6 \end{bmatrix} \begin{pmatrix} g_1 \\ g_2 \\ g_3 \end{pmatrix} = \begin{pmatrix} 0 \\ 0 \\ 0 \end{pmatrix}$$

Determine: The definiteness of the matrix of coefficients.

Method: Triangular factorization

5:6. Gauss-Seidel iteration.

Given: The following matrix equation and solution.

$$\begin{bmatrix} 1 & -1 & 0 \\ -1 & 2 & -1 \\ 0 & -1 & 2 \end{bmatrix} \begin{pmatrix} g_1 \\ g_2 \\ g_3 \end{pmatrix} = \begin{pmatrix} 1 \\ 0 \\ 0 \end{pmatrix}, \quad \mathbf{g} = \begin{pmatrix} 3 \\ 2 \\ 1 \end{pmatrix}$$

Find: The Gauss-Seidel solution estimate and error estimate at the end of three successive cycles of iteration under the following assumptions.

(a) The starting guess is $\mathbf{g}^T = (1.0, 0, 0)$.

(b) The overrelaxation factor is 1.5.

(c) Calculations are to be performed with four digits of accuracy.

(d) The error measures are the change of **g** for two successive iterates compared with the exact **g** and the change of **g** compared with the current estimate of **g**.

5.7 COMPUTERWORK

5-1. One-equation validation of solver.

Given: The rod element of the figure.

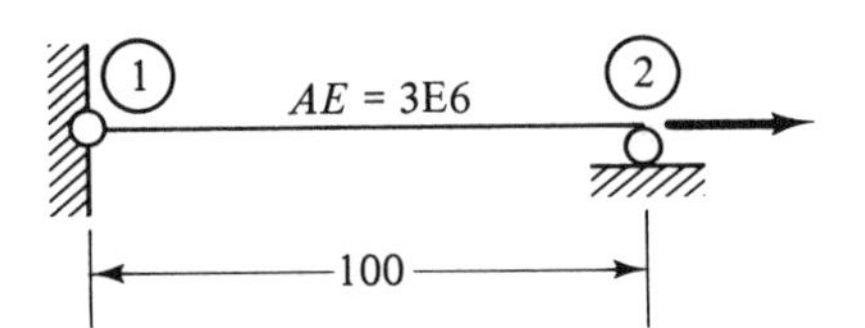

Computerwork Problem 5-1

Find:

(a) The exact deflection at node 2.

(b) The computer-calculated deflection of node 2.

***5-2.** Fully populated matrix validation of solver.

Given: The truss structure of the figure.

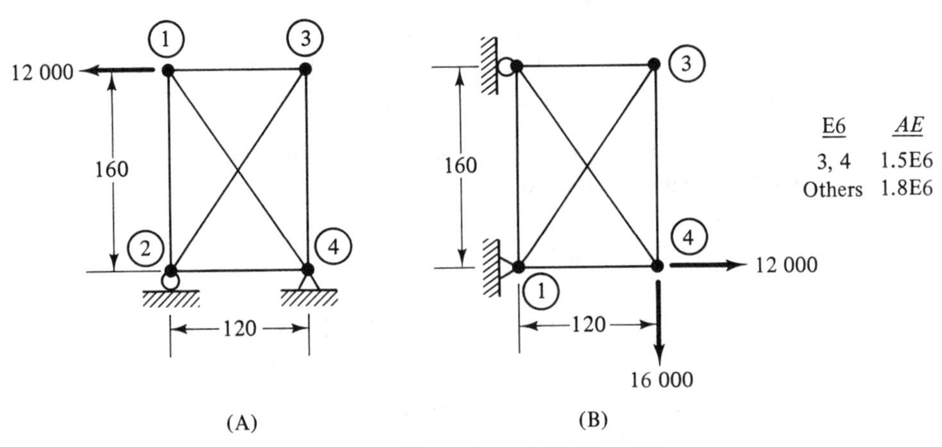

Computerwork Problem 5-2

Determine: If the equation solver can get the correct solution for a fully populated stiffness matrix.

Method: Predict the deflections of the left structure and make a self-consistency check with the solution of the right structure for rod forces.

5-3. Validation for sparsely populated equation solution.

Given: The cantilevered beam of the figure.

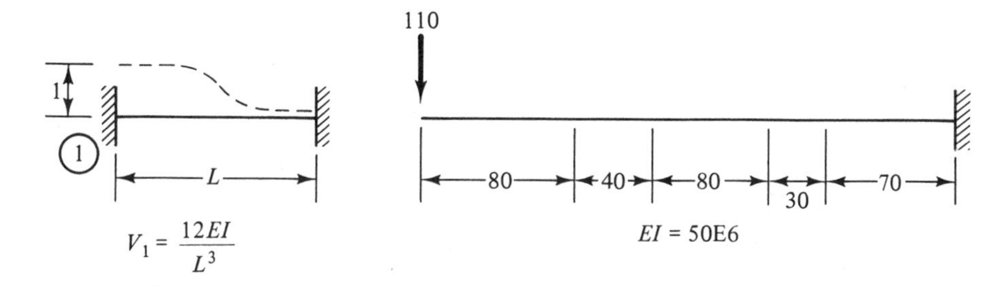

Computerwork Problem 5-3

Validate: The value of the 2nd, 5th, and 11th diagonals of **D**.

Method: Compare diagonals calculated by the computer with values calculated from the physical interpretation of D_{ii}.

5-4. Validation of positive definite logic.

Given: The bent of the figure.

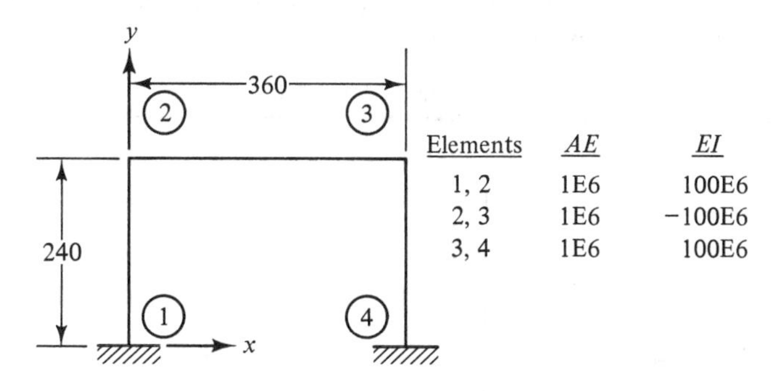

Elements	AE	EI
1, 2	1E6	100E6
2, 3	1E6	−100E6
3, 4	1E6	100E6

Computerwork Problem 5-4

Determine: If the equation solver can distinguish that the constrained stiffness matrix is not positive definite.

***5-5.** Validation of equation consistency evaluation logic.

Given: The five-legged truss of the figure.

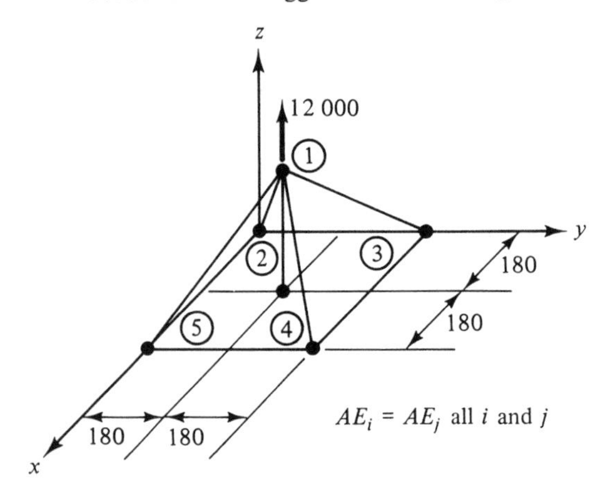

Fixities: 2, 1; 2, 2; 3, 1; 3, 2; 4, 1; 4, 2; 5, 1; −5, 2

Computerwork Problem 5-5

Determine: The force in each rod.

5-6. Validation of double-precision equation solver.

Given: The cantilevered beam of Prob. 5.3.

Validate: The double-precision equation solver.

Method: Compare computer-predicted moments at each node with their exact values.

5-7. Validation of solver using a superposition check.

Given: The structure of part (A) of the figure loaded by the wind and earthquake loadings of parts (A) and (B).

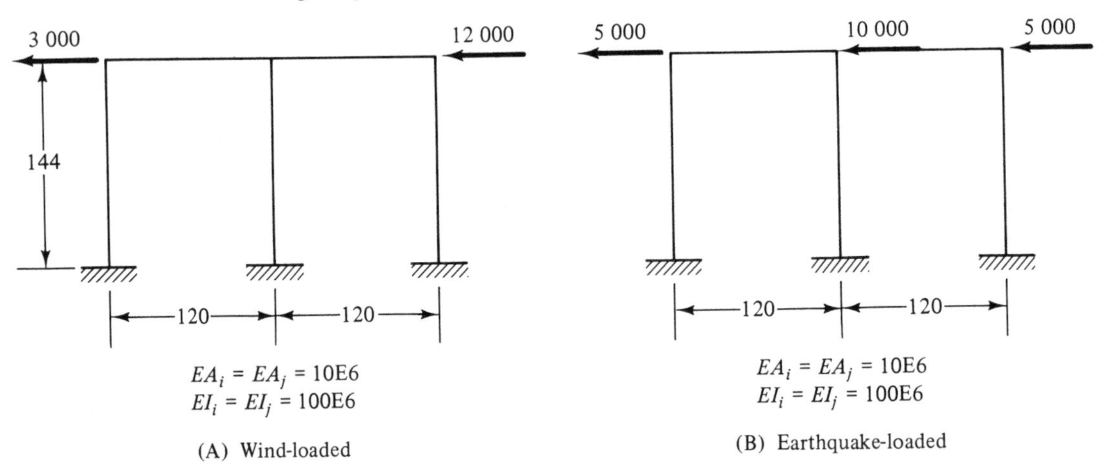

Computerwork Problem 5-7

Do the following:

(a) Generate the computer solution for deflections under the loading of part (A).

(b) Generate the computer solution for deflections under the loading of part (B).

(c) Generate the computer solution for deflections under the sum of the loadings of parts (A) and (B).
(d) Compare the deflections found in step (c) with the sum of the deflections from steps (a) and (b).

5-8. Validation of Gauss-Seidel iteration.

Given: The cantilevered beam of Prob. 5-3.

Validate: The ability of the computer configuration to find the moment at each node using accelerated Gauss-Seidel iteration with an acceleration factor of 1.0 and, separately, 1.9.

5.8 REFERENCES

1. Klyuyev, V.V.; Kokovkin-Shcherbak, N.I., "On the Minimization of the Number of Operations for the Solution of Linear Algebraic Systems of Equations," Tech. Report CS24, Computer Science Dept., Stanford University, 1965.
2. Melosh, R.J.; Bamford, R.M., "Efficient Solution of Load-Deflection Equations," *J. Struct. Div. ASCE,* vol. 95, no. ST4, Apr. 1969, pp. 661–676.
3. Irons, B.M., "A Frontal Solution Program for Finite Element Analysis," *Int. J. Num. Meth. Engineering*, vol. 2, no. 1, 1970, pp. 5–32.
4. Wilson, E.L.; Bathe, K.J; Doherty, W.P., "Direct Solution of Large Systems of Linear Equations," *Computers and Structures*, vol. 4, no. 2, 1974, pp. 363–372.
5. Cuthill, E.H., "Several Strategies for Reducing the Bandwidth of Matrices," *Sparse Matrices and Their Applications*, ed. D.J. Rose, R.A. Willoughby, New York: Plenum Press, 1972.
6. Gibbs, N.E.; Poole, W.G.; Stockmeyer, P.K., "An Algorithm for Reducing Bandwidth and Profile of a Sparse Matrix," *SIAM J. Num. Anal.*, vol. 13, no. 2, 1976, pp. 236–250.
7. Utku, S.; Melosh, R.J.; Salam, M., "Mathematical Characterization of Concurrent Processing: Machines, Jobs, and Stratagems," *3rd Conf. in Computing in Civil Engineering*, San Diego, Calif., Apr. 1984.
8. Melosh, R.J.; Utku, S.; Salama, M., "Direct Finite Element Equation Solving Algorithms," *Computer and Structures*, vol. 20, no. 1-3, 1985, pp. 99–105.
9. Hayashi, A.; Melosh, R.J.; Utku, S.; Salama, M., "Variation in Efficiency of Parallel Algorithms," *Computers and Structures*, vol. 21, no. 5, 1985, pp. 1025–1034.
10. Melosh, R.J., "Seidel-Type Iteration for Solving Equilibrium Equations of Mechanical Systems," *Proc. Comp. Meth. in Nucl. Engineering*, Conf. 750413, vol. 2, Charleston, S.C., Apr. 1975, pp. IV-1 to IV-22.
11. Hestenes, M.R.; Stiefel, E., "Methods of Conjugate Gradients for Solving Linear Systems," *J. Res. National Bureau of Standards*, vol. 49, no. 6, December 1962.
12. Fletcher, R.; Powell, M.J.D., "A Rapidly Convergent Descent Method for Minimization," *Computer Journal*, vol. 6, no. 2, 1963.
13. Fletcher, R.; Reeves, C.M., "Function Minimization by Conjugate Gradients," *Computer Journal*, vol. 7, no. 2, 1964.
14. Melosh, R.J.; Smith, H.A.; Ghattas, O., "A Technology in Transition: Finite Element Analysis," *Proc. SECTAM XIV Conf.*, Biloxi, Miss., Apr. 1988.

Chapter 6

Controlling Round-off Error

Round-off error is the error caused by using a limited number of digits in the computer's representation of a number. It includes errors in converting numbers to and from the computer representation and errors in computer arithmetic.

Its definition contains its cure. Whenever round-off error is excessive, it is always possible to increase computer accuracy by increasing the number of digits used in the number model. While effective, use of this panacea is usually unnecessarily inefficient.

The magnitude of round-off error varies with the problem, the problem description, and the computer configuration. Estimates of the maximum error tend to increase for the same number of calculations when the lattice structure changes from three-dimensional to two, from a single material to two or more dissimilar materials, from prismatic members to tapered members, and from indeterminate supports to determinate.

Round-off error changes with the number, numbering sequence, spacing, and coordinate reference system for the nodes and with the finite element model type, shape, size, and disposition over the structural geometry. It is affected by the computerized implementation of number conversion and arithmetic, by the algorithms used in the analysis, and by the effectiveness of ancillary logic that directs calculations to minimize the error.

Round-off errors sometimes seem to defy intuition. We can produce deflection predictions of unacceptable accuracy while internal force predictions are accurate. Meaningless results can associate with uniform nodal spacing while nonuniform spacing yields accurate responses. Accurate behavior estimates may affiliate with analyses of a 1000-node structure and inaccurate with a 3-node model of a cantilevered beam.

Regardless of this masquerade, round-off errors are repeatable, systematic, and easily controlled. Before performing the FEA, we can estimate the maximum error. During FEA, we can measure error magnitudes from each source, increasing data processing costs by only a few percentage points. Using these error magnitudes, we can determine the accuracy of analysis results and take appropriate action to improve inaccurate results.[1]

6.1 NUMBER REPRESENTATION

The computer hardware represents each real variable as a floating-binary-point number. Figure 6.1 shows the basic formats of floating-point number representations in accord with the IEEE standard.[2] The sign bit indicates whether the number is plus or minus. The coefficient bits define the value of the number in base 2, with the implied binary point defining a binary fraction. The exponent is the power of 2 multiplier that scales the coefficient to its value.

The precision of the number model is the number of binary places allocated to the coefficient. Because the precision limits the accuracy, it is the most important feature of the number model. Figure 6.1 shows conversion of the binary precision to define the number of decimal digits of precision D_p: the maximum number of decimal digits of accuracy of computer calculations.

The importance of precision reflects in the availability of double-precision arithmetic number models. In double precision, about twice as many bits are allocated to the real variable representation as for single precision. Since the number of bits for the exponent may not be doubled, the double-precision model can actually have more than twice the precision of the single-precision format. Figure 6.1(C) illustrates that IEEE double precision has 15.6 digits, single-precision only 6.92.

When the computer coefficient bits are too few to represent a number, one of several options is selected to approximate the number. The usual option requires rounding the last binary position to the nearest binary representation. Then the result differs from the exact result by no more than one-half of the least significant binary place in the coefficient.

Usually, the coefficient is normalized so the bit immediately following the binary point has a value of 1. When this happens, the exponent is adjusted to preserve the value of the number.

When the range of the exponent is exceeded, the immediate results, at least, are inaccurate. When the excess requires representing the exponent that is too small for the number model, a state of "underflow" exists; when too large, "overflow."

The processing action upon exponent underflow or overflow varies with the

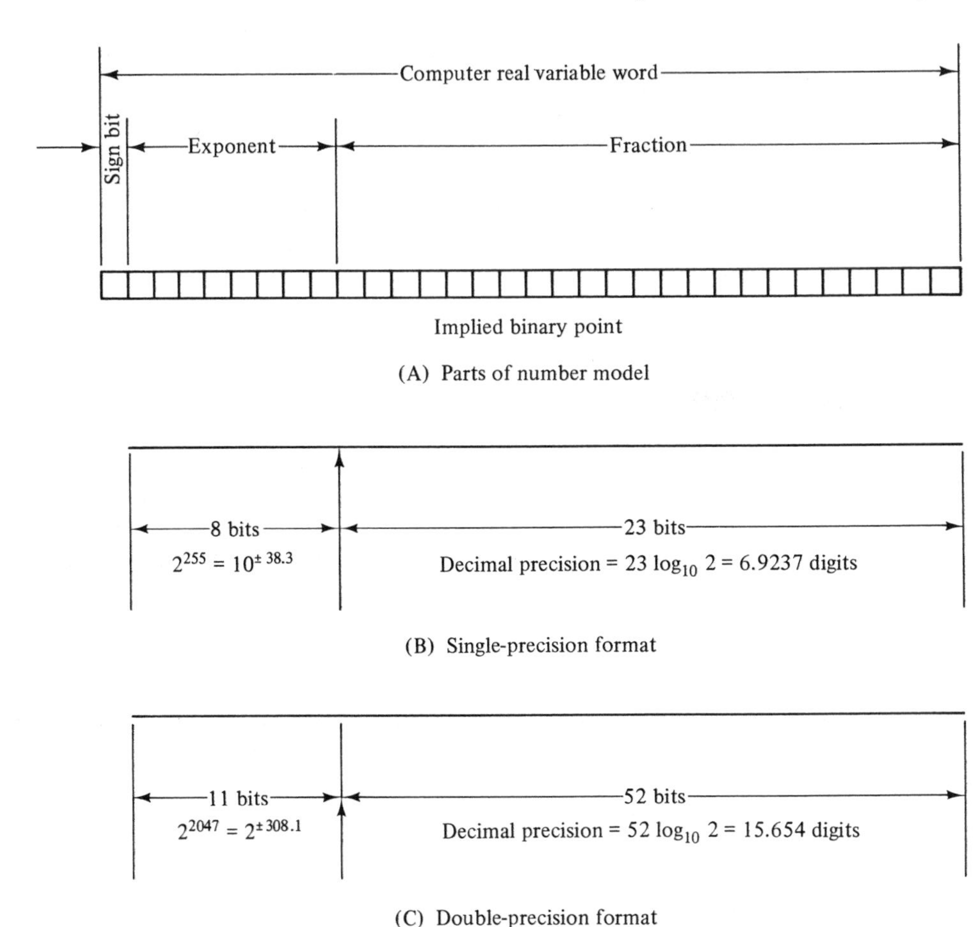

Figure 6.1 Basic computer formats for floating-point numbers.

computer software. Typically, when underflow occurs, the fraction and exponent of the model are set to zero—representing the variable as an absolute zero. When overflow occurs, an error message is produced. When too many overflows occur, as defined by the developer of the computer language compiler being used, calculations are stopped. With some language compilers, such as C and Pascal, exponent underflow and overflow are simply ignored.

The IEEE code requires that the hardware or software be capable of supporting three other rounding modes, including round to zero (truncation), and permits "extended formats" that differ from those of Fig. 6.1. Some existing computer configurations do not comply with the standard. Thus the arithmetic mode may change when only the implementing hardware or software changes and may vary from one version of the language compiler to the next.

The best course of action to establish details of the arithmetic mode is to code directly and run tests in the language of the FEA code to be used. Because of errors in converting numbers between decimal and binary number bases, accurate assessment of the arithmetic mode must involve binary arithmetic and internal testing of number models rather than reliance on the accuracy of number conversion to decimal form.

6.2 ARITHMETIC ERROR AND ACCURACY

Real variables that can be represented without approximation on a computer configuration are "perfect" reals. When the basic single-precision format of Fig. 6.1(B) is in use, perfect reals include all integers from zero to $2^{24} - 1 = 16\,777\,215$ and up to about seven 2-multiples of those integers implied by changing the value of the exponent. In addition, all of these integers are perfect reals when their exponent implies they have decimal parts. Thus the fractions ½, ¼, ⅛, . . . and many of their sums are perfect reals.

In a perfect computer world, all input and arithmetic operations would involve perfect reals. Then there would be no round-off error even though precision is limited. Though we can exploit perfect numbers in finite element implementation validation, using finite element problems that involve only perfect numbers to strengthen tests, practical problems always incur round-off error.

In accordance with the IEEE standard, the result of each addition, subtraction, multiplication, and division must be developed exactly before rounding. Thus with the usual rounding, the result of an arithmetic operation must have an error that is less than one-half of the last binary place of the unnormalized result. That is,

$$e \leq 2^{-t} \tag{6.1}$$

where

e is the absolute error and
t is the number of binary places in the coefficient.

When two accurate and nearly equal numbers are subtracted, the error magnitude does not reflect the fact that the accuracy of the solution is reduced to a maximum of one-half of one binary position. This fact is reflected better by the relative error: the absolute error divided by the exact value of the arithmetic operation. The relative error in subtracting c_2 from c_1 is

$$\epsilon \leq \frac{2^{-t}}{c_1 - c_2}$$

where ϵ is the relative error. This traditional definition of relative error disintegrates when $c_1 = c_2$, so we define round-off relative error by

$$\epsilon \leq \frac{2^{-t}}{\max.\ (|c_1|, |c_2|)} \tag{6.2}$$

where the value in the denominator is the maximum of the absolute value of the numbers c_1 and c_2.

In practical applications, we are interested in the number of decimal digits lost in the computer calculations. This number is found by

$$D_L = |\log_{10}(\epsilon)| \tag{6.3}$$

where

D_L is the number of digits lost and
$\log_{10}(\ )$ means the logarithm of the value in parentheses is to the base 10.

Given D_L, the number of digits retained is simply

$$D_R = D_p - D_L \tag{6.4}$$

where

D_R is the number of digits retained and
D_p is the number of digits of precision of the computer configuration in use.

These definitions emphasize the concept that the number of digits lost depends on the problem being solved while the number retained involves direct consideration of the arithmetic precision of the computer configuration employed.

An arithmetic operation that loses most of the accuracy of its input in generating its result is "critical arithmetic." When it associates with exponent underflow or overflow, the analyst may be alerted to its occurrence by the computer configuration. When the exponent is not exceeded, it is standard arithmetic management for hardware and software to ignore the occurrence of critical arithmetic in subtraction.

To explore the error and accuracy in a series of arithmetic operations, we will assume IEEE standard arithmetic. To simplify study, we will use rounding to the nearest binary number.

Figure 6.2 facilitates comparing the growth of relative error in three series arithmetic operations:

1. **Differencing:** $x = \Sigma(c_1 - c_2)$ summed over $i = 1, 2, \ldots, N$, where $c_1 = 2^{24} - 1$, $c_2 = 2^{24} - 2$
2. **Addition/subtraction:** $x = \Sigma c_3$ summed over $i = 1, 2, \ldots, N$, where $c_3 = 3$
3. **Multiplication/division:** $x = \Pi c_4$ for N products, where $c_4 = (2^{24} - 1)/2^{24}$.

Here the constants c_j have been chosen to aggravate round-off errors per calculation.

The curves of Fig. 6.2 reemphasize that the loss of accuracy in differencing is much more significant than errors in adding or multiplying numbers. They exhibit the fact that the maximum error increases with the number of arithmetic operations. Since the 2 million calculations of the figure would be required for a system of 200 equations with a bandwidth of 50, this data suggests that round-off errors may be destructive even when calculations involve few equations.

The round-off errors of Fig. 6.2 can be largely avoided by using different arithmetic algorithms. In differencing, the error will be zero if all the subtractions are performed first and then the results are summed. The relative error in serial addition will not exceed one digit if the additions are "pyramided": adding pairs

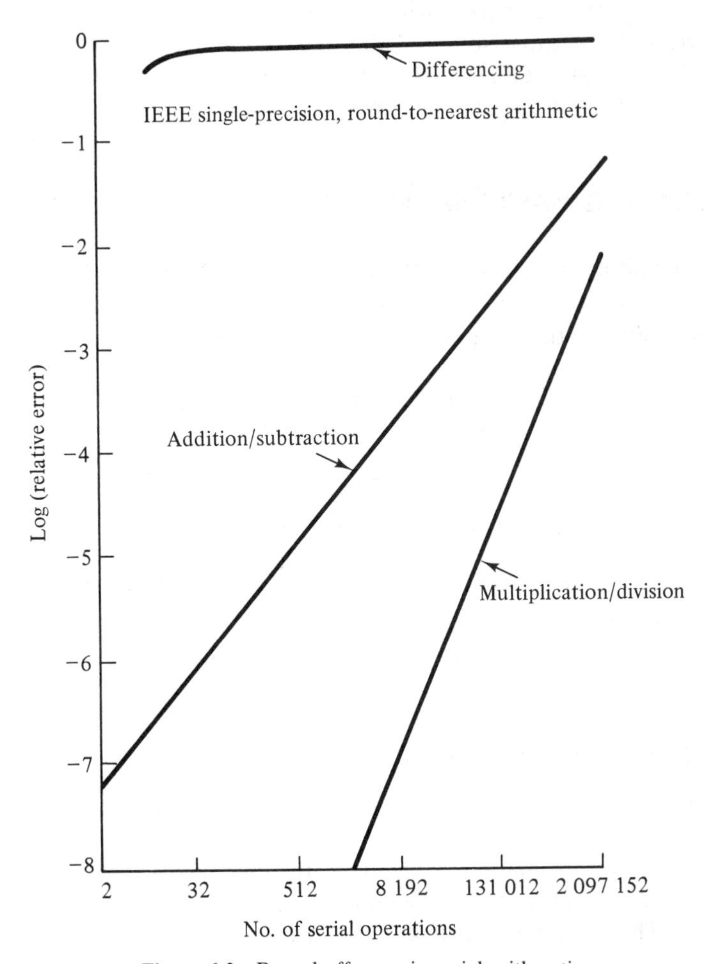

Figure 6.2 Round-off error in serial arithmetic.

of numbers, adding the pair sums, pairs of sums of the sums, and so on until all additions have been made. Similarly, when multiplications are pyramided, the relative error reaches only three digits lost rather than five digits when N reaches 2 million.

Changing the sequence of arithmetic operations can change the relative error drastically. Likewise, changing how the arithmetic is implemented changes round-off error. Replacing the serial addition by a single multiplication reduces the relative error to less than one digit lost. Furthermore, zero errors are realizable when these serial arithmetic operations involve perfect numbers.

We conclude that the relative round-off error is sensitive to the number representation, type of arithmetic, sequence of the arithmetic operations, and values of the numbers involved. Errors of differencing are most important because they can involve critical arithmetic without providing any signal that this has occurred. By comparison, in addition or subtraction and in multiplication or division, such errors are benign. Even in the worst case, millions of operations are needed to lose all accuracy in these operations.

6.3 DETERMINING MAXIMUM ERRORS

By combining matrix operations and the definition of a matrix norm, we can formulate equations for the maximum relative error induced by rounding.
The mathematical requirements of a norm are as follows:

1. The norm of a matrix must be greater than zero unless all the coefficients of the matrix are zero; that is,

$$\|\mathbf{A}\| > 0, \text{ unless all } A_{ij} = 0 \tag{6.5}$$

2. The norm of a scalar times a matrix must satisfy

$$\|\alpha\mathbf{A}\| = \|\alpha\| \, \|\mathbf{A}\| = |\alpha| \cdot \|\mathbf{A}\| \tag{6.6}$$

   where α is the scalar.

3. The norm of the sum of two matrices must satisfy

$$\|\mathbf{A} + \mathbf{B}\| \leq \|\mathbf{A}\| + \|\mathbf{B}\| \tag{6.7}$$

4. The norm of the product of two matrices must satisfy

$$\|\mathbf{A}\,\mathbf{B}\| \leq \|\mathbf{A}\| \, \|\mathbf{B}\| \tag{6.8}$$

There are many particular norms that satisfy Eqs. (6.5) through (6.8). We limit our attention to the Euclidean, one, infinity, and energy norms.

The Euclidean norm is defined by

$$\|\mathbf{A}\|_U = [\Sigma\, \Sigma\, (A_{ij})^2]^{0.5} \tag{6.9}$$

where summing extends over all coefficients of the matrix. This norm is the generalization of vector length to N-dimensional space.

The one and infinity norms are

$$\|\mathbf{A}\|_O = \max._j \Sigma_i |A_{ij}| \tag{6.10}$$

where the maximum is selected from among all the column sums of **A**, and

$$\|\mathbf{A}\|_I = \max._i \Sigma_j |A_{ij}| \tag{6.11}$$

where the maximum is selected from among all the row sums of **A**.

The Euclidean, one, and infinity norms are of special interest because they apply to any matrix and require few calculations to evaluate. When the matrix is a single vector, the infinity norm is interpreted as the vector coefficient of maximum absolute value, the one norm as the sum of the absolute value of the coefficients. In taking the norm of a product of a matrix times a vector, the infinity matrix norm of the matrix must be used with the infinity norm of the vector to satisfy the requirements of a norm. For a vector times a matrix, the one norm of the vector must be used with the one norm of the matrix. Since each of these norms must be equal to or greater than any coefficient of the matrix, the norm is a measure of the largest coefficient.

The energy norm, as previously noted, is defined by

$$\|\mathbf{A}\|_E = 0.5\, \mathbf{g}^T\, \mathbf{A}\, \mathbf{g} \tag{6.12}$$

where

A is a symmetric, positive definite matrix and
g is any non-null vector.

This norm plays an important role in both the theories of structures and finite elements.

The requirements of the norm define a basis for a norm algebra. Given a matrix equation, we can take the norm of both sides, arriving at an inequality relation that may be useful in bounding round-off error.

Table 6.1 illustrates formulation of an error bound for multiplication of a nonsingular matrix by a vector using the norm algebra. The final result, Eq. (f), expresses the relative error in the matrix product as a function of the norms of the errors in the input matrices. We arrive at this result following Wilkinson's approach of defining errors in the output in terms of hypothesized errors in the input.[3] We use the matrix algebra to express the absolute error in the solution vector in terms of the input matrices. We use the norm rules as the basis for developing inequalities in the norms of the solution vector and the error in the solution. Taking the ratio of these norms yields the desired result: an equation for evaluating the maximum relative error expressed in terms of norms of the input matrices.

Equation (f) expresses the bound as the product of the "condition number"

TABLE 6.1 FINDING AN ERROR BOUND

Given: $\mathbf{x} = \mathbf{Ab}$, $\mathbf{A}$ nonsingular (a)

Find: An equation bounding the relative error in $\mathbf{x}$.

1. We suppose that the error in $\mathbf{x}$, $\Delta\mathbf{x}$, is caused by an error in $\mathbf{A}$ and $\mathbf{B}$; that is,

$$(\mathbf{x} + \Delta\mathbf{x}) = (\mathbf{A} + \Delta\mathbf{A})(\mathbf{b} + \Delta\mathbf{b}) \tag{b}$$

2. Subtracting Eq. (a) from Eq. (b) gives

$$\Delta\mathbf{x} = \mathbf{A}\,\Delta\mathbf{b} + \Delta\mathbf{A}\,\mathbf{b} + \Delta\mathbf{A}\,\Delta\mathbf{b} \tag{c}$$

3. Taking the norms of both sides of Eq. (c) gives

$$\|\Delta\mathbf{x}\| \le \|\mathbf{A}\| \times \|\Delta\mathbf{b}\| + \|\Delta\mathbf{A}\| \times \|\mathbf{b}\| + \|\Delta\mathbf{A}\| \times \|\Delta\mathbf{b}\| \tag{d}$$

4. Multiplying Eq. (a) by $\mathbf{A}^{-1}$ and taking the norm of the equation gives

$$\|\mathbf{x}\| \ge \frac{\|\mathbf{b}\|}{\|\mathbf{A}^{-1}\|} \tag{e}$$

5. Dividing Eq. (d) by Eq. (e) gives the desired result,

$$\frac{\|\Delta\mathbf{x}\|}{\|\mathbf{x}\|} \le \underbrace{\|\mathbf{A}^{-1}\| \times \|\mathbf{A}\|}_{\text{condition number}} \underbrace{\left(\frac{\|\Delta\mathbf{b}\|}{\|\mathbf{b}\|} + \frac{\|\Delta\mathbf{A}\|}{\|\mathbf{A}\|} + \frac{\|\Delta\mathbf{b}\|}{\|\mathbf{b}\|} \times \frac{\|\Delta\mathbf{A}\|}{\|\mathbf{A}\|}\right)}_{\text{relative input error}} \tag{f}$$

of matrix $\mathbf{A}$, $\|A\|\ \|A^{-1}\|$, and the relative input error. The largest value of the relative input error is 3×2^{-t}. The condition number magnifies this error by a factor that has a minimum value of 1 but no upper bound. Thus this bound on relative error in the matrix product reflects the fact that large round-off errors can arise with some 2×2 matrices and little with other very large order matrices, depending on the magnitude of the condition number.

Equation (f) determines the error bound as the sum of norm error contributions from each of the participating matrices and a norm product term: the last term in the equation. Since the product term is at most 2^{-t} times the other terms in the relative input error, it can be disregarded for engineering purposes.

Following the process of Table 6.1 for solving linear simultaneous equations, Wilkinson developed the bound for relative error:[3]

$$\frac{\|\Delta\mathbf{g}\|}{\|\mathbf{g}\|} \le \|\mathbf{K}\|\ \|\mathbf{K}^{-1}\| \left[\frac{\|\Delta\mathbf{p}\|}{\|\mathbf{p}\|} + \frac{\|\Delta\mathbf{K}\|}{\|\mathbf{K}\|}\left(1 - \|\mathbf{K}\|\ \|\mathbf{K}^{-1}\| \frac{\|\Delta\mathbf{K}\|}{\|\mathbf{K}\|}\right)^{-1}\right] \tag{6.13}$$

Again, the matrix condition number plays the role of magnifying the small errors of the input matrices. Again, the error bound adds the norm error contributions of the input matrices.

The error bound, Eq. (6.13), is useful for evaluating preanalysis estimates of round-off error. The maximum stiffness and influence matrix diagonal values are easily estimated to arrive at an estimate of the condition number. Multiplying the condition number by 2^{-t} then provides a worst-case estimate of the relative round-off error and, by Eq. (6.3), the maximum number of digits lost in solution of the simultaneous equations.

6.4 DIRECT MEASUREMENT OF ERRORS

Though the norm algebra leads to rigorous upper-bound error equations like Eq. (6.13), these expressions furnish bounds that are much greater than realizable errors.[4] Much sharper estimates can be generated using direct measurements of round-off errors on the actual analysis of interest. Because direct measurements of errors reflect the actual arithmetic, they can be used to choose among solution processes and modeling options.

The direct approach to estimating round-off errors considers FEA as a sequence of matrix operations: assembly, decomposition, forward substitution, back substitution, and stress recovery. A direct measurement of round-off error is made or implied for each operation. To calculate the minimum number of digits retained, the maximum number of digits of accuracy lost in each operation are summed and subtracted from the digits of precision of arithmetic.[4] This approach assumes that the error inherited from previous operations is not canceled in the current operation. The error internal to the current operation adds to the previous error. The approach is consistent with the error accumulation of matrix products and the interpretation of error bound summing of Table 6.1.

To crystallize the ideas, we will consider an appropriate set of measures and detail their use. The entries of Table 6.2 summarizes the set and some properties of each measure.

Two measures span triangularization of the stiffness matrix: the numerical singularity and the translation measures. The numerical singularity measure senses the loss of accuracy in the determinant of the stiffness matrix. The translation measure senses the attrition error in the triangularization.

The numerical singularity measure is available in most production computer codes. It is based on the relative minimum factor in the determinant of the constrained stiffness matrix. Since for a positive definite matrix the diagonal coefficients must be reduced from their original value by the decoupling calculations of factorization, the measure determines directly how many of the significant digits can have been lost in the differencing calculations.

The numerical singularity error can induce unbounded errors in FEA response predictions as singularity is approached. In addition, the measure reflects realizable errors. Therefore, the measure can serve as the basis for analytical evaluation of limiting problem sizes as a function of the precision of computer arithmetic. Such an analysis indicates, for example, that with IEEE single precision, singularity error can limit the number of equal-span elements of a prismatic cantilevered beam to 322.[5] With unequal span elements, the limit can be two elements or several million.

The u translation measure senses how well the norm of the nodal forces

TABLE 6.2 SET OF DIRECT ERROR MEASURES

Error Measure	Operation	Ratio	Ideal Value	Digits Lost
1. Numerical singularity	Decomposition	$SR = \min\left(\frac{\mathbf{d}_{ii}}{\mathbf{K}_{ii}}\right)$	1	$\lvert\log_{10}(SR)\rvert$
2. Translations in x_i $i = 1, 2, 3$	Decomposition	$TR = \frac{\mathbf{r}^T\mathbf{K}\mathbf{r} - \mathbf{r}^T\mathbf{L}\mathbf{D}\mathbf{L}^T\mathbf{r}}{\mathbf{r}^T\mathbf{K}\mathbf{r}}$	0	$\log 2^t - \log TR$
3. Forward substitution	Forward substitution	$FR = \frac{\lVert\mathbf{p} - \mathbf{L}^{-1}\mathbf{p}\rVert}{\lVert\mathbf{p}\rVert}$	0	$\log 2^t - \log FR$
4. Energy	Back substitution	$BR = \frac{EW - 2SE}{EW}$	0	$\log 2^t - \log BR$
5. Displacement vector	Stress recovery	$DR = \frac{\lVert\mathbf{d}\rVert - \lVert\mathbf{R}\mathbf{d}\rVert}{\lVert\mathbf{d}\rVert}$	0	$\log 2^t - \log DR$
6. Vector differencing	Stress recovery	$VR = \frac{u_i - u_j}{\min(\lvert u_i\rvert, \lvert u_j\rvert)}$	0	$\log(VR)$

needed to move every node a unit distance in the x direction is preserved through the triangularization calculations. Here $\mathbf{r}^T$ is taken to be a vector of ones and zeros when a single rectangular coordinate system defines the global reference axes. The sensor provides a comprehensive measure of errors in all terms of the matrix, rather than just monitoring the diagonal coefficients. The check is made more complete by considering separately u, v, and w translations and unit rotations about each axis.

When the structure is an unbraced lattice, calculations of $\mathbf{r}^T\mathbf{K}_c\mathbf{r}$ can be drastically reduced by exploiting the knowledge that $\mathbf{r}^T\mathbf{K}\mathbf{r} = 0$. Then the value of the triple product can be determined by involving only the constrained degrees of freedom or those associated with unconstrained degrees of freedom, whichever engenders the least calculations. Pursuing this implementation, the measure also provides a partial check of element stiffness matrix generation and assembly accuracy.

The forward substitution measure is a direct calculation of the error in forward substitution. By calculating the residual associated with the forward substitution, the measure provides a necessary and sufficient check on the numerical process.

Similarly, the energy measure provides a direct check on the backward substitution process. The strain energy calculation is effected by summing the element energies; that is,

$$SE = 0.5 \sum \mathbf{g}_e^T \mathbf{K}_e \mathbf{g}_e \tag{6.14}$$

where summation extends over all element stiffness matrices, and comparing the result with the external work calculated by

$$EW = \mathbf{d}^T \mathbf{D}^{-1} \mathbf{d} \tag{6.15}$$

where

d is the vector evaluated by forward substitution and

D is the diagonal matrix of the triangular factorization.

Accordingly, the energy measure embraces errors of the back substitution process.

The stress recovery check recognizes stress evaluation as a two-step process. In the first step, the nodal displacements of an element are transformed by axis rotations to the element's axes; in the second, the displacements are differenced to determine strains. Since the cogradient transformation preserves the Euclidean length of the vector, comparison of the vector lengths before and after transformation senses the round-off. The differencing test compares the numbers involved in differencing to establish the loss in accuracy corresponding to the number of leading digits with the same value using Eq. (6.2). The sum of the digits lost in each of the two steps supplies the maximum number of digits losable in stress recovery.

The calculations associated with these measures are a small fraction of the calculations of an FEA. When many equations are involved, the analysis calculations are of the order of Nb^2, while the calculations for the set of measures add less than $5Nb$. The percentage increase in calculations to provide round-off error sensing is $500/b$. Therefore, when the bandwidth is greater than 100, the increase in calculations is less than 5 percent.

The error set includes no direct measures of round-off error in element matrix generation or assembly. This is consistent with a philosophy of element model tests independent of production analyses and experience confirming that round-off error measures of assembly sense negligible errors.

Note that round-off error measurements involve calculations that are themselves subject to round-off error. Usually, this means that round-off error calculations must be performed in higher-precision arithmetic than the structural analysis calculations. In any event, the careful analyst will include assessment of the logic and accuracy of error measures in validating the equation-solving and stress recovery calculations.

6.5 DETERMINING ACTUAL ERRORS

A facility for higher-precision arithmetic is a facility for evaluating the actual round-off errors experienced in an analysis.

If the lower-precision analysis is sufficient to produce a solution with some accurate binary places, higher-precision arithmetic will enhance the accuracy by the difference in the number of digits represented in the coefficient of the number representations. Accordingly, the number of digits that match in the two analysis results is the number of digits of the first analysis, untainted by round-off error. If no digits match, we know that the lower-precision analysis results are meaningless due to excessive round-off error.

A work of caution: To ensure that accuracy is increased by the precision difference, both analyses must involve the same input data and arithmetic mode. Some computer codes will direct use of double precision for only some of the analysis steps, thereby voiding the facility for exact round-off error assessment based on the precision change.

Availability of a higher-precision equation solver is a valuable adjunct for validating error measures and is useful in determining actual round-off errors in production analyses.

The round-off self-consistency test[6] can provide estimates of round-off error magnitudes when a double-precision implementation is not available. This test involves the following steps:

1. Predict the responses of the structure of interest by a numerical analysis. (We will refer to this as the "baseline" analysis.)

2. Multiply Young's modulus and the applied loads by a factor $f = (1 + 2^{-t+3})$, and perform the computer analysis of the revised structure. (The second term in parentheses perturbs the last digit of the computer representation of the modulus and loading.)

3. Compare response predictions of the baseline and revised problems. If there is no round-off error, results will be identical. The number of digits, counting left to right, that match in the two analysis results provides a measure of the actual round-off error.

6.6 ILLUSTRATIVE PROBLEM

We introduce error analysis for the structure of Fig. 6.3 to clarify the ideas of round-off error analysis. The structure is a two-story building frame-loaded with uniformly distributed roof, floor, and wind loads.

The structural model consists of three roof beam elements, eight floor beam elements, and four column elements. This discretization was the analyst's choice based on computer program limitations—EASE, the program used, evaluates displacements only at the nodes. The equivalent nodal forces and moments are used. Thus the model is a lattice model. Table 6.3 presents the results of IEEE single-precision computer analysis of the structure of Fig. 6.3.

Table 6.4 lists the results of round-off error measure evaluations for this structure. This table furnishes the maximum number of digits lost in evaluating internal forces for each element and for each analysis step.

Using the data of Table 6.4, we conclude that the maximum error in deflection predictions is less than or equal to 5.094 digits. To arrive at this maximum, we sum the value of the maximum digits lost for each operation, factorization, forward substitution, and back substitution: 2.280 + 1.476 + 1.339 = 5.094. Subtracting the maximum from the computer's digits of precision, we conclude that the displacement predictions of Table 6.3 will have at least 6.924 − 5.094 = 1.830 digits of accuracy.

To arrive at the probable digits of accuracy retained, we start by calculating the root mean square digits lost for the three error sources. This calculation is consistent with the premise that the error from each source is statistically independent and a representation of a binomially distributed population. The result of this calculation and use of Eq. (6.3) furnishes the fact that the analysis has an expected minimum of at least 3.70 digits of accuracy.

To determine the maximum number of digits losable through the evaluation of internal loads, we add the digits lost in stress recovery. From the data of Table 6.4, this leads to the conclusion that in the worst case, internal load predictions have no guaranteed digits of accuracy, but an accuracy of at least 2.76 digits can be expected.

Table 6.5 lists response predictions from a double-precision analysis of the structure of Fig. 6.3. Compaing displacement predictions of Tables 6.5 and 6.3

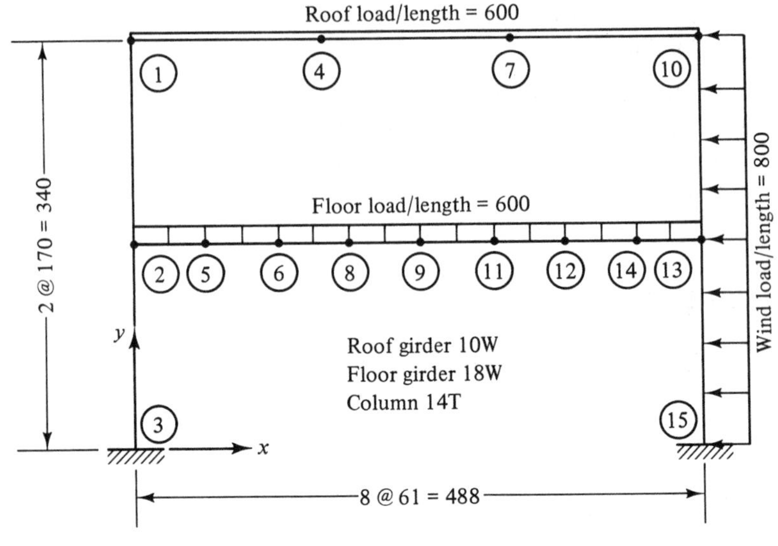

Element	EA	EI	Node	x	y
1, 4	10E7	$5.72\overline{2}$E7	1	0	28.25
4, 7	10E7	$5.72\overline{2}$E7	2	0	$14.16\overline{6}$
7, 10	10E7	$5.72\overline{2}$E7	3	0	0
2, 5	24E7	1.422917E8	4	$13.5\overline{5}$	28.25
5, 6	24E7	1.422917E8	5	$5.08\overline{3}$	$14.1\overline{6}$
6, 8	24E7	1.422917E8	6	$10.16\overline{6}$	14.16
8, 9	24E7	1.422917E8	7	$27.1\overline{1}$	28.25
9, 11	24E7	1.422917E8	8	15.25	$14.1\overline{6}$
11, 12	24E7	1.422917E8	9	$20.3\overline{3}$	$14.1\overline{6}$
12, 14	24E7	1.422917E8	10	$40.6\overline{6}$	28.25
14, 13	24E7	1.422917E8	11	$25.41\overline{6}$	$14.1\overline{6}$
1, 2	48E7	$8.16\overline{6}$E7	12	30.5	$14.1\overline{6}$
2, 3	48E7	$8.16\overline{6}$E7	13	$40.6\overline{6}$	$14.1\overline{6}$
10, 13	48E7	$8.16\overline{6}$E7	14	$35.58\overline{3}$	$14.1\overline{6}$
13, 15	48E7	$8.16\overline{6}$E7	15	$40.6\overline{6}$	0

Fixity: 3, 1; 3, 2; 3, 3; 15, 1; 15, 2; −15, 3

Loads: Wind: 10, 1, $-5666.6\overline{6}$; 13, 1 −11.33; 10, 3, −13, 379.63
Roof: 1, 2, $-4066.\overline{6}$; 4, 2, $-8133.\overline{3}$; 7, 2, $-8133.\overline{3}$, 10, 2, $-4066.\overline{6}$; 1, 3, −9187.646; 10, 3, 9187.646
Floor: 2, 2, $-4066.\overline{6}$; 5, 2, $-8133.\overline{3}$; 6, 2, $-8133.\overline{3}$; 8, 2, $-8133.\overline{3}$; 9, 2, $-8133.\overline{3}$; 11, 2, $-8133.\overline{3}$; 12, 2, $-8133.\overline{3}$; 14, 2, $-8133.\overline{3}$; 13, 2, $-4066.\overline{6}$; 2, 1, −3445.370; 13, 1, 3445.370

Figure 6.3 Illustrative frames.

TABLE 6.3 SINGLE-PRECISION ANALYSIS RESULTS

Analysis Results: Nodal Displacements

Node	*x* Direction	*y* Direction	Direction 3
1	-8.35971E - 02	-1.81310E - 03	+6.38192E - 04
2	-4.41414E - 02	-1.43005E - 03	-1.15326E - 03
3	+0.00000E + 00	+0.00000E + 00	+0.00000E + 00
4	-8.56461E - 02	-6.74113E - 02	-5.50577E - 03
5	-4.42032E - 02	-2.49671E - 02	-7.15894E - 03
6	-4.42649E - 02	-6.58229E - 02	-8.21334E - 03
7	-8.76950E - 02	-7.96605E - 02	+4.15637E - 03
8	-4.43267E - 02	-1.02582E - 01	-5.79347E - 03
9	-4.43884E - 02	-1.21339E - 01	-1.37635E - 03
10	-8.97440E - 02	-1.32161E - 03	+3.50670E - 03
11	-4.44502E - 02	-1.15694E - 01	+3.56095E - 03
12	-4.45119E - 02	-8.67572E - 02	+7.54143E - 03
13	-4.46355E - 02	-9.88759E - 04	+6.72378E - 03
14	-4.45737E - 02	-4.31467E - 02	+9.08806E - 03
15	+0.00000E + 00	+0.00000E + 00	+0.00000E + 00

Analysis Results: Member End Forces

Member	Axial	Shear	Node A Moment	Node B Moment
1	-1.51151E + 04	+8.98882E + 03	+8.68598E + 04	+3.49886E + 04
2	-1.51151E + 04	+8.55487E + 02	-3.49886E + 04	+4.65852E + 04
3	-1.51152E + 04	-7.27785E + 03	-4.65852E + 04	-5.20701E + 04
4	-2.91543E + 03	+3.13313E + 04	+2.47744E + 05	-8.84761E + 04
5	-2.91562E + 03	+2.31979E + 04	+8.84760E + 04	+2.94473E + 04
6	-2.91525E + 03	+1.50646E + 04	-2.94472E + 04	+1.06026E + 05
7	-2.91561E + 03	+6.93128E + 03	-1.06026E + 05	+1.41260E + 05
8	-2.91560E + 03	-1.20225E + 03	-1.41260E + 05	+1.35149E + 05
9	-2.91560E + 03	-9.33550E + 03	-1.35149E + 05	+8.76933E + 04
10	-2.91596E + 03	-1.74690E + 04	-8.76928E + 04	-1.10719E + 03
11	-2.91613E + 03	-2.56023E + 04	-1.31253E + 05	+1.10769E + 03
12	-1.30555E + 04	-1.51151E + 04	-9.60474E + 04	-1.16824E + 05
13	-4.84535E + 04	-1.80306E + 04	-1.34365E + 05	-1.21069E + 05
14	-1.13445E + 04	+9.44853E + 03	+4.78781E + 04	+8.51887E + 04
15	-3.35015E + 04	+1.03107E + 03	+4.60641E + 04	-3.14572E + 04

Axial internal force is tension.
Let the vector from node A to node B define the *x*-axis vector. Let the *y*-axis vector be the *x* rotated counterclockwise 90 degrees. Then,
1. Positive moment is a counterclockwise external moment.
2. Positive shear is an A node, *y*-direction external force.

TABLE 6.4 SINGLE-PRECISION ANALYSIS ERROR REPORTS

Element Member	Relative Energies and Stretching	Bending	Stress Error Digits Lost
1	+0.00458	+0.06687	+1.62118
2	+0.00458	+0.05863	+1.63145
3	+0.00458	+0.02866	+1.64147
4	+0.00003	+0.16037	+2.85481
5	+0.00003	+0.01072	+2.85539
6	+0.00003	+0.02680	+2.85605
7	+0.00003	+0.08126	+2.85660
8	+0.00003	+0.10086	+2.85721
9	+0.00003	+0.06654	+2.85781
10	+0.00003	+0.01337	+2.85836
11	+0.00003	+0.03058	+2.85894
12	+0.00074	+0.09899	+0.67517
13	+0.01024	+0.14053	+0.00000
14	+0.00056	+0.04648	+0.59885
15	+0.00490	+0.03897	+0.00000

```
u translation digits lost = 2.062955
v translation digits lost = 0.2968063
Forward substitution digits lost = -1.476379
Singularity ratio = 5.244894E - 03 at node 14,
component 1
Numerical singularity digits lost = 2.280263
Energy digits lost = 1.338307
```

yields the conclusion that the least accuracy is 4.50 digits rather than the bound estimate of 1.83. Similarly, comparing member end force predictions of double- and single-precision analyses, we find that internal loads are predicted by single-precision arithmetic to at least 2.91 digits rather than the bound estimate of zero digits.

The following table summarizes the round-off error results for this problem. The expected maximum errors provide a conservative estimate of the actual errors, but excellent estimates of error are also produced, for less computer resources, by the round-off consistency test.

Digits Lost Due to Round-Off Error

ITEM	DISPLACEMENT	STRESSES
Error bound	5.09	> 6.92
Expected maximum	3.02	4.16
Actual maximum	2.42	4.01
Self-consistency maximum	2+	4+

TABLE 6.5 DOUBLE-PRECISION ANALYSIS RESULTS

Frame.C6 is the problem data set name.
The sum of (AEL), the weight factor, = 4.094667E + 10
Singularity ratio = 5.244788E - 03 at node 14, component 1

Analysis Results: Nodal Displacements

Node	x Direction	y Direction	Direction 3
1	-8.35990D - 02	-1.81310D - 03	+6.38217D - 04
2	-4.41425D - 02	-1.43005D - 03	-1.15317D - 03
3	+0.00000D + 00	+0.00000D + 00	+0.00000D + 00
4	-8.56479D - 02	-6.74113D - 02	-5.50579D - 03
5	-4.42042D - 02	-2.49667D - 02	-7.15886D - 03
6	-4.42660D - 02	-6.58222D - 02	-8.21328D - 03
7	-8.76969D - 02	-7.96608D - 02	+4.15636D - 03
8	-4.43277D - 02	-1.02581D - 01	-5.79345D - 03
9	-4.43895D - 02	-1.21338D - 01	-1.37638D - 03
10	-8.97458D - 02	-1.32160D - 03	+3.50675D - 03
11	-4.44513D - 02	-1.15693D - 01	+3.56090D - 03
12	-4.45130D - 02	-8.67568D - 02	+7.54137D - 03
13	-4.46365D - 02	-9.88751D - 04	+6.72381D - 03
14	-4.45748D - 02	-4.31467D - 02	+9.08801D - 03
15	+0.00000D + 00	+0.00000D + 00	+0.00000D + 00

Analysis Results: Member End Forces

Member	Axial	Shear	Node A Moment	Node B Moment
1	-1.51151E + 04	+8.98883D + 03	+8.68601D + 04	+3.49886D + 04
2	-1.51151E + 04	+8.55501D + 02	-3.49886D + 04	+4.65853D + 04
3	-1.51151E + 04	-7.27783D + 03	-4.65853D + 04	-5.20698D + 04
4	-2.91565E + 03	+3.13313D + 04	+2.47744D + 05	-8.84766D + 04
5	-2.91565E + 03	+2.31979D + 04	+8.84766D + 04	+2.94464D + 04
6	-2.91565E + 03	+1.50646D + 04	-2.94464D + 04	+1.06025D + 05
7	-2.91564E + 03	+6.93126D + 03	-1.06025D + 05	+1.41259D + 05
8	-2.91565E + 03	-1.20208D + 03	-1.41259D + 05	+1.35148D + 05
9	-2.91565E + 03	-9.33541D + 03	-1.35148D + 05	+8.76930D + 04
10	-2.91564E + 03	-1.74687D + 04	-8.76930D + 04	-1.10633D + 03
11	-2.91564E + 03	-2.56021D + 04	-1.31250D + 05	+1.10633D + 03
12	-1.30555E + 04	-1.51151D + 04	-9.60477D + 04	-1.16824D + 05
13	-4.84534E + 04	-1.80308D + 04	-1.34366D + 05	-1.21070D + 05
14	-1.13445E + 04	+9.44846D + 03	+4.78778D + 04	+8.51880D + 04
15	-3.35012E + 04	+1.03077D + 03	+4.60621D + 04	-3.14595D + 04

Axial internal force is tension.
Let the vector from node A to node B define the x-axis vector. Let the y-axis vector be the x rotated counterclockwise 90 degrees. Then,
1. Positive moment is a counterclockwise external moment.
2. Positive shear is an A node, y-direction external force.

For both the displacements and the stresses, the statistical estimates of expected error are closer to the actual error than the error bound. In general, although the error bound is a realizable bound, the statistical estimates of the maximum error are rarely exceeded. Both actual errors and maximum error analyses indicate that stress recovery is the largest source of round-off error in this problem.

6.7 MINIMIZING ERRORS

Choices the analyst makes in describing the structural problem for FEA influence the round-off error magnitude. By biasing these choices, the analyst can drastically reduce round-off errors.

The most influential choices and the desirable bias are as follows:

1. *Choice of dimensional units.* This decision can change the input data from truncated to perfect numbers. Insofar as calculations with this data result in perfect numbers, the decision will eliminate round-off error.

In the United States, the best choice of the length unit is inches. The architect's scale is still the most commonly used basis for detailing and construction. Thus use of inches and the usual inch fractions (½, ¼, ⅛, 1/16, 1/32 means that all geometry data will be perfect numbers. If centimeters and decimal fractions of a centimeter are used, numbers with fractional parts will be imperfect. For this reason alone, U.S. engineers will tend to incur less round-off error than their counterparts in most of the rest of the world.

In the United States, pounds are the best unit of force for minimizing round-off error. Here most structural design codes specify design loadings in terms of an integer number of pounds of force or force per unit length.

2. *Choice of constraints.* Maximizing the number of displacement constraints not only decreases the number of calculations in equation solving but also can only increase the stability of the structure, as measured by its numerical singularity. Therefore, maximizing the number of constraints reduces the potential for destructive round-off.

If the structure is symmetric, with lines or planes of symmetric or antisymmetric response, round-off is reduced by representing only part of the structure and introducing constraints to imply the rest of the structure. Similarly, where measured settlements are available, these should be used to reduce round-off error.

When it is the analyst's intent to represent part of the structure as infinitely stiff compared with another part, round-off is virtually avoided by introducing appropriate constraints rather than choosing a relatively large number to imply infinite stiffness. Conversely, when a member is idealized as offering no resistance to deformation, that member should be represented by zero stiffness rather than a relatively small stiffness value.

3. *Choice of element models.* To explore the relation between round-off error and element model selection, consider the Bernouilli-Euler and Timoshenko beam alternatives. When the beam is long relative to its depth, shear deformations will be negligible. If this beam is modeled by many short segments and the Timoshenko model, excessive round-off occurs due to the dominance of the shear flexibility on the main diagonal of the stiffness matrix. In this case, the Bernouilli-Euler model is a better choice.

The same situation arises with respect to plate and shell analysis when shear deformation models are used. When the shell can be idealized as having infinite shear resistance, round-off error is reduced by using the bending models that associate with the Germain-Lagrange plate-bending equations. When shear flexibility must be included, higher round-off errors are expected.

4. *Number of nodes.* A given lattice structure discretized with some number of nodes will almost always evoke more round-off error when the number of nodes is increased. Both the numerical singularity measure and the matrix condition number will define higher error bounds. Since the range of the error increases with the number of calculations, the assumption of a common population distribution leads to the conclusion that the variance of the error must increase. Thus we expect the errors to increase with the number of nodes.

5. *Node sequencing.* Algebraic studies of truss and beam problems prove that there is a nodal sequence that minimizes the numerical singularity error measure. To define this optimum, the number 1 node is taken to be the node of the greatest flexibility—the node with maximum movement when a unit load is applied there. The number 2 node is assigned to the next most flexible point, and so on.

For the cantilevered beam, this sequencing numbers the nodes from the tip to the root of the cantilever. For a simply supported beam, optimum numbering starts near midspan and moves toward the supports, alternately approaching left and right supports.

The importance of sequencing on numerical singularity is dramatized by the cantilevered beam. Numbering from the root to the tip limits the number of equal span elements to 332 under IEEE single-precision arithmetic.[1] Numbering from tip to root means, theoretically, that there is no limit on the number of elements—the singularity ratio is never smaller than ¼.

6. *Nodal spacing.* Since stiffness coefficients vary with element length, nodal spacing plays a role in reducing round-off error. With the optimum nodal spacing, all stiffnesses added at a node will have the same values.

7. *Scale factor of the loading vector.* When addressing linear static analysis, we are free to scale the loading vector by a factor and scale the resulting displacements and element internal loads by the reciprocal of the factor. This device should be used if the nominal loading induces exponent underflow or overflow.

8. *Choice of coordinate axes.* The choice of coordinate axes affects round-off error in stress calculations. If the axes can be aligned with element axes, the transformation of displacements from the global to the local axes will be free of round-off error.

The best problem description with respect to round-off errors will usually produce results of the desired accuracy. It will be a compromise of the optimum for each of the particular choices.

6.8 INCREASING SOLUTION ACCURACY

When lattice analysis results are not accurate enough, accuracy can be improved by reanalysis with higher-precision arithmetic or problem reformulation.

Higher-precision arithmetic is the direct way of increasing the accuracy. Higher-precision reanalysis will usually more than double the number of digits retained. Because a more complex second analysis is activated, the total computer resources needed for the analysis will be more than twice those of single-precision analysis. Hard-wired double-precision arithmetic consumes 15 to 30 percent more computer time than single-precision. Storage and data transfer data volume requirements of the double-precision analysis will be twice those of the single. Therefore, double-precision analysis will require between 1.15 and 2.00 times the computer time of single-precision.

Reformulation may involve changing the problem description to improve its optimality with respect to round-off or changing the basis for effecting the solution. In either case, data of the error sensors will indicate the most advantageous reformulation.

Table 6.6 summarizes reformulation indicated by each of the error sensors. The third column lists problem redescription indicated by each sensor. The fourth column lists indicated changes in the analysis basis. Items in this column merit further discussion.

Unfortunately, changing the solution algorithm does not guarantee reduced round-off error. Because an alternate algorithm will involve different numbers in the arithmetic, a different arithmetic sequence, or both, it will usually result in different round-off.

It is a popular misconception that an iteration process will produce results with lower round-off error than a direct method. Suppose that the iteration cutoff criteria require that the process error, the error associated with cutting off the iterations, be less than the round-off error. There is no guarantee that the iteration will produce a converging sequence of solution estimates. Even if convergence occurs, round-off error in the error calculation may result in erroneous conclusions about solution accuracy. An iteration algorithm can only be expected to have less round-off than a direct process when it uses higher-precision arithmetic than the direct.

Usually, a large error in the substitutions is caused by a loading that changes sign each time the node number increases: a flip-flop loading. This type of loading

TABLE 6.6 REFORMULATION INDICATED BY ERROR SENSORS

Dominant Error Measure	Likely Error Source	Problem Reformulation Action	Problem Basis Change Action
Numerical singularity	Disparate stiffness	Change node sequencing Change node spacing Increase constraints	Change solution algorithm
Translations	Disparate stiffness	Change node spacing Change element model	Change solution algorithm
Forward substitution	Flip-flop loading	None	Superimpose flip and flop solutions
Energy balance	Flip-flop strains	None	Superimpose flip and flop solutions
Displacement vector	Disparate strains	Change reference axes	None
Vector differencing	Large rigid motions	None	Change to displacement/strain variables

converts substitutions into differencing calculations. When this type of loading occurs, separating plus and minus load components into independent loading vectors reduces the differencing to differencing the solution—calculations that are easily monitored to determine which response predictions are accurate and which inaccurate.

Large errors in the vector differencing of stress recovery usually arise when the deformations are small relative to rigid body motion of the element. The "Digits Lost" column of Table 6.4 reflects this characteristic for the two-story bent of Fig. 6.3. This data confirms that elements with maximum deflections, elements generally far removed from the supports, are the prime candidates for large stress recovery differencing errors.

The differencing error can be significantly reduced by changing from displacement unknowns to a mixture of displacement and element strain variables. This transformation will replace displacement differencing with differencing of element stiffness coefficients, a calculation that is intrinsically better with respect to round-off error because it involves local rather than system behavior.

6.9 ROUND-OFF ERROR CONTROL

Figure 6.4 is characteristic of the growth of worst error as a function of the number of equations. This graph shows the growth of the triangularization and stress recovery round-off error, the two operations that most frequently indicate signif-

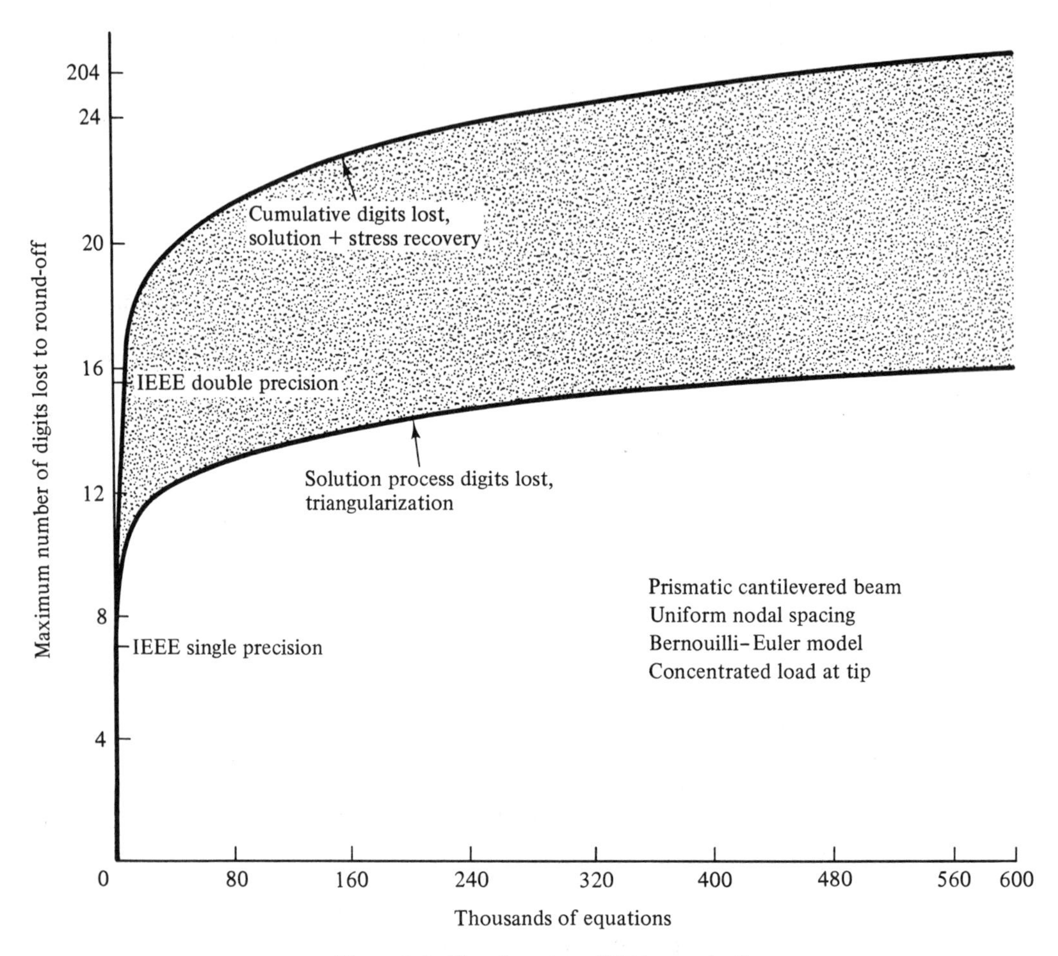

Figure 6.4 The character of FEA round-off error.

icant error. The data indicates that stress recovery errors, though not as large as triangularization, can involve ruinous errors when only a few thousand equations are treated. The data suggests that IEEE single precision will be inadequate in dealing with more than 1000 equations. Because most production analyses involve less than 1000 equations, application experiences may lull the analyst into a sense of false security with respect to round-off. To be realistic, the analyst should always be alert to the signals of excessive round-off errors and be prepared to control them.

Engineers can control computer-induced round-off error by the practices they adopt for describing structural problems for computer analysis, by monitoring error data produced in the course of the analysis, and, when necessary, by directing a reanalysis using higher-precision arithmetic or redefining the problem-descriptive data. Control is facilitated by a computer configuration that includes a complete set of error sensors and equation-solving capability for at least two levels of arithmetic precision.

6.10 HOMEWORK

***6:1.** Numerical singularity of a beam stiffness matrix.

Given: A prismatic cantilevered beam discretized into n equal-length elements.

Develop: An equation of the numerical singularity ratio.

(a) With nodal numbering from tip to root

(b) With nodal numbering from root to tip

Method: Use the physical interpretation of the diagonals of the **D** matrix.

6:2. Computer arithmetic.

Given:

(a) A hypothetical computer with a floating-decimal arithmetic of three digits of accuracy.

(b) The following matrix.

$$\begin{bmatrix} 3.0000 & 7.2000 & -3.900000 \\ & 17.2801 & -9.359840 \\ \text{Sym.} & & 6.070256 \end{bmatrix}$$

Determine: The computer's value of the determinant of the stiffness matrix calculated by triangular factorization. (The exact value of the determinant is 0.0003.)

***6:3.** The norm algebra.

Given: The matrix equation $\mathbf{C} = \mathbf{A}\,\mathbf{B}$ where **A** and **B** are square, nonsingular matrices.

Find: An inequality relationship between the relative error in **C** and errors $\Delta\mathbf{A}$ and $\Delta\mathbf{B}$ in **A** and **B**.

Method: Use the norm algebra and Wilkinson's method.

6:4. Bounding equation-solving error.

Given: The equations $\mathbf{K}\,\mathbf{g} = \mathbf{p}$.

Prove: If we hypothesize inherited errors $\Delta\mathbf{K}$ and $\Delta\mathbf{p}$ in **K** and **p**, respectively, the relative error norm measure is that given by Eq. (6.13).

Method: Use the norm algebra and Wilkinson's method.

6:5. Maximum equation-solving error.

Given: A diagonal stiffness matrix of order 50 with each diagonal having a value of 0.5.

Do the following:

(a) Calculate the condition number of this matrix using the maximum value norm.

(b) Calculate the maximum error using Eq. (6.13).

(c) Determine the actual maximum error in matrix inversion.

6:6. Digits of accuracy.

Given: An arithmetic result is given exactly by 3.172594 and from computer calculation by 3.171983.

Find: The number of digits of accuracy of the computer result.

***6:7.** Wilson's notorious matrix.

Given: Wilson's matrix and its inverse. (The matrix is notorious because though both the coefficients of the matrix and its inverse are integers, the matrix is ill conditioned. T. S. Wilson included the matrix in a paper published in 1946.)

$$\mathbf{K} = \begin{bmatrix} 5 & 7 & 6 & 5 \\ 7 & 10 & 8 & 7 \\ 6 & 8 & 10 & 9 \\ 5 & 7 & 9 & 10 \end{bmatrix} \qquad \mathbf{K}^{-1} = \begin{bmatrix} 68 & -41 & -17 & 10 \\ -41 & 25 & 10 & -6 \\ -17 & 10 & 5 & -3 \\ 10 & -6 & -3 & 2 \end{bmatrix}$$

Find:

(a) The condition number of the matrix using the zero, infinity, or Euclidean norm.

(b) The maximum error in equation solution assuming IEEE single-precision arithmetic.

6.11 COMPUTERWORK

The computerwork problems from 6.2 on have been designed so that round-off errors are measurable for the EASE computer configuration. For these problems, do not scale input data.

6-1. Computer arithmetic mode.

Given: A particular computer and a computer language.

Determine:

(a) The number of digits of precision, to five digits of accuracy, using single-precision arithmetic.

(b) The maximum value of the exponent in floating-point arithmetic and computer action when the exponent overflows.

(c) The minimum value of the exponent in floating-point arithmetic and computer action when the exponent underflows.

(d) Whether results of multiplication are rounded or truncated.

Method: Develop and run small computer programs that test, internally, for each feature and report results.

***6-2.** Numerical singularity error.

Given: The structure of the figure.

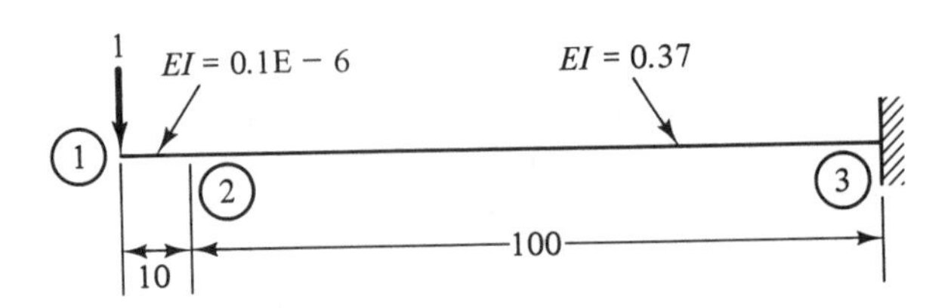

Computerwork Problem 6-2

Do the following:

(a) Find the exact value of the singularity ratio and the tip deflection.

(b) Find the value of the singularity ratio and tip deflection calculated by the computer configuration.
(c) Compare the results of activities (a) and (b).

6-3. Low-rise truss.

Given: The truss of the figure.

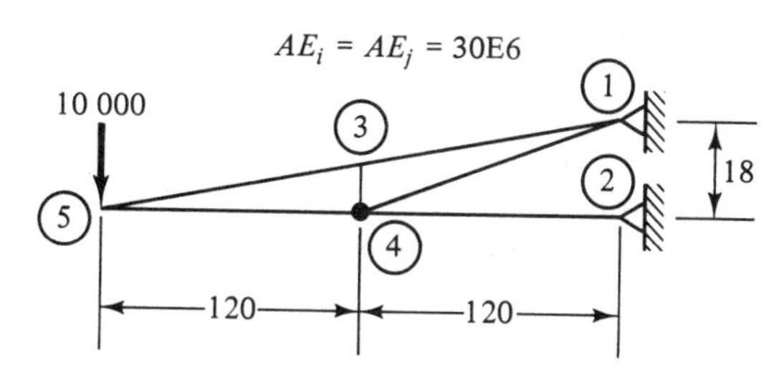

Computerwork Problem 6-3

Find:

(a) The actual error in prediction of the force in member 3, 5.
(b) The maximum possible error as predicted by the error measures.
(c) The expected error as predicted by the error measures.

6-4. Equation-sequencing error.

Given: The structures of parts (A) and (B) of the figure.

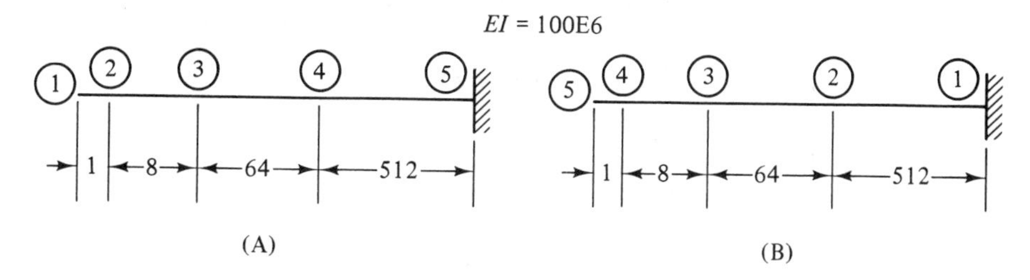

Computerwork Problem 6-4

Do the following:

(a) Analyze each structure using single-precision arithmetic.
(b) Analyze the structure in part (B) using double-precision arithmetic.
(c) Compare the single- and double-precision analysis results to determine the actual round-off error.
(d) Compare the deduction of actual round-off error with the exact solution from $v_{tip} = PL^3/(3EI)$.
(e) Define the effect of changing the nodal numbering sequence.

***6-5.** Round-off self-consistency test.

Given: The structure of part (A) of Prob. 6-4 with a lateral load of 1000 at node 1 and node 4 fixed.

Do the following:

(a) Analyze the baseline structure.

(b) Define the Young's modulus and loading factor $f = 1.00007$ and analyze the revised structure.

(c) Compare the results of the baseline and revised structure element bending moments to estimate the maximum round-off error in bending moments caused by round-off errors.

(d) Analyze the baseline structure using double-precision arithmetic and determine actual bending-moment round-off errors by comparing results with those of task (a).

(e) Compare estimated errors of the task (c) analysis with the actual errors of the task (d) result.

6-6. Fieldhouse strip.

Given: The beam strip model of a fieldhouse shell of the figure.

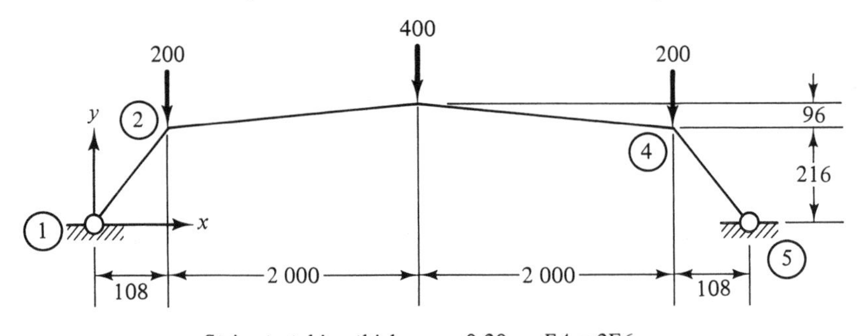

Strip stretching thickness = 0.20 $EA = 2E6$
Strip bending thickness = 0.93 $EI = 6.67E6$

Computerwork Problem 6-6

Do the following:

(a) Find the vertical reactions at the supports.

(b) Compare reactions with their exact values.

6-7. King post truss.

Given: The king post truss of the figure.

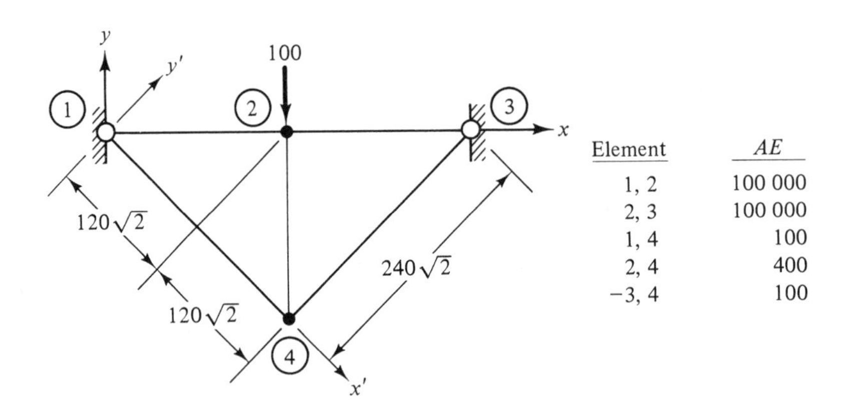

Element	AE
1, 2	100 000
2, 3	100 000
1, 4	100
2, 4	400
−3, 4	100

Computerwork Problem 6-7

Do the following:

(a) Analyze the truss using the xy coordinate system and the computer configuration.
(b) Analyze the truss using the $x'y'$ coordinate system and the computer configuration.
(c) Compare the actual errors in the prediction of the force in member 2, 4 with the maximum errors possible, using the digits-lost data.

6.12 REFERENCES

1. Melosh, R.J., "Characteristics of Manipulation Errors in Solving Load-Deflection Equations," ASME PVP Meeting, New York, Nov. 1970.
2. IEEE Standard 754, "Binary Floating Point Arithmetic," Mar. 1985.
3. Wilkinson, J.H., *Rounding Errors in Algebraic Processes*, Englewood Cliffs, N.J.: Prentice-Hall, 1963.
4. Melosh, R.J.; Utku, S., "Estimating Manipulation Errors in Finite Element Analysis," *Finite Elements in Analysis and Design*, vol. 3, no. 3, Oct. 1987.
5. Melosh, R.J., "Manipulation Errors in Finite Element Analysis," in *Recent Advances in Matrix Methods of Structural Analysis and Design*, Huntsville: University of Alabama Press, 1971, pp. 857–877.
6. Melosh, R.J.; Utku, S., "Verification Tests for Computer-aided structural Analysis," ASCE Spring National Conference, Nashville, Tenn., May 1988.

Chapter 7

Continuum Element Models

All finite element models that are not lattice models are continuum element models. The continuum class includes all elements with three or more nodes. Since continuum element models evoke discretization error, these models also encompass some two-noded elements.

Most of the finite element models in a given computer code, and in the engineering literature, are continuum element models. In NASTRAN, for example, more than 84 percent of the total of 26 element models are continuum models.[1] In Fredriksson and Mackerle's review, more than 92 percent of the element models are continuum models.[2]

Continuum elements model membranes, plates, shells, and three-dimensional structure of virtually any geometry and material composition. Their use is essential when assumptions of the structural theory are unrealistic: in the analysis of joints of a frame, surfaces that are thick compared with other dimensions, and massive structures like gravity dams where there is no other recourse than the theory of elasticity.

Continuum element models are preferred by analysts when biaxial and triaxial effects govern behavior due to material or geometry effects. These include Poisson ratio effects such as shear lag and near incompressibility, stress failures due to multiaxial stress states, and complex interactions induced by material anisotropy. They include geometry effects of two-directional bending and buckling under multiaxial stress states.

This chapter classifies existing continuum models, presents some element models in closed form, describes the potential energy and other methods of deriving model components, and identifies the characteristics of the continuum element models.

7.1 TYPES OF CONTINUUM ELEMENT MODELS

Figure 7.1 relates the terms that identify the types of continuum models. The two highest levels of the tree characterize all element types. Field and boundary terms for continuum elements parallel the terms unbraced and braced elements for lattice models. Geometry terms for continuum elements embrace line, surface, and solid components. The number of nodes, the structural equation model, and the formulation basis are the remaining distinguishing features between element models.

Figures 7.2 through 7.4 provide a listing of some of the element models available. Many of these models are detailed by Fredriksson and Mackerle.[2] Membrane models include three- and four-sided geometries with the shape of sides defined by up to third-order polynomials in the midplane coordinates. There are elements with nodes only on the sides and elements with nodes both on the sides and in the interior. Semi-infinite elements are available. Nodal variables, in almost all cases, are the u and v displacements of the midsurface of the element. The mathematical model for all the membrane elements is the theory of elasticity reduced by plane stress or plane strain assumptions to two-dimensional elasticity.

The plate models of Fig. 7.3 include three- and four-sided geometries but with sides defined by up to second-order polynomials of the midplane coordinates. Nodal variables include displacements normal to the surface, the first derivatives of normal displacements with respect to midplane coordinates, and in some cases higher derivatives. The mathematical model for most of the plate elements is the Germain-Lagrange plate equations developed by constraining solutions of the three-dimensional theory of elasticity by the Kirchhoff hypothesis (plane sections normal to the midsurface remain plane) and Love's hypothesis (the stress normal to the midsurface is negligible).

Shell models are only indirectly represented by the figures. The flat shell element is represented by assembling the membrane stiffnesses and plate thicknesses in a single matrix. Similarly, curved shell element models include both midplane stretching and shearing and out-of-lane bending and shearing, but the curved geometry results in direct coupling of in-surface and normal-to-surface resistance to applied loads. Elements could be based on at least a dozen shell theories, although the Fredriksson and Mackerle[2] survey cites popular use of only the Koiter-Sanders and Kirchhoff theories.

Three-dimensional solid models include general and axisymmetric models, as Fig. 7.4 suggests. Both tetrahedral and "brick" geometries are supported, with sides represented by up to third-order polynomials. Whenever geometry and boundary conditions imply symmetric responses, use of the axisymmetric elements offers major economies in the computer resources needed.

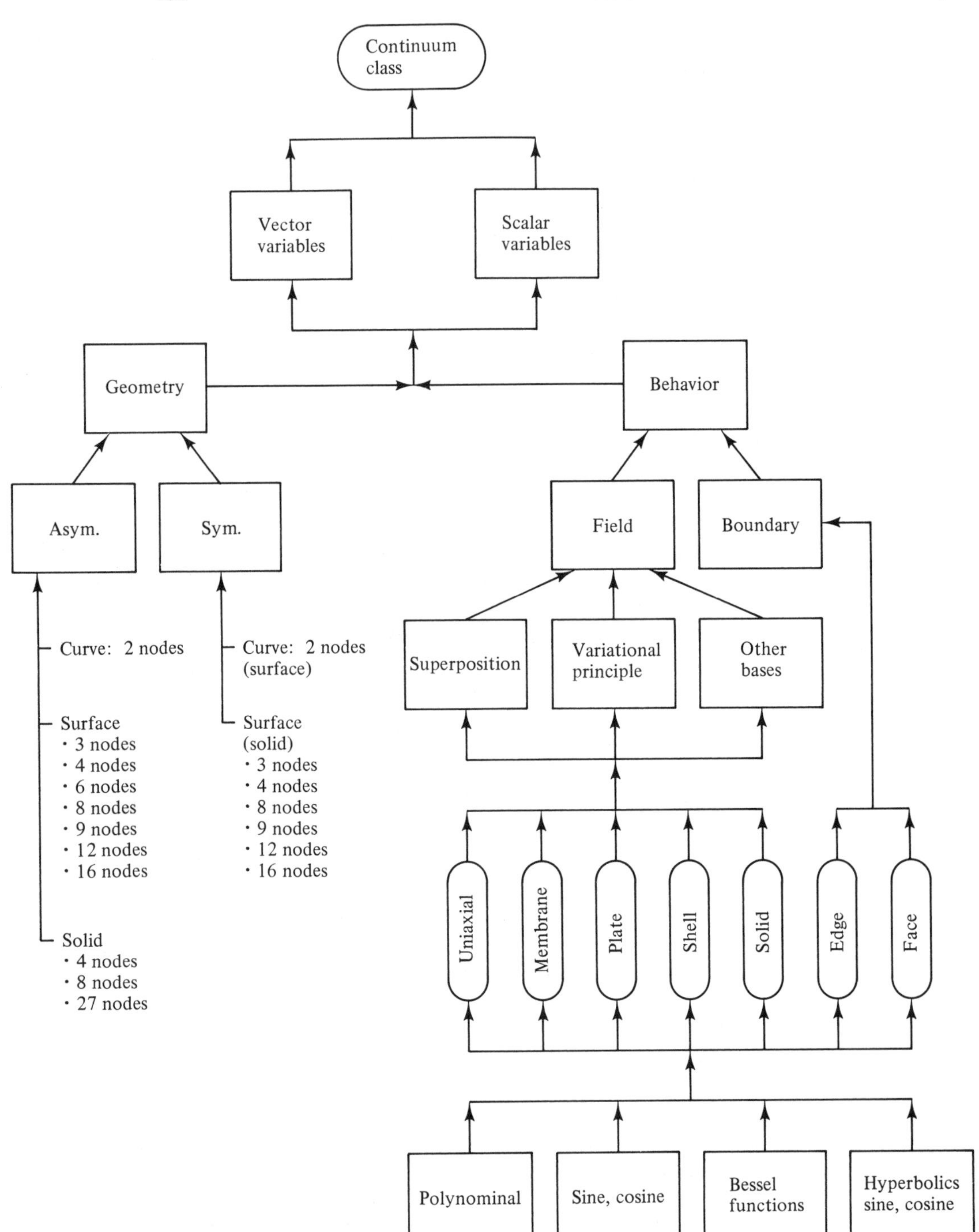

Figure 7.1 Types of continuum element models.

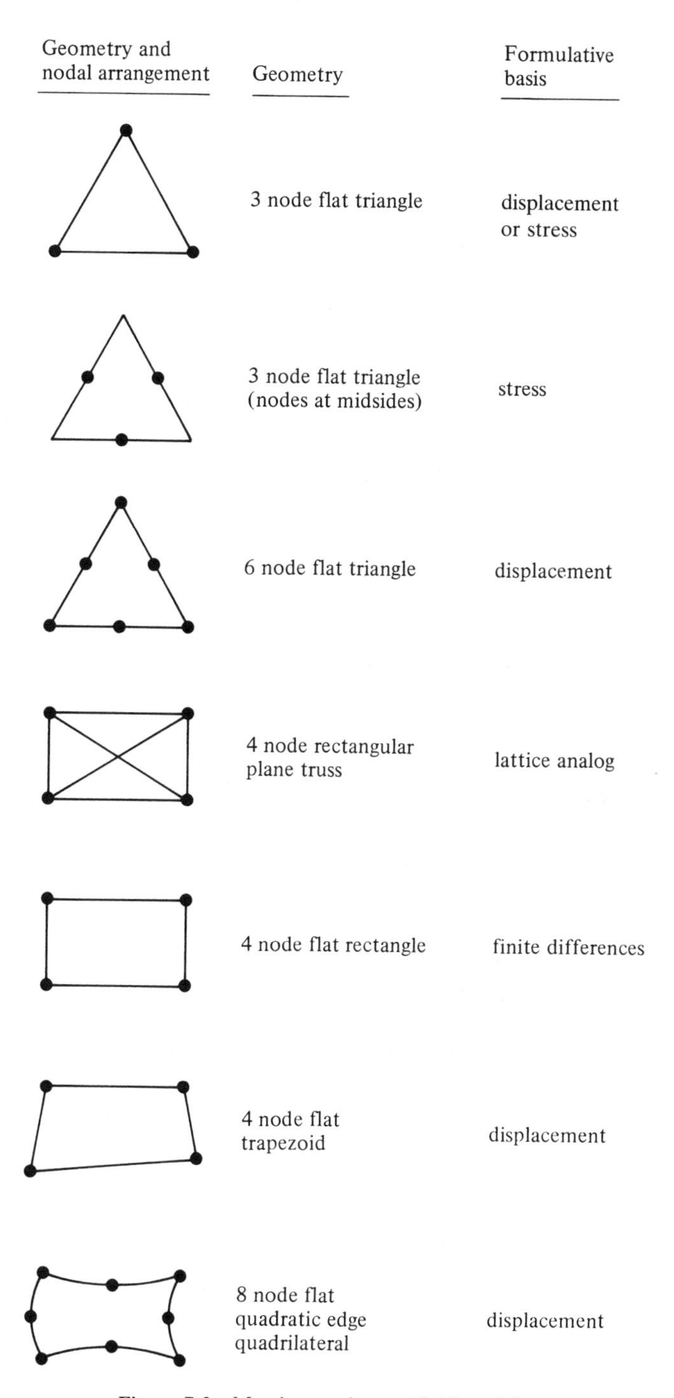

Geometry and nodal arrangement
Geometry
Formulative basis
3 node flat triangle
displacement or stress
3 node flat triangle (nodes at midsides)
stress
6 node flat triangle
displacement
4 node rectangular plane truss
lattice analog
4 node flat rectangle
finite differences
4 node flat trapezoid
displacement
8 node flat quadratic edge quadrilateral
displacement

Figure 7.2 Membrane element field models.

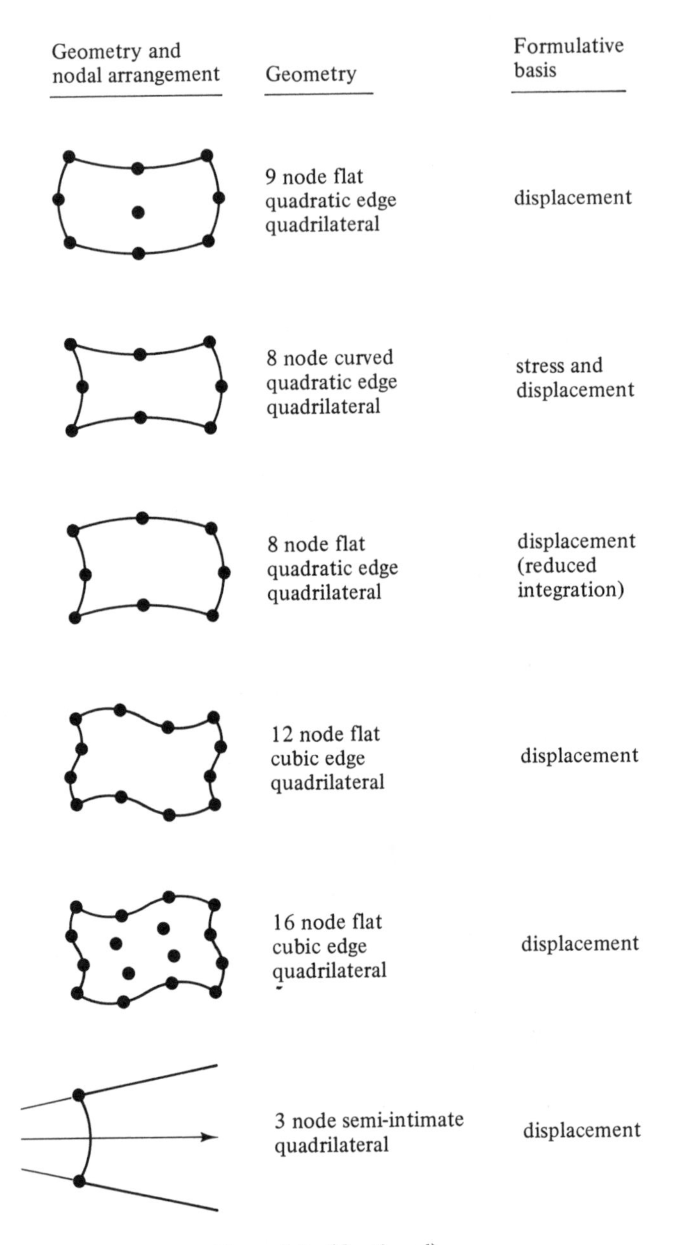
Geometry and nodal arrangement
Geometry
Formulative basis
9 node flat quadratic edge quadrilateral
displacement
8 node curved quadratic edge quadrilateral
stress and displacement
8 node flat quadratic edge quadrilateral
displacement (reduced integration)
12 node flat cubic edge quadrilateral
displacement
16 node flat cubic edge quadrilateral
displacement
3 node semi-intimate quadrilateral
displacement

Figure 7.2 (Continued)

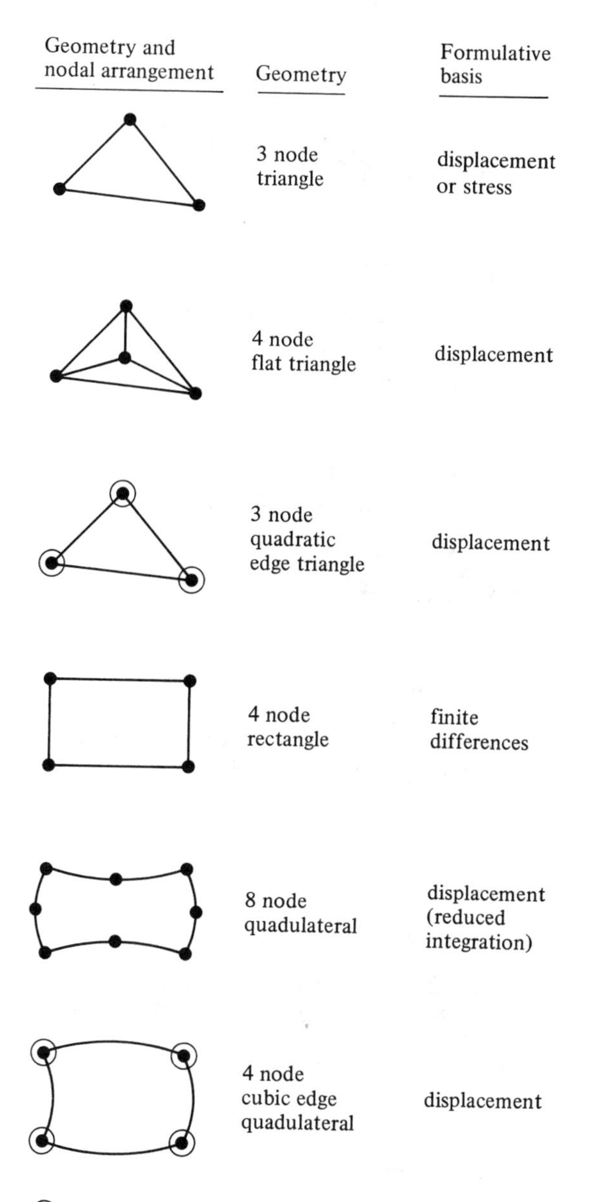

Geometry and nodal arrangement
Geometry
Formulative basis
3 node triangle
displacement or stress
4 node flat triangle
displacement
3 node quadratic edge triangle
displacement
4 node rectangle
finite differences
8 node quadulateral
displacement (reduced integration)
4 node cubic edge quadulateral
displacement
indicates that displacements, slopes, and higher derivatives of displacements are used as nodal variables.

Figure 7.3 Plate element field models.

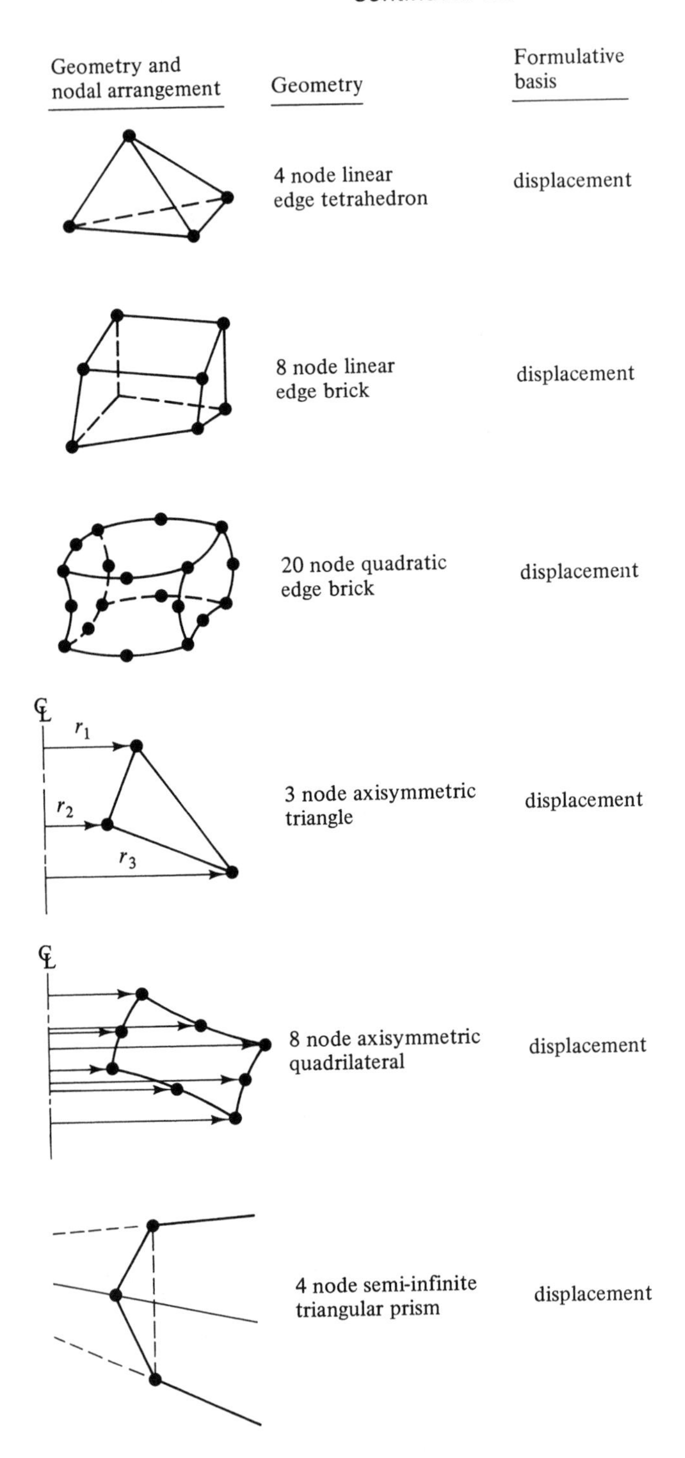

Geometry and nodal arrangement
Geometry
Formulative basis
4 node linear edge tetrahedron
displacement
8 node linear edge brick
displacement
20 node quadratic edge brick
displacement
℄
r_1
r_2
r_3
3 node axisymmetric triangle
displacement
℄
8 node axisymmetric quadrilateral
displacement
4 node semi-infinite triangular prism
displacement

Figure 7.4 Solid element field models.

7.2 MATHEMATICAL BASIS FOR CONTINUUM MODELS

The basis of finite element continuum models for structural behavior is the three-dimensional theory of elasticity. The fundamental differential equations of this theory are the strain displacement, equilibrium, and compatability equations.

The strain displacement equations define the strains in terms of the deformations:

$$\begin{aligned} e_{xx} &= \delta u/\delta x \\ e_{yy} &= \delta v/\delta y \\ e_{zz} &= \delta w/\delta z \\ e_{xy} &= \delta u/\delta y + \delta v/\delta \mathrm{x} \\ e_{xz} &= \delta u/\delta z + \delta w/\delta x \\ e_{yz} &= \delta v/\delta z + \delta w/\delta y \end{aligned} \tag{7.1}$$

or

$$\mathbf{e} = \mathbf{B}\ \mathbf{u} \tag{7.2}$$

where

u, v, and w are displacements measured parallel to the x, y, and z rectangular Cartesian axes,

δ denotes partial differentiation,

e is the vector of e_{ij} strains,

B is a matrix whose coefficients are partial differentiation operators, and

u is the vector u, w, and w displacements.

The strains are the nondimensional components of the microscopic changes in length. The deformations are differential measures of geometry change. The strain displacement equations imply that displacements are continuous functions of the coordinates that map an undeformed point uniquely to a deformed position.

For strains small compared to 1 and stresses below the proportional limit, the strains are related to stresses by the generalized Hooke's law:

$$\begin{pmatrix} \sigma_{xx} \\ \sigma_{yy} \\ \sigma_{zz} \\ \sigma_{xy} \\ \sigma_{xz} \\ \sigma_{yz} \end{pmatrix} = \begin{bmatrix} E_{11} & E_{12} & E_{13} & E_{14} & E_{15} & E_{16} \\ & E_{22} & E_{23} & E_{24} & E_{25} & E_{26} \\ & & E_{33} & E_{34} & E_{35} & E_{36} \\ & & & E_{44} & E_{45} & E_{46} \\ & & & & E_{55} & E_{56} \\ \text{Sym.} & & & & & E_{66} \end{bmatrix} \begin{pmatrix} e_{xx} \\ e_{yy} \\ e_{zz} \\ e_{xy} \\ e_{xz} \\ e_{yz} \end{pmatrix} \tag{7.3}$$

or

$$\sigma = \mathbf{E}\ \mathbf{e} \tag{7.4}$$

where

σ_{xy} is the component of the stress vector at the point of interest, acting in the y direction on the plane perpendicular to the x axis,

σ is the vector of stress components, and

E is the matrix of material properties.

We arrive at the requirement that **E** be symmetric by limiting concern to materials for which a strain energy potential exists. We define the potential as

$$\phi = 0.5\, \mathbf{e}^T \mathbf{E}\, \mathbf{e} - \mathbf{e}^T \sigma \tag{7.5}$$

where ϕ is the material potential. Since ϕ is a scalar, we can, without loss of generality, adopt the convenience of requiring that **E** be symmetric.

This definition of the potential means that we can recover the stress-strain equations by partial differentiation with respect to the strains, which is to say that Castigliano's theorem applies to our materials.

Green's materials are those for which the material potential is defined by a positive definite **E** matrix. The definiteness in turn implies that a nonzero stress state cannot exist without a nonzero strain vector. Furthermore, strains define stresses uniquely and reversibly.

Assuming that stresses and strains are unknowns, enforcing the stress-strain equations implies that only 6 of the 12 stress and strain components are independent variables.

We can reduce the stress-strain relations to two-dimensional space by assumptions about some strains and stresses or both. Most common are the assumptions of plane stress and plane strain.

Let us assume that $e_{zz} = e_{xz} = e_{yz} = 0$ in Eqs. (7.3), the plane strain assumption with respect to the z axis. All the stresses are thus defined in terms of the e_{xx}, e_{yy}, e_{xy} strain components. We can then express the material strain energy, Eq. (7.5), in terms of the three unknown strains alone.

We note that the plane strain assumption does *not* require that $\sigma_{zz} = \sigma_{xz} = \sigma_{yz} = 0$. The value of these stresses may be recovered from the three-dimensional stress-strain equations. They may, but need not, be zero.

Assume that $\sigma_{zz} = \sigma_{xz} = \sigma_{yz} = 0$ in Eqs. (7.3), the plane stress assumption with respect to the z axis. We can then use selective Gauss elimination to decouple the σ_{xx}, σ_{yy}, σ_{xy} equations from the e_{zz}, e_{xz}, e_{yz} variables. We thereby arrive an expression for the material strain energy that is a function of only e_{xx}, e_{yy}, and e_{xy}. This reduction does not imply that e_{zz}, e_{xz}, and e_{yz} are necessarily zero.

In general, we can assume any linear constraint relationship between e_{zz}, e_{xz}, e_{yz} and e_{xx}, e_{yy}, e_{zz} or, alternately, between σ_{zz}, σ_{xz}, σ_{yz} and σ_{xx}, σ_{yy}, σ_{xy} and transform Eqs. (7.3) to three equations in three stress and three strain variables. The transformation of the **E** matrix will be a congruent transformation.

If the stress-strain relation is required to be independent of the sense, plus or minus, of the z axis, the material is said to imply one plane of symmetry. Then some of the **E** coefficients must be zero. In particular, the stress-strain relation must take the form

$$\begin{pmatrix} \sigma_{xx} \\ \sigma_{yy} \\ \sigma_{zz} \\ \sigma_{xy} \\ \sigma_{xz} \\ \sigma_{yz} \end{pmatrix} = \begin{bmatrix} E_{11} & E_{12} & E_{13} & 0 & 0 & E_{16} \\ & E_{22} & E_{23} & 0 & 0 & E_{26} \\ & & E_{33} & 0 & 0 & E_{36} \\ & & & E_{44} & E_{45} & 0 \\ & & & & E_{55} & 0 \\ \text{Sym.} & & & & & E_{66} \end{bmatrix} \begin{pmatrix} e_{xx} \\ e_{yy} \\ e_{zz} \\ e_{xy} \\ e_{xz} \\ e_{yz} \end{pmatrix} \tag{7.6}$$

Assuming that the coefficients of the elastic properties matrix are independent of the sense of the x and y axes as well produces the "orthotropic" stress-strain form,

$$\begin{pmatrix} \sigma_{xx} \\ \sigma_{yy} \\ \sigma_{zz} \\ \sigma_{xy} \\ \sigma_{xz} \\ \sigma_{yz} \end{pmatrix} = \begin{bmatrix} E_{11} & E_{12} & E_{13} & 0 & 0 & 0 \\ & E_{22} & E_{23} & 0 & 0 & 0 \\ & & E_{33} & 0 & 0 & 0 \\ & & & E_{44} & 0 & 0 \\ & & & & E_{55} & 0 \\ \text{Sym.} & & & & & E_{66} \end{bmatrix} \begin{pmatrix} e_{xx} \\ e_{yy} \\ e_{zz} \\ e_{xy} \\ e_{xz} \\ e_{yz} \end{pmatrix} \tag{7.7}$$

The orthotropic material model of Eqs. (7.7) is simplified to the isotropic case by imposing the requirement that the coefficients of the elastic properties matrix be independent of both the direction of the x and y axes and independent of rotations about any axis. These requirements limit the matrix coefficients to being a function of two parameters,

$$\begin{pmatrix} \sigma_{xx} \\ \sigma_{yy} \\ \sigma_{zz} \\ \sigma_{xy} \\ \sigma_{xz} \\ \sigma_{yz} \end{pmatrix} = \begin{bmatrix} \lambda(1-\nu) & & & 0 & 0 & 0 \\ & \lambda(1-\nu) & & 0 & 0 & 0 \\ & & \lambda(1-\nu) & 0 & 0 & 0 \\ & & & G & 0 & 0 \\ & & & & G & 0 \\ \text{Sym.} & & & & & G \end{bmatrix} \begin{pmatrix} e_{xx} \\ e_{yy} \\ e_{zz} \\ e_{xy} \\ e_{xz} \\ e_{yz} \end{pmatrix} \tag{7.8}$$

where

$\lambda = E/((1 + \nu)(1 - 2\nu))$,
E is Young's modulus,
ν is Poisson's ratio, and
$G = E/(2(1 + \nu))$, the shear modulus.

The isotropic stress-strain relations for axisymmetric three-dimensional solids are the stress-strain equations of Eq. (7.8) expressed in cylindrical coordinates.

The stress-strain equations, (7.5) through (7.8), are reduced to plane strain and plane stress form by following the same reasoning by which the anisotropic equations, (7.3), are reduced. We thereby arrive at the plane stress case for the isotropic material by operating on Eq. (7.8):

$$\begin{pmatrix} \sigma_{xx} \\ \sigma_{yy} \\ \sigma_{xy} \end{pmatrix} = \begin{matrix} E/(1 - \nu^2) \\ \\ \\ \end{matrix} \begin{bmatrix} 1\ \nu & 1 & 0 \\ & 1 & 1 & 0 \\ \text{Sym.} & & (1 - \nu)/2 \end{bmatrix} \begin{pmatrix} e_{xx} \\ e_{yy} \\ e_{xy} \end{pmatrix} \tag{7.9}$$

Similarly, Eq. (7.8) yields the plane strain equations

$$\begin{pmatrix} \sigma_{xx} \\ \sigma_{yy} \\ \sigma_{xy} \end{pmatrix} = \begin{matrix} E/((1 + \nu)(1 - 2\nu)) \\ \\ \\ \end{matrix} \begin{bmatrix} 1 - \nu & \nu & 0 \\ & 1 - \nu & 0 \\ \text{Sym.} & & (1 - 2\nu)/2 \end{bmatrix} \begin{pmatrix} e_{xx} \\ e_{yy} \\ e_{xy} \end{pmatrix} \tag{7.10}$$

The plane stress-strain relations are changed to the plate structural stress-strain relations by integrating the stresses over the plate thickness to replace stress with generalized forces and using the Kirchhoff assumption to replace strains with curvatures. Accordingly, we define the generalized forces by

$$M_{xx} = \int \sigma_{xx}\, z\, dz,\ M_{yy} = \int \sigma_{yy}\, z\, dz,\ M_{xy} = \int \sigma_{xy}\, z\, dz \tag{7.11}$$

where

M_{xx} is the moment about the x axis,
M_{yy} is the moment about the y axis,
M_{xy} is the moment about the x axis induced by shear stress, and
the limits of integration are $-t/2$ to $+t/2$, with t the thickness of the plate.

The strains are assumed to be

$$\begin{aligned} e_{xx} &= e_{xx0} + z\, \delta^2 w/\delta x^2,\ e_{yy} = e_{yy0} + z\, \delta^2 w/\delta y^2, \\ e_{xy} &= e_{xy0} + z\, \delta^2 w/(\delta x \delta y) \end{aligned} \tag{7.12}$$

where the subscript 0 denotes strains in the middle surface of the plate. Using Eqs. (7.11) and (7.12) in Eq. (7.9) and integrating the right-hand side over the thickness limits yields the structural stress-strain relations for the plate:

$$\begin{pmatrix} M_{xx} \\ M_{yy} \\ M_{xy} \end{pmatrix} = \frac{Et^3}{12(1 - \nu^2)} \begin{bmatrix} 1 & \nu & 0 \\ & 1 & 0 \\ \text{Sym.} & & (1 - \nu)/2 \end{bmatrix} \begin{pmatrix} \delta^2 w/\delta x^2 \\ \delta^2 w/\delta y^2 \\ \delta^2 w/(\delta x \delta y) \end{pmatrix} \tag{7.13}$$

Here we omit the shear forces normal to the midplane because their strain energy, in the Germain-Lagrange theory, is assumed to be negligible.

The stress-strain relations for the shell are the combination of the membrane and plate equations at each point of the midsurface of the shell.

The requirement that equilibrium be satisfied at a point takes the form

$$\mathbf{B}^T \sigma = \mathbf{0} \tag{7.14}$$

Under the plane stress assumption, these three equations reduce directly to two equations by dropping the zero stress terms. Under the plane strain assumption, the stresses σ_{zz}, σ_{xz}, and σ_{yz} are expressed in terms of the strains e_{xx}, e_{yy}, and e_{xy} and these in terms of σ_{xx}, σ_{yy}, and σ_{xy}, hence modifying the equations of equilibrium. It is often hypothesized that the σ_{zz}, σ_{xz}, and σ_{yz} stresses are small compared with other stresses, and their terms neglected in the microscopic equations.

The compatibility equations of elasticity require that if we are given the deformations, they are integrable to obtain a unique set of displacements. Since we intend to develop element models by assuming displacement functions, the compatibility equations will always be satisfied if we restrict attention to functions that are continuous with respect to the spatial coordinates, at least through the first derivatives.

From the viewpoint of the theory of elasticity, there are 15 unknowns at a point: 6 stresses, 6 strains, and 3 displacements. There are 15 equations: 6 strain displacement, 6 stress-strain, and 3 equilibrium. The fact that there will be only one solution of these equations is ensured by the condition that the elastic properties matrix E be positive definite.

7.3 ILLUSTRATIVE ELEMENT MODELS

Tables 7.1 through 7.6 cite formulas for stiffness coefficients for nine continuum elements: five field elements and four boundary elements.[3-7] The field elements include two membrane elements, two plate elements, and a model for an eight-node brick. The boundary elements deal with Winkler (linear) foundations and up to cubic displacements.

In general, these matrices are fully populated and symmetric. Since they are fully populated, nodal bandedness is a practical basis for planning efficient equation solution. Symmetry ensures that only a triangular representation of all stiffness matrices need be generated or stored by the computer configuration.

The field elements are distinguished by the fact that the forces of any column of the stiffness matrix satisfy the appropriate macroscopic equilibrium requirements. Thus any column of a membrane model satisfies the requirement that the sum of the force components in any direction be zero. Any column of a plate model stiffness matrix satisfies the requirements that both the sum of the forces and the moment of the forces be zero. Accordingly, the rank of a field element stiffness matrix is lower than its order by at least the number of linearly independent equilibrium equations satisfied by each and every column of the matrix.

Conversely, a boundary element stiffness matrix is distinguished by the fact that at least one of the appropriate macroscopic equilibrium equations will not be satisfied by each and every column of the stiffness matrix. The rank of the boundary element stiffness matrix may be equal to its order.

Axisymmetric models are a special subclass of boundary element models. Whether they represent shells or solids, the axisymmetric response constraints limit

TABLE 7.1 TRIANGULAR MEMBRANE STIFFNESS MATRIX

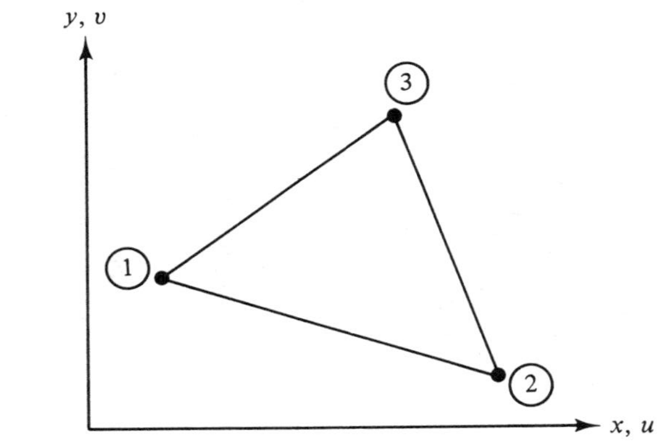

$$\left.\begin{aligned} u &= a_0 + a_1 x + a_2 y \\ v &= b_0 + b_1 x + b_2 y \end{aligned}\right\} \quad \text{(a)}$$

t = uniform thickness

$$\begin{pmatrix} \sigma_{xx} \\ \sigma_{yy} \\ \sigma_{xy} \end{pmatrix} = \frac{E}{(1-\nu^2)} \begin{bmatrix} 1 & \nu & 0 \\ \nu & 1 & 0 \\ 0 & 0 & \dfrac{1-\nu}{2} \end{bmatrix} \begin{pmatrix} e_{xx} \\ e_{yy} \\ e_{xy} \end{pmatrix} \quad \text{(b)}$$

$$\sigma_{zz} = \sigma_{yz} = \sigma_{xz} = 0$$

$$\frac{tE}{2(G_0)(1-\nu^2)} \quad K_e g_e = \begin{bmatrix} y_{23}^2 + \gamma x_{32}^2 & & & \text{Sym.} & & \\ y_{23}y_{31} + \gamma x_{13}x_{32} & y_{31}^2 + \gamma x_{13}^2 & & & & \\ y_{23}y_{31} + \gamma x_{13}x_{32} & y_{12}y_{31} + \gamma x_{21}x_{13} & y_{12}^2 + \gamma x_{21}^2 & & & \\ \nu x_{32}y_{23} + \gamma x_{32}y_{23} & \nu x_{32}y_{31} + \gamma x_{13}y_{23} & \nu x_{32}y_{12} + \gamma x_{21}y_{23} & x_{32}^2 + \gamma y_{23}^2 & & \\ \nu x_{13}y_{23} + \gamma x_{32}y_{31} & \nu x_{13}y_{31} + \gamma x_{13}y_{31} & \nu x_{13}y_{12} + \gamma x_{21}y_{31} & x_{13}x_{32} + \gamma y_{31}y_{23} & x_{13}^2 + \gamma y_{31}^2 & \\ \nu x_{21}y_{23} + \gamma x_{32}y_{12} & \nu x_{21}y_{31} + \gamma x_{13}y_{12} & \nu x_{21}y_{12} + \gamma x_{21}y_{12} & x_{21}x_{32} + \gamma y_{12}y_{23} & x_{21}x_{13} + \gamma y_{12}y_{31} & x_{21}^2 + \gamma y_{12}^2 \end{bmatrix} \begin{Bmatrix} u_1 \\ u_2 \\ u_3 \\ v_1 \\ v_2 \\ v_3 \end{Bmatrix} \quad \text{(c)}$$

$$x_{ij} = x_i - x_j; \quad y_{ij} = y_i - y_j$$

$$\gamma = \frac{1-\nu}{2}; \quad -1 < \nu < 0.5$$

$$G_0 = |x_2 y_3 + x_1 y_2 + y_1 x_3 - x_2 y_1 - \chi_3 y_2 - x_1 y_3| = 2 \times \text{plan area}$$

TABLE 7.2 RECTANGULAR MEMBRANE STIFFNESS MATRIX

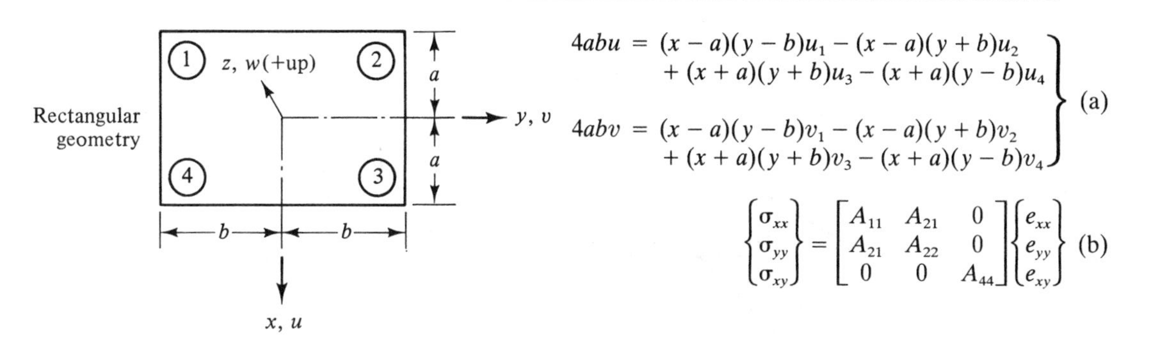

$$\left.\begin{aligned} 4abu &= (x-a)(y-b)u_1 - (x-a)(y+b)u_2 \\ &\quad + (x+a)(y+b)u_3 - (x+a)(y-b)u_4 \\ 4abv &= (x-a)(y-b)v_1 - (x-a)(y+b)v_2 \\ &\quad + (x+a)(y+b)v_3 - (x+a)(y-b)v_4 \end{aligned}\right\} \quad \text{(a)}$$

$$\begin{Bmatrix} \sigma_{xx} \\ \sigma_{yy} \\ \sigma_{xy} \end{Bmatrix} = \begin{bmatrix} A_{11} & A_{21} & 0 \\ A_{21} & A_{22} & 0 \\ 0 & 0 & A_{44} \end{bmatrix} \begin{Bmatrix} e_{xx} \\ e_{yy} \\ e_{xy} \end{Bmatrix} \quad \text{(b)}$$

Rectangular slice stiffness matrix

$$[K] = \begin{bmatrix} K_{11} & K_{21} \\ K_{21} & K_{22} \end{bmatrix} \quad (\text{columns } u_1, v_1)$$

$$a_{11} = tA_{11}b/6a \qquad a_{22} = tA_{22}a/6b \qquad a_{12} = tA_{21}/4$$

t = slice thickness

$$\bar{a}_{44} = tA_{44}a/6b \qquad \dot{a}_{44} = tA_{44}b/6a \qquad a_{44} = tA_{44}/4$$

$$[K_{11}] = \begin{bmatrix} 2a_{11}+2\bar{a}_{44} & & & \text{symmetric} \\ a_{11}-2\bar{a}_{44} & 2a_{11}+2\bar{a}_{44} & & \\ -a_{11}-\bar{a}_{44} & -2a_{11}+\bar{a}_{44} & 2a_{11}+2\bar{a}_{44} & \\ -2a_{11}+\bar{a}_{44} & -a_{11}-\bar{a}_{44} & a_{11}-2\bar{a}_{44} & 2a_{11}+2\bar{a}_{44} \end{bmatrix} \quad (u_1, u_2, u_3, u_4)$$

$$[K_{22}] = \begin{bmatrix} 2a_{22}+2\dot{a}_{44} & & & \text{symmetric} \\ -2a_{22}+\dot{a}_{44} & 2a_{22}+2\dot{a}_{44} & & \\ -a_{22}-\dot{a}_{44} & a_{22}-2\dot{a}_{44} & 2a_{22}+2\dot{a}_{44} & \\ a_{22}-2\dot{a}_{44} & -a_{22}-\dot{a}_{44} & -2a_{22}+\dot{a}_{44} & 2a_{22}+2\dot{a}_{44} \end{bmatrix} \quad (v_1, v_2, v_3, v_4)$$

$$[K_{21}] = \begin{bmatrix} a_{12}+a_{44} & a_{12}-a_{44} & -a_{12}-a_{44} & -a_{12}+a_{44} \\ -a_{12}+a_{44} & -a_{12}-a_{44} & a_{12}-a_{44} & a_{12}+a_{44} \\ -a_{12}-a_{44} & -a_{12}+a_{44} & a_{12}+a_{44} & a_{12}-a_{44} \\ a_{12}-a_{44} & a_{12}+a_{44} & -a_{12}+a_{44} & -a_{12}-a_{44} \end{bmatrix} \quad (u_1, u_2, u_3, u_4)$$

all rigid body movement except movement along the axis. Therefore, despite the fact that these models emulate three-dimensional structures, only one of the macroscopic equilibrium equations is satisfied by the forces in any column of the stiffness matrix.

Forces in a column of a continuum element stiffness matrix may contradict intuition developed in using lattice element models. For the rectangular membrane, for example, we would expect that all x-direction forces of the first column

TABLE 7.3 FLAT TRIANGULAR SHELL ELEMENT STIFFNESS MATRIX

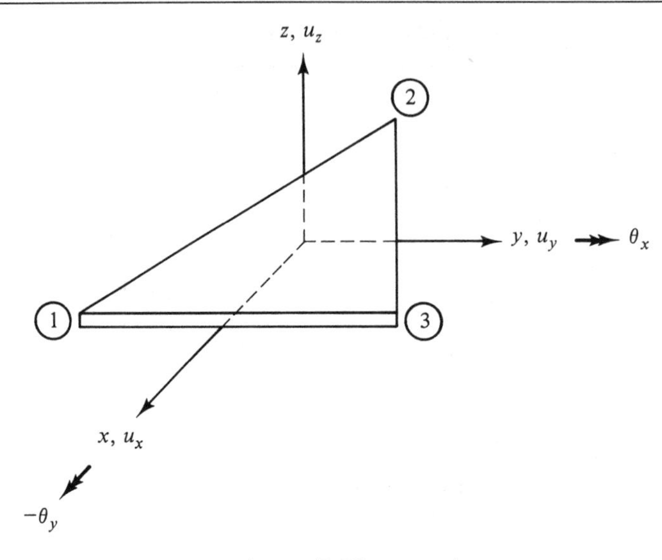

$$u_i = \lfloor 1\ x\ y \rfloor [\mathbf{H}]\{v_{ji} + z\theta_{ji}\}$$
$$u_z = \lfloor 1\ x\ y \rfloor [\mathbf{H}]\{v_{jz} + f(z)\} \tag{a}$$

where

$$i = x, y; \quad j = 1, 2, 3; \quad x_{KL} = x_K - x_L;$$

$$[\mathbf{H}] = \frac{1}{x_{21}y_{31} - x_{31}y_{21}} \begin{bmatrix} x_2y_3 - x_3y_2 & x_3y_1 - x_1y_3 & x_1y_2 - y_1x_2 \\ y_{23} & y_{31} & y_{12} \\ x_{32} & x_{13} & x_{21} \end{bmatrix}$$

and $f(z)$ is some undefined function of z.

$$\sigma = \mathbf{E}\, e \tag{b}$$

$$\sigma = \begin{bmatrix} \sigma_{xx} \\ \sigma_{yy} \\ \sigma_{xy} \\ \sigma_{zz} \\ \sigma_{xz} \\ \sigma_{yz} \end{bmatrix} \qquad e = \begin{bmatrix} e_{xx} \\ e_{yy} \\ e_{xy} \\ e_{zz} \\ e_{xz} \\ e_{yz} \end{bmatrix}$$

$$\mathbf{E} = \begin{bmatrix} \mathbf{E}_{11} - \mathbf{E}_{41}^2/\mathbf{E}_{44} & & & \text{Sym.} & \\ \mathbf{E}_{21} - \mathbf{E}_{41}\mathbf{E}_{42}/\mathbf{E}_{44} & \mathbf{E}_{22} - \mathbf{E}_{42}^2/\mathbf{E}_{44} & & & \\ \mathbf{E}_{31} - \mathbf{E}_{41}\mathbf{E}_{43}/\mathbf{E}_{44} & \mathbf{E}_{32} - \mathbf{E}_{42}\mathbf{E}_{43}/\mathbf{E}_{44} & \mathbf{E}_{33} - \mathbf{E}_{43}^2/\mathbf{E}_{44} & & \\ 0 & 0 & 0 & \mathbf{E}_{55} & \\ 0 & 0 & 0 & \mathbf{E}_{65} & \mathbf{E}_{66} \end{bmatrix} \tag{c}$$

$$\begin{bmatrix} V_{jx} \\ \Theta_{jx} \\ V_{jy} \\ \Theta_{jy} \\ V_{jz} \end{bmatrix} = \begin{bmatrix} \mathbf{K}_{11} & & & & \text{Sym.} \\ 0 & \mathbf{K}_{22} & & & \\ \mathbf{K}_{31} & 0 & \mathbf{K}_{33} & & \\ 0 & \mathbf{K}_{42} & 0 & \mathbf{K}_{44} & \\ 0 & \mathbf{K}_{52} & 0 & \mathbf{K}_{54} & \mathbf{K}_{55} \end{bmatrix} \begin{bmatrix} v_{jx} \\ \theta_{jx} \\ v_{jy} \\ \theta_{jz} \\ v_{jz} \end{bmatrix} \tag{d}$$

TABLE 7.3 (Continued)

where V_{jx} and Θ_{fx} are the generalized loads associated with $v_{j\kappa}$ and θ_{jx} and

$$\mathrm{K}_{11} = \overline{D}_{11}\mathbf{Y}^T\mathbf{Y} + \overline{D}_{31}\mathbf{X}^T\mathbf{Y} + \overline{D}_{31}\mathbf{Y}^T\mathbf{X} + \overline{D}_{33}\mathbf{X}^T\mathbf{X}$$

$$\mathrm{K}_{22} = \tilde{D}_{11}\mathbf{Y}^T\mathbf{Y} + \tilde{D}_{31}\mathbf{X}^T\mathbf{Y} + \tilde{D}_{31}\mathbf{Y}^T\mathbf{X} + \tilde{D}_{33}\mathbf{X}^T\mathbf{X}$$

$$\mathrm{K}_{31} = \overline{D}_{21}\mathbf{X}^T\mathbf{Y} + \overline{D}_{31}\mathbf{Y}^T\mathbf{Y} + \overline{D}_{32}\mathbf{X}^T\mathbf{X} + \overline{D}_{33}\mathbf{Y}^T\mathbf{X}$$

$$\mathrm{K}_{33} = \overline{D}_{22}\mathbf{X}^T\mathbf{X} + \overline{D}_{32}\mathbf{X}^T\mathbf{Y} + \overline{D}_{32}\mathbf{Y}^T\mathbf{X} + \overline{D}_{33}\mathbf{Y}^T\mathbf{Y}$$

$$\mathrm{K}_{42} = \tilde{D}_{21}\mathbf{X}^T\mathbf{Y} + \tilde{D}_{31}\mathbf{Y}^T\mathbf{Y} + \tilde{D}_{32}\mathbf{X}^T\mathbf{X} + \tilde{D}_{33}\mathbf{Y}^T\mathbf{X}$$

$$\mathrm{K}_{44} = \tilde{D}_{22}\mathbf{X}^T\mathbf{X} + \tilde{D}_{32}\mathbf{X}^T\mathbf{Y} + \tilde{D}_{32}\mathbf{Y}^T\mathbf{X} + \tilde{D}_{33}\mathbf{Y}^T\mathbf{Y}$$

$$\overline{D}_{ij} = t \cdot A \cdot (E_{ij} - E_{4i}\,E_{4j}/E_{44}); \qquad \tilde{D}_{ij} = (t^2/12)\overline{D}_{ij}$$

$$\mathbf{X} = s\lfloor x_{32}, x_{13}, x_{21} \rfloor; \qquad \mathbf{Y} = s\lfloor y_{23}, y_{31}, y_{12} \rfloor$$

$$\mathbf{s} = (x_{21}y_{31} - x_{31}y_{21})^{-1}; \qquad \mathbf{x}_{KL} = x_K - x_L$$

$$\mathbf{K}_{22_s} = \beta_{12}x_{21}^2\lfloor 1,1,0\rfloor^T\lfloor 1,1,0\rfloor + \beta_{13}x_{31}^2\lfloor 1,0,1\rfloor^T\lfloor 1,0,1\rfloor + \beta_{23}x_{32}^2\lfloor 0,1,1\rfloor^T\lfloor 0,1,1\rfloor$$

$$\mathbf{K}_{42_s} = \beta_{12}x_{21}y_{21}\lfloor 1,1,0\rfloor^T\lfloor 1,1,0\rfloor + \beta_{13}x_{31}y_{31}\lfloor 1,0,1\rfloor^T\lfloor 1,0,1\rfloor + \beta_{23}x_{32}y_{32}\lfloor 0,1,1\rfloor^T\lfloor 0,1,1\rfloor$$

$$\mathbf{K}_{44_s} = \beta_{12}y_{21}^2\lfloor 1,1,0\rfloor^T\lfloor 1,1,0\rfloor + \beta_{13}y_{31}^2\lfloor 1,0,1\rfloor^T\lfloor 1,0,1\rfloor + \beta_{23}y_{32}^2\lfloor 0,-1,1\rfloor^T\lfloor 0,1,1\rfloor$$

$$\mathbf{K}_{52_s} = 2\beta_{12}x_{21}\lfloor -1,1,0\rfloor^T\lfloor 1,1,0\rfloor + 2\beta_{13}x_{31}\lfloor -1,0,1\rfloor^T\lfloor 1,0,1\rfloor + 2\beta_{23}x_{32}\lfloor 0,-1,1\rfloor^T\lfloor 0,1,1\rfloor \qquad \text{(e)}$$

$$\mathbf{K}_{54_s} = 2\beta_{12}y_{21}\lfloor -1,1,0\rfloor^T\lfloor 1,1,0\rfloor + 2\beta_{13}y_{31}\lfloor -1,0,1\rfloor^T\lfloor 1-,0,-1\rfloor + 2\beta_{23}y_{32}\lfloor 0,-1,-1\rfloor^T\lfloor 0,1,-1\rfloor$$

$$\mathbf{K}_{55_s} = 4\beta_{12}\lfloor 1,-1,0\rfloor^T\lfloor 1,-1,0\rfloor + 4\beta_{13}\lfloor 1,0,-1\rfloor^T\lfloor 1,0,-1\rfloor + 4\beta_{23}\lfloor 0,1,-1\rfloor^T\lfloor 0,1,-1\rfloor$$

$\beta_{ij} = 6A_{ij}/2L_{ij}$

where A_{ij} – = area resisting shear along side ij

L_{ij} = length of side ij

$$\beta_{12} = |y_{23}y_{31}\rho_1 + x_{31}x_{13}\rho_2 + (x_{13}y_{23} + x_{32}y_{31})\rho_3|; \quad \rho_1 = \frac{E_{55}t}{4A}$$

$$\beta_{13} = |y_{23}y_{12}\rho_1 + x_{32}x_{21}\rho_2 + (x_{21}y_{23} + x_{32}y_{12})\rho_3|; \quad \rho_2 = \frac{E_{66}t}{4A} \qquad \text{(f)}$$

$$\beta_{23} = |y_{12}y_{31}\rho_1 + x_{21}x_{13}\rho_2 + (x_{21}y_{31} + x_{13}y_{12})\rho_3|; \quad \rho_3 = \frac{E_{56}t}{4A}$$

with A = plan form area

TABLE 7.4 RECTANGULAR PLATE STIFFNESS MATRIX

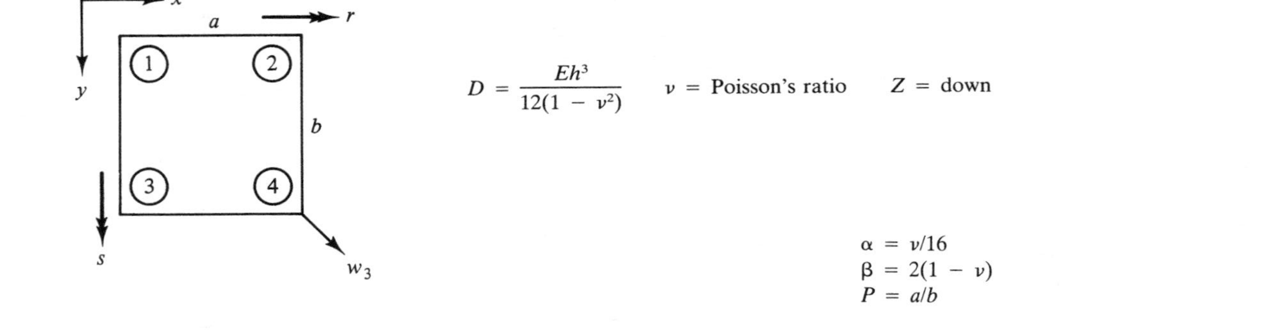

$$D = \frac{Eh^3}{12(1 - \nu^2)} \qquad \nu = \text{Poisson's ratio} \qquad Z = \text{down}$$

$\alpha = \nu/16$
$\beta = 2(1 - \nu)$
$P = a/b$

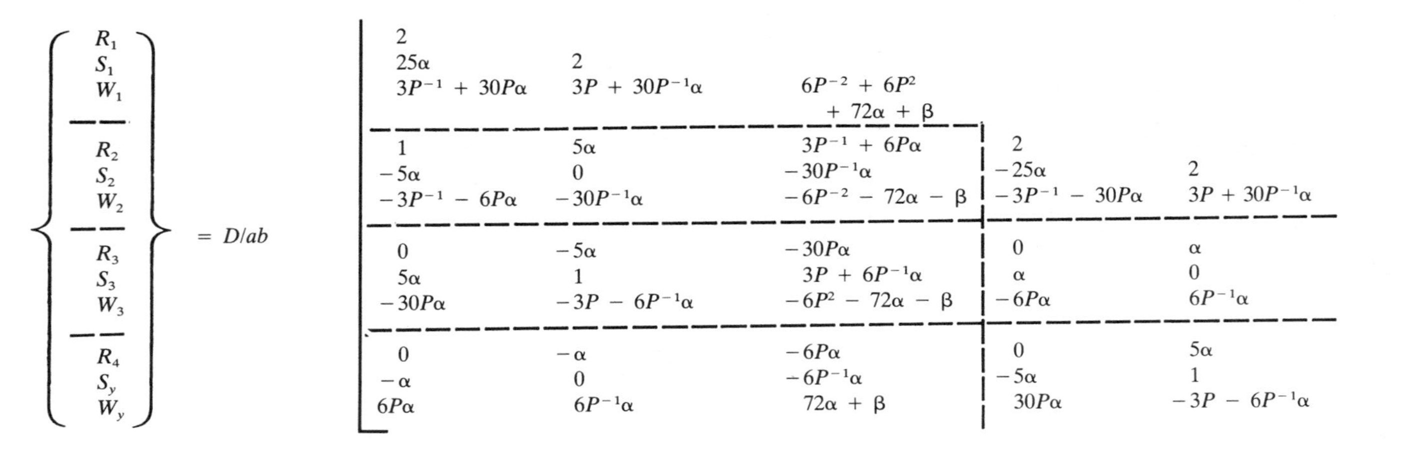

$$\begin{Bmatrix} R_1 \\ S_1 \\ W_1 \\ \hline R_2 \\ S_2 \\ W_2 \\ \hline R_3 \\ S_3 \\ W_3 \\ \hline R_4 \\ S_y \\ W_y \end{Bmatrix} = D/ab \left[\begin{array}{ccc|cc}
2 & & & & \\
25\alpha & 2 & & & \\
3P^{-1} + 30P\alpha & 3P + 30P^{-1}\alpha & 6P^{-2} + 6P^2 + 72\alpha + \beta & & \\
\hline
1 & 5\alpha & 3P^{-1} + 6P\alpha & 2 & \\
-5\alpha & 0 & -30P^{-1}\alpha & -25\alpha & 2 \\
-3P^{-1} - 6P\alpha & -30P^{-1}\alpha & -6P^{-2} - 72\alpha - \beta & -3P^{-1} - 30P\alpha & 3P + 30P^{-1}\alpha \\
\hline
0 & -5\alpha & -30P\alpha & 0 & \alpha \\
5\alpha & 1 & 3P + 6P^{-1}\alpha & \alpha & 0 \\
-30P\alpha & -3P - 6P^{-1}\alpha & -6P^2 - 72\alpha - \beta & -6P\alpha & 6P^{-1}\alpha \\
\hline
0 & -\alpha & -6P\alpha & 0 & 5\alpha \\
-\alpha & 0 & -6P^{-1}\alpha & -5\alpha & 1 \\
6P\alpha & 6P^{-1}\alpha & 72\alpha + \beta & 30P\alpha & -3P - 6P^{-1}\alpha
\end{array}\right]$$

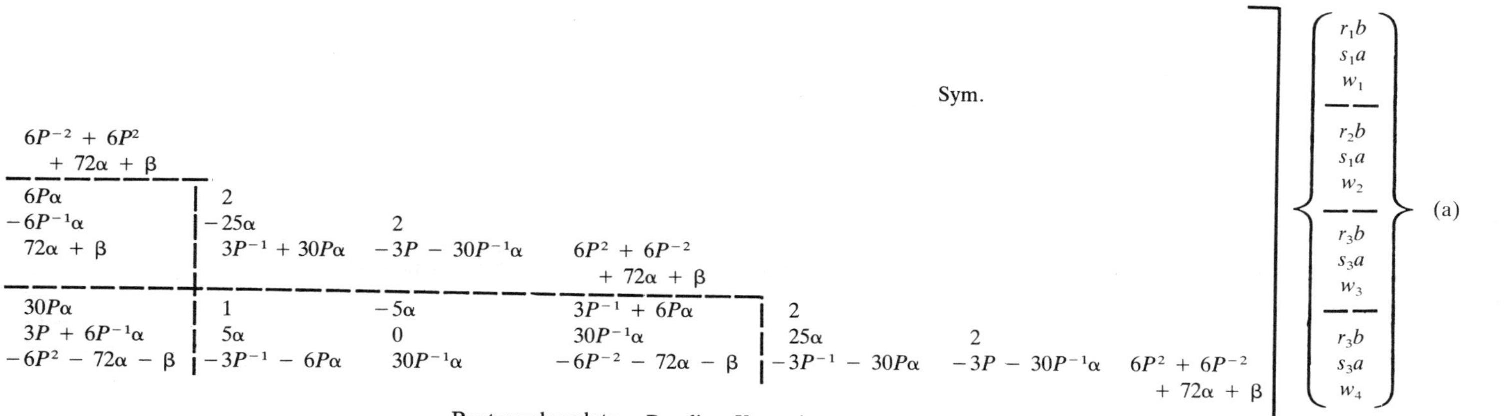

$$
\begin{bmatrix}
6P^{-2} + 6P^{2} + 72\alpha + \beta & & & & & & \\
6P\alpha & 2 & & & & & \\
-6P^{-1}\alpha & -25\alpha & 2 & & & \text{Sym.} & \\
72\alpha + \beta & 3P^{-1} + 30P\alpha & -3P - 30P^{-1}\alpha & 6P^{2} + 6P^{-2} + 72\alpha + \beta & & & \\
30P\alpha & 1 & -5\alpha & 3P^{-1} + 6P\alpha & 2 & & \\
3P + 6P^{-1}\alpha & 5\alpha & 0 & 30P^{-1}\alpha & 25\alpha & 2 & \\
-6P^{2} - 72\alpha - \beta & -3P^{-1} - 6P\alpha & 30P^{-1}\alpha & -6P^{-2} - 72\alpha - \beta & -3P^{-1} - 30P\alpha & -3P - 30P^{-1}\alpha & 6P^{2} + 6P^{-2} + 72\alpha + \beta
\end{bmatrix}
\begin{Bmatrix}
r_1 b \\ s_1 a \\ w_1 \\ r_2 b \\ s_1 a \\ w_2 \\ r_3 b \\ s_3 a \\ w_3 \\ r_3 b \\ s_3 a \\ w_4
\end{Bmatrix}
\quad \text{(a)}
$$

Rectangular plate Bending K matrix

TABLE 7.5 RIGHT PRISM STIFFNESS MATRIX

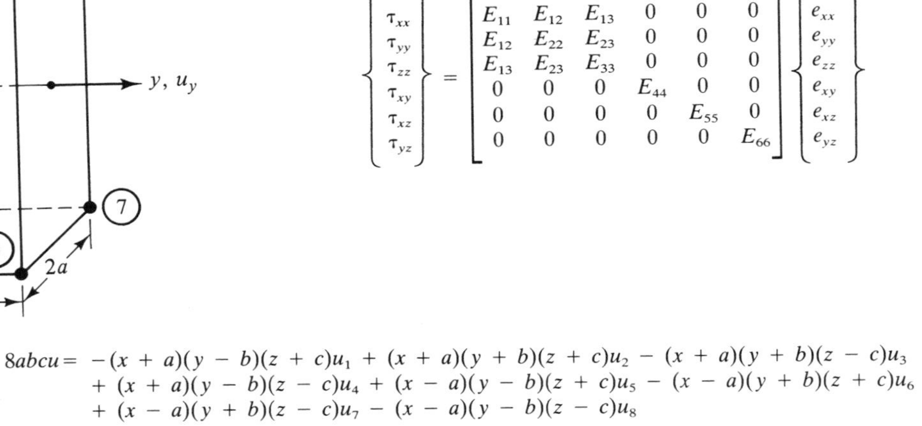

$$\begin{Bmatrix} \tau_{xx} \\ \tau_{yy} \\ \tau_{zz} \\ \tau_{xy} \\ \tau_{xz} \\ \tau_{yz} \end{Bmatrix} = \begin{bmatrix} E_{11} & E_{12} & E_{13} & 0 & 0 & 0 \\ E_{12} & E_{22} & E_{23} & 0 & 0 & 0 \\ E_{13} & E_{23} & E_{33} & 0 & 0 & 0 \\ 0 & 0 & 0 & E_{44} & 0 & 0 \\ 0 & 0 & 0 & 0 & E_{55} & 0 \\ 0 & 0 & 0 & 0 & 0 & E_{66} \end{bmatrix} \begin{Bmatrix} e_{xx} \\ e_{yy} \\ e_{zz} \\ e_{xy} \\ e_{xz} \\ e_{yz} \end{Bmatrix}$$

$$\begin{aligned} 8abcu = & -(x+a)(y-b)(z+c)u_1 + (x+a)(y+b)(z+c)u_2 - (x+a)(y+b)(z-c)u_3 \\ & + (x+a)(y-b)(z-c)u_4 + (x-a)(y-b)(z+c)u_5 - (x-a)(y+b)(z+c)u_6 \\ & + (x-a)(y+b)(z-c)u_7 - (x-a)(y-b)(z-c)u_8 \end{aligned}$$

$\vdots$

Stiffness Matrix K_L

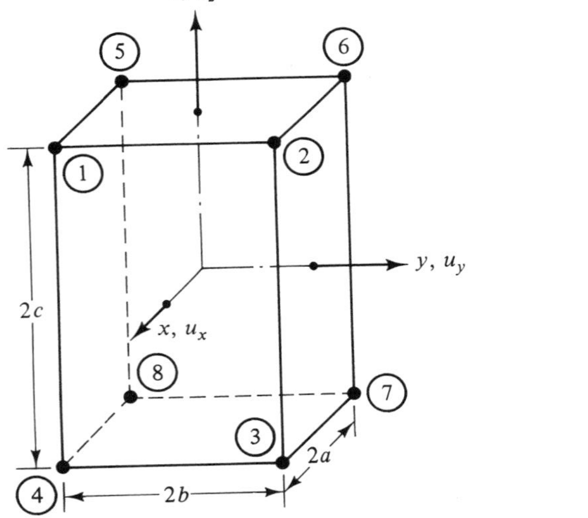

$$K_{11} = \frac{1}{18abc} \begin{bmatrix} \overset{u_{x1}}{4d_{11} + 4d_{44} + 4d_{55}} & \overset{u_{x2}\cdots}{} & & & \text{Sym.} \\ 2d_{11} - 4d'_{44} + 2d_{55} & 4d_{11} + 4d'_{44} + 4d_{55} & & & \\ d_{11} - 2d'_{44} - 2d_{55} & 2d_{11} + 2d'_{44} - 4d_{55} & 4d_{11} + 4d'_{44} + 4d_{55} & & \\ 2d_{11} + 2d'_{44} - 4d_{55} & d_{11} - 2d'_{44} - 2d_{55} & 2d_{11} - 4d'_{44} + 2d_{55} & 4d_{11} + 4d'_{44} + 4d_{55} & \\ -4d_{11} + 2d'_{44} + 2d_{55} & -2d_{11} - 2d'_{44} + d_{55} & -d_{11} - d'_{44} - d_{55} & -2d_{11} + d'_{44} - 2d_{55} & \\ -2d_{11} - 2d'_{44} + d_{55} & -4d_{11} + 2d'_{44} + 2d_{55} & -2d_{11} + d'_{44} - 2d_{55} & -d_{11} - d'_{44} - d_{55} & \\ -d_{11} - d'_{44} - d_{55} & -2d_{11} + d'_{44} - 2d_{55} & -4d_{11} + 2d'_{44} + 2d_{55} & -2d_{11} - 2d'_{44} + d_{55} & \\ -2d_{11} + d'_{44} - 2d_{55} & -d_{11} - d'_{44} - d_{55} & -2d_{11} - 2d'_{44} + d_{55} & -4d_{11} + 2d'_{44} + 2d_{55} & \end{bmatrix}$$

u_{x8}

$$
\begin{array}{llll}
4d_{11} + 4d'_{44} + 4d_{55} & & & \\
2d_{11} - 4d'_{44} + 2d_{55} & 4d_{11} + 4d'_{44} + 4d_{55} & & \\
d_{11} - 2d'_{44} - 2d_{55} & 2d_{11} + 2d'_{44} - 4d_{55} & 4d_{11} + 4d'_{44} + 4d_{55} & \\
2d_{11} + 2d'_{44} - 4d_{55} & d_{11} - 2d'_{44} - 2d_{55} & 2d_{11} - 4d'_{44} + 2d_{55} & 4d_{11} + 4d'_{44} + 4d_{55}
\end{array}
\Bigg]
$$

$u_{y1} \qquad u_{y2} \; . \; . \; .$

$$
K_{22} = \frac{1}{18abc}
\left[
\begin{array}{llll}
4d_{22} + 4d_{44} + 4d_{66} & & & \\
-4d_{22} + 2d_{44} + 2d_{66} & 4d_{22} + 4d_{44} + 4d_{66} & & \\
-2d_{22} - d_{44} - 2d_{66} & 2d_{22} + 2d_{44} - 4d_{66} & 4d_{22} + 4d_{44} + 4d_{66} & \\
2d_{22} + 2d_{44} - 4d_{66} & -2d_{22} + d_{44} - 2d_{66} & -4d_{22} + 2d_{44} + 2d_{66} & 4d_{22} + 4d_{44} + 4d_{66} \\
2d_{22} - 4d_{44} + 2d_{66} & -2d_{22} - 2d_{44} + d_{66} & -d_{22} - d_{44} - d_{66} & d_{22} - 2d_{44} - 2d_{66} \\
-2d_{22} - 2d_{44} + d_{66} & 2d_{22} - 4d_{44} + 2d_{66} & d_{22} - 2d_{44} - 2d_{66} & -d_{22} - d_{44} - d_{66} \\
-d_{22} - d_{44} - d_{66} & d_{22} - 2d_{44} - 2d_{66} & 2d_{22} - 4d_{44} + 2d_{66} & -2d_{22} - 2d_{44} + d_{66} \\
d_{22} - 2d_{44} - 2d_{66} & -d_{22} - d_{44} - d_{66} & -2d_{22} - 2d_{44} + d_{66} & 2d_{22} - 4d_{44} + 2d_{66}
\end{array}
\right.
$$

u_{y8}

$$
\begin{array}{llll}
4d_{22} + 4d_{44} + 4d_{66} & & & \\
-4d_{22} + 2d_{44} + 2d_{66} & 4d_{22} + 4d_{44} + 4d_{66} & & \\
-2d_{22} + d_{44} - 2d_{66} & 2d_{22} + 2d_{44} - 4d_{66} & 4d_{22} + 4d_{44} + 4d_{66} & \\
2d_{22} + 2d_{44} - 4d_{66} & -2d_{22} + d_{44} - 2d_{66} & -4d_{22} + 2d_{44} + 2d_{66} & 4d_{22} + 4d_{44} - 4d_{66}
\end{array}
\Bigg]
$$

$u_{z1} \qquad u_{z2} \; . \; . \; .$

$$
K_{33} = \frac{1}{18abc}
\left[
\begin{array}{llll}
4d_{33} + 4d'_{55} + 4d'_{66} & & & \\
2d_{33} + 2d'_{55} - 4d'_{66} & 4d_{33} + 4d'_{55} + 4d'_{66} & & \\
-2d_{33} + d'_{55} - 2d'_{66} & -4d_{33} + 2d'_{55} + 2d'_{66} & 4d_{33} + 4d'_{55} + 4d'_{66} & \\
-4d_{33} + 2d'_{55} + 2d'_{66} & -2d_{33} + d'_{55} - 2d'_{66} & 2d_{33} + 2d'_{55} - 4d'_{66} & 4d_{33} + 4d'_{55} + 4d'_{66} \\
2d_{33} - 4d'_{55} + 2d'_{66} & d_{33} - 2d'_{55} - 2d'_{66} & -d_{33} - d'_{55} - d'_{66} & -2d_{33} - 2d'_{55} + d'_{66} \\
d_{33} - 2d'_{55} - 2d'_{66} & 2d_{33} - 4d'_{55} + 2d'_{66} & -2d_{33} - 2d'_{55} + d'_{66} & -d_{33} - d'_{55} - d'_{66} \\
-d_{33} - d'_{55} - d'_{66} & -2d_{33} - 2d'_{55} + d'_{66} & 2d_{33} - 4d'_{55} + 2d'_{66} & d_{33} - 2d'_{55} - 2d'_{66} \\
-2d_{33} - 2d'_{55} + d'_{66} & -d_{33} - d'_{55} - d'_{66} & d_{33} - 2d'_{55} - 2d'_{66} & 2d_{33} - 4d'_{55} + 2d'_{66}
\end{array}
\right.
$$

(Continued)

TABLE 7.5 (Continued)

$$
\begin{array}{llll|}
 & & & u_{z8} \\
4d_{33} + 4d'_{55} + 4d'_{66} & & & \\
2d_{33} + 2d'_{55} - 4d'_{66} & 4d_{33} + 4d'_{55} + 4d'_{66} & & \\
-2d_{33} + d'_{55} - 2d'_{66} & -4d_{33} + 2d'_{55} + 2d'_{66} & 4d_{33} + 4d'_{55} + 4d'_{66} & \\
-4d_{33} + 2d'_{55} + 2d'_{66} & -2d_{33} + d'_{55} - 2d'_{66} & 2d_{33} + 2d'_{55} - 4d'_{66} & 4d_{33} + 4d'_{55} + 4d'_{66}
\end{array}
$$

$$
\mathbf{K}_{21} = \frac{d_{12}}{12}\begin{bmatrix}
-2 & -2 & -1 & -1 & 2 & 2 & 1 & 1 \\
2 & 2 & 1 & 1 & -2 & -2 & -1 & -1 \\
1 & 1 & 2 & 2 & -1 & -1 & -2 & -2 \\
-1 & -1 & -2 & -2 & 1 & 1 & 2 & 2 \\
-2 & -2 & -1 & -1 & 2 & 2 & 1 & 1 \\
2 & 2 & 1 & 1 & -2 & -2 & -1 & -1 \\
1 & 1 & 2 & 2 & -1 & -1 & -2 & -2 \\
-1 & -1 & -2 & -2 & 1 & 1 & 2 & 2
\end{bmatrix} + \frac{d_{44''}}{12}\begin{bmatrix}
-2 & 2 & 1 & -1 & -2 & 2 & 1 & -1 \\
-2 & 2 & 1 & -1 & -2 & 2 & 1 & -1 \\
-1 & 1 & 2 & -2 & -1 & 1 & 2 & -2 \\
-1 & 1 & 2 & -2 & -1 & 1 & 2 & -2 \\
2 & -2 & -1 & 1 & 2 & -2 & -1 & 1 \\
2 & -2 & -1 & 1 & 2 & -2 & -1 & 1 \\
1 & -1 & -2 & 2 & 1 & -1 & -2 & 2 \\
1 & -1 & -2 & 2 & 1 & -1 & -2 & 2
\end{bmatrix}
$$

$$
\mathbf{K} = \left[\begin{array}{c|c|c}
u_{xi} & u_{yi} & u_{zi} \\
\mathrm{K}_{11} & & \text{Sym.} \\
\mathrm{K}_{21} & \mathrm{K}_{22} & \\
\mathrm{K}_{31} & \mathrm{K}_{32} & \mathrm{K}_{33}
\end{array}\right]
$$

$d_{11} = E_{11}b^2c^2$
$d_{22} = E_{22}a^2c^2$
$d_{33} = E_{33}a^2b^2$
$d_{44} = E_{44}b^2c^2$

$$
\mathbf{K}_{31} = \frac{d_{13}}{12}\begin{bmatrix}
2 & 1 & 1 & 2 & -2 & -1 & -1 & -2 \\
1 & 2 & 2 & 1 & -1 & -2 & -2 & -1 \\
-1 & -2 & -2 & -1 & 1 & 2 & 2 & 1 \\
-2 & -1 & -1 & -2 & 2 & 1 & 1 & 2 \\
2 & 1 & 1 & 2 & -2 & -1 & -1 & -2 \\
1 & 2 & 2 & 1 & -1 & -2 & -2 & -1 \\
-1 & -2 & -2 & -1 & 1 & 2 & 2 & 1 \\
-2 & -1 & -1 & -2 & 2 & 1 & 1 & 2
\end{bmatrix} + \frac{d_{13}}{12}\begin{bmatrix}
2 & 1 & -1 & -2 & 2 & 1 & -1 & -2 \\
1 & 2 & -2 & -1 & 1 & 2 & -2 & -1 \\
1 & 2 & -2 & -1 & 1 & 2 & -2 & -1 \\
2 & 1 & -1 & -2 & 2 & 1 & -1 & -2 \\
-2 & -1 & 1 & 2 & -2 & -1 & 1 & 2 \\
-1 & -2 & 2 & 1 & -1 & -2 & 2 & 1 \\
-1 & -2 & 2 & 1 & -1 & -2 & 2 & 1 \\
-2 & -1 & 1 & 2 & -2 & -1 & 1 & 2
\end{bmatrix}
$$

$d_{55} = E_{55}a^2b^2$
$d_{66} = E_{66}a^2b^2$

$d'_{44} = E_{44}a^2c^2$
$d'_{55} = E_{55}b^2c^2$
$d'_{66} = E_{66}a^2c^2$

$$
\mathbf{K}_{32} = \frac{d_{23}}{12}\begin{bmatrix}
-2 & 2 & 2 & -2 & -1 & 1 & 1 & -1 \\
-2 & 2 & 2 & -2 & -1 & 1 & 1 & -1 \\
2 & -2 & -2 & 2 & 1 & -1 & -1 & 1 \\
2 & -2 & -2 & 2 & 1 & -1 & -1 & 1 \\
-1 & 1 & 1 & -1 & -2 & 2 & 2 & -2 \\
-1 & 1 & 1 & -1 & -2 & 2 & 2 & -2 \\
1 & -1 & -1 & 1 & 2 & -2 & -2 & 2 \\
1 & -1 & -1 & 1 & 2 & -2 & -2 & 2
\end{bmatrix} + \frac{d_{66''}}{12}\begin{bmatrix}
-2 & -2 & 2 & 2 & -1 & -1 & 1 & 1 \\
2 & 2 & -2 & -2 & 1 & 1 & -1 & -1 \\
2 & 2 & -2 & -2 & 1 & 1 & -1 & -1 \\
-2 & -2 & 2 & 2 & -1 & -1 & 1 & 1 \\
-1 & -1 & 1 & 1 & -2 & -2 & 2 & 2 \\
1 & 1 & -1 & -1 & 2 & 2 & -2 & -2 \\
1 & 1 & -1 & -1 & 2 & 2 & -2 & -2 \\
-1 & -1 & 1 & 1 & -2 & -2 & 2 & 2
\end{bmatrix}
$$

$d''_{44} = c\ E_{44}$
$d''_{55} = b\ E_{55}$
$d''_{66} = a\ E_{66}$

$d_{12} = c\ E_{12}$
$d_{13} = b\ E_{13}$
$d_{23} = a\ E_{23}$

TABLE 7.6 BOUNDARY FINITE ELEMENTS

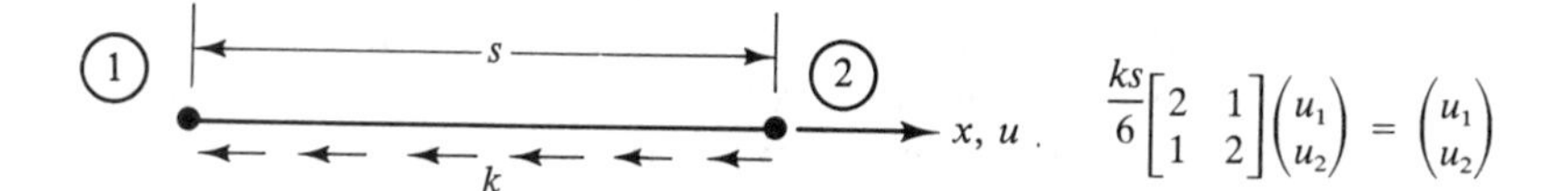

$$\frac{ks}{6}\begin{bmatrix} 2 & 1 \\ 1 & 2 \end{bmatrix}\begin{pmatrix} u_1 \\ u_2 \end{pmatrix} = \begin{pmatrix} u_1 \\ u_2 \end{pmatrix}$$

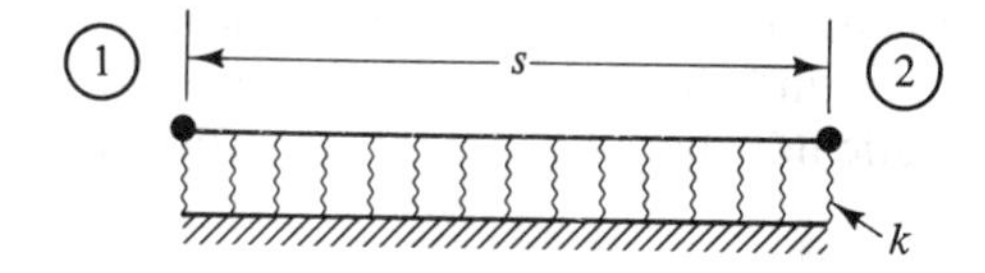

$$\frac{ks}{6}\begin{bmatrix} 2 & 1 \\ 1 & 2 \end{bmatrix}\begin{pmatrix} u_1 \\ u_2 \end{pmatrix} = \begin{pmatrix} u_1 \\ u_2 \end{pmatrix}$$

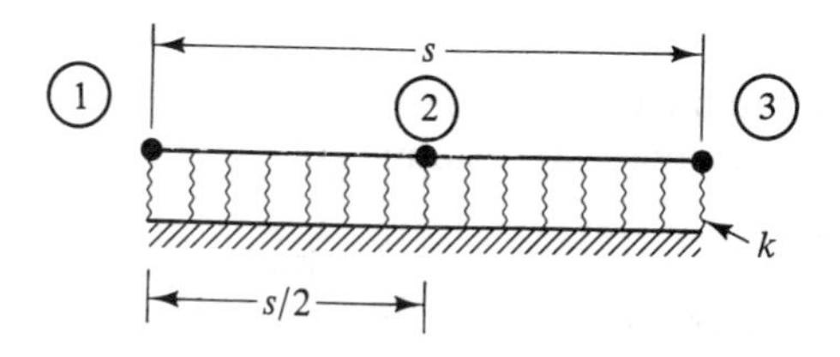

$$\frac{ks}{30}\begin{bmatrix} 4 & 2 & -1 \\ 2 & 16 & 2 \\ -1 & 2 & 4 \end{bmatrix}\begin{pmatrix} u_1 \\ u_2 \\ u_3 \end{pmatrix} = \begin{pmatrix} u_1 \\ u_2 \\ u_3 \end{pmatrix}$$

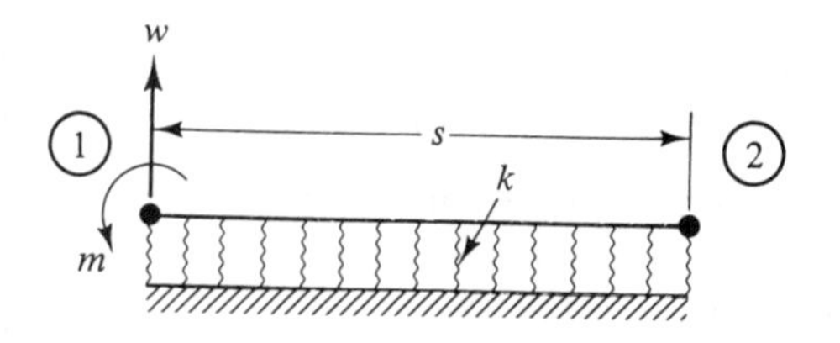

$$\frac{k}{420}\begin{bmatrix} 156 & & & \text{Sym.} \\ 22s & 4s^2 & & \\ 54 & 13s & 156 & \\ -13s & -3s^2 & -22s & 4s \end{bmatrix}\begin{pmatrix} w_1 \\ m_1 \\ w_2 \\ m_2 \end{pmatrix} = \begin{pmatrix} W_1 \\ M_1 \\ W_2 \\ M_2 \end{pmatrix}$$

of the matrix would be negative except the first. The formulas of Table 7.3 show this is not the case. For the rectangular plate, we would expect x-moment reactions at each of the other nodes due to a unit x rotation at the first node. The formulas of Table 7.5 show that this is not the case.

Our intuition founders on the fact that continuum element models are necessarily approximate. We cannot measure responses of a physical model and deduce stiffness matrix coefficients without making some approximation. Given the measured influence coefficients for a continuum structure, we will find a system stiffness matrix that is not nearly as sparse as that associated with lattice finite element models.

The coefficients of the stiffness matrix depend on the material stiffness of the **E** matrix. Mathematically, at least one coefficient of the stiffness matrices of Figs. 7.2 through 7.5 becomes infinite in the isotropic case when Poisson's ratio is plus or minus 1. Since no known structural material has such a Poisson's ratio, the associated infinities never arise in practice. When plane strain assumptions are made, however, as Eqs. (7.10) suggest, infinite stiffnesses also occur when Poisson's ratio is 0.5. Since rubber and some plastics exhibit Poisson's ratios near 0.5, their plane strain analyses often involve badly conditioned stiffness matrices.

The membrane models of Figs. 7.1, 7.2, and 7.3 have the special property that their stiffness matrices can be expressed by a dimensional scalar multiplying a matrix of nondimensional stiffness matrix coefficients. This property must associate with any membrane model that relates only u and v nodal variables, based on arguments of dimensional consistency. The independence of the stiffness matrix from the size of the element is a characteristic not shared by plate, shell, or three-dimensional solid elements.

7.4 DERIVATION OF MODELS BY POTENTIAL ENERGY

Since more than 88 percent of the continuum element models are derived by minimizing the potential energy over the element,[2] we examine this approach in detail.

Table 7.7 illustrates the derivation of the stiffness matrix for a triangular membrane element in plane stress. The steps of the derivation and their particularization for the triangular membrane element are as follows:

1. Choose nodes and nodal variables for the element. (Table 7.7 indicates the apexes of the triangle at the midsurface as the sites of the three nodes. The nodal variables are chosen to be the u and v displacement of each node.)

2. Choose continuous functions defining the displacements everywhere over the element in terms of as many arbitrary constants as there are nodal displacement variables. This step casts nodal displacements into the form

$$\mathbf{g}(x, y, z) = \mathbf{N}^{-1}\boldsymbol{\alpha} \tag{7.15}$$

where

$\mathbf{g}$ (x, y, z,) is the vector of nodal generalized displacements expressed as a function of coordinates of points in the element,

$\mathbf{N}^{-1}$ is a nonsingular matrix whose coefficients are functions of the coordinates, and

$\boldsymbol{\alpha}$ is the vector of arbitrary constants.

(In Table 7.7 Eqs. (a) are the assumed displacement functions. Since these functions are planes in three-dimensional space, the resulting element model is often referred to as a "simplex" element.)

3. Fit the functions to the nodal displacements. Let $\mathbf{g}_e = \mathbf{N}^{-1}\boldsymbol{\alpha}_e$; so

$$\boldsymbol{\alpha} = \mathbf{N}\mathbf{g}_e \tag{7.16}$$

TABLE 7.7 DERIVATION OF SIMPLEX MEMBRANE STIFFNESS

Step	Simplex Element
1. Choose nodes and nodal variables for the element.	1. 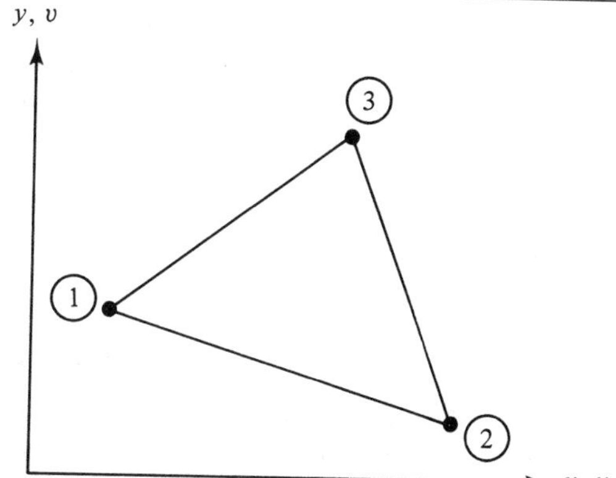
2. Choose interpolation function for displacements (linear in arbitrary constants).	2. Let $u = a_0 + a_1 x + a_z y$ $v = b_0 + b_1 x + b_2 y$ (a)
3. Fit interpolation function to nodal displacements.	3. $\begin{pmatrix} u_1 \\ u_2 \\ u_3 \end{pmatrix} = \begin{bmatrix} 1 & x_1 & y_1 \\ 1 & x_2 & y_2 \\ 1 & x_3 & y_3 \end{bmatrix} \begin{pmatrix} a_0 \\ a_1 \\ a_2 \end{pmatrix}$ (b) $\therefore \begin{pmatrix} a_0 \\ a_1 \\ a_2 \end{pmatrix} = \frac{1}{\text{Det.}} \begin{bmatrix} x_2 y_3 - y_2 x_3 & x_3 y_1 - x_1 y_3 & x_1 y_2 - x_2 y_1 \\ y_2 - y_3 & y_3 - y_1 & y_1 - y_2 \\ x_3 - x_2 & x_1 - x_3 & x_2 - x_1 \end{bmatrix} \begin{pmatrix} u_1 \\ u_2 \\ u_3 \end{pmatrix}$ (c)
4. Use strain displacement equations to evaluate strains in terms of interpolation constants.	4. $\begin{pmatrix} e_{xx} \\ e_{yy} \\ e_{xy} \end{pmatrix} = \begin{pmatrix} \frac{\partial u}{\partial x} \\ \frac{\partial v}{\partial y} \\ \frac{\partial u}{\partial y} + \frac{\partial v}{\partial x} \end{pmatrix} = \begin{pmatrix} a_1 \\ b_2 \\ (a_2 + b_1) \end{pmatrix}$ (d)
5. Express stresses in terms of strains using σ-e equation.	5. $\begin{pmatrix} \sigma_{xx} \\ \sigma_{yy} \\ \sigma_{xy} \end{pmatrix} = \frac{E}{1 - \nu^2} \begin{bmatrix} 1 & \nu & 0 \\ \nu & 1 & 0 \\ 0 & 0 & \frac{1-\nu}{2} \end{bmatrix} \begin{pmatrix} e_{xx} \\ e_{yy} \\ e_{xy} \end{pmatrix} = \mathbf{De}$ (e)
6. Form SE.	6. $SE = \frac{1}{2}\mathbf{e}^T\mathbf{E}\mathbf{e}$
7. Differentiate SE.	$\begin{bmatrix} y_{23} & 0 & x_{32} \\ y_{31} & 0 & x_{13} \\ y_{12} & 0 & x_{12} \\ 0 & x_{32} & y_{23} \\ 0 & x_{13} & y_{31} \\ 0 & x_{12} & y_{12} \end{bmatrix} \begin{bmatrix} y_{23} & y_{31} & y_{12} & \nu x_{32} & \nu x_{13} & \nu x_{21} \\ \nu y_{23} & \nu y_{31} & \nu y_{12} & x_{32} & x_{13} & x_{21} \\ \gamma x_{32} & \gamma x_{13} & \gamma x_{21} & \gamma y_{23} & \gamma y_{31} & \gamma y_{12} \end{bmatrix}$ (f)

where the subscript e indicates particularization of the coefficients of the vectors to the nodal points of the finite element. (Eqs. (b) of Table 7.7 lists the fitting equations. The matrix of coefficients, $\mathbf{N}^{-1}$, is the Vandermode matrix. The Vandermode matrix must be nonsingular so that the arbitrary constants can be expressed uniquely in terms of the nodal displacements producing Eqs. (c).)

4. Using the strain displacement equations and the assumed displacement functions, evaluate the strains in terms of the nodal displacements:

$$\mathbf{e} = \mathbf{B}\,\mathbf{N}^{-1}\,\mathbf{N}\mathbf{g}_e \tag{7.17}$$

(This step produces Eqs. (d) of the table.)

5. Integrate the terms of the potential energy functional for the external work and the internal strain energy; that is, evaluate

$$PE = \iint \mathbf{g}_e^T\,\mathbf{B}^T\,\sigma\,(\mathrm{x},\,\mathrm{y},\,\mathrm{z})\,dS - 0.5\iiint \mathbf{g}_e^T\,\mathbf{N}^T\,\mathbf{B}^T\,\mathbf{E}\,\mathbf{B}\,\mathbf{N}\,\mathbf{g}_e\,dV \tag{7.18}$$

where

PE is the potential energy of the element,
$\sigma(\mathrm{x}, \mathrm{y}, \mathrm{z})$ reflects the variation of applied tractions,
dS is the differential surface area over which tractions are applied, and
dV is the differential volume of the element.

(As Eqs. (f) suggest, the integration of the external work requires knowledge of the distribution of the applied boundary stresses. Integration of the strain energy proceeds by expressing the stresses in terms of the strains using the stress-strain equations of the material of interest.)

6. Differentiate the potential with respect to the nodal displacements. The external work term yields the equivalent loads. The strain energy term yields the stiffness matrix.

$$\mathbf{q}_e = \frac{\delta}{\delta \mathbf{g}_{ei}} \iint \mathbf{B}^T\,\sigma\,(x,\,y,\,z)\,dS \tag{7.19}$$

$$\mathbf{K}_e = \iiint \mathbf{B}^T\,\mathbf{N}^T\,\mathbf{E}\,\mathbf{B}\,\mathbf{N}\,dV \tag{7.20}$$

where $\delta\mathbf{g}_{ei}$ flags differentiation with respect to generalized nodal displacement i. (For the triangular membrane, this step produces the stiffness matrix of Table 7.2.)

Eqs. (7.20) specify that the stiffness matrix be developed by a congruent transformation of the material properties matrix. Therefore, the stiffness matrix developed by the potential energy approach will always be symmetric because the **E** matrix is symmetric.

The key to development of the macro equation coefficients is the selection of the generalized displacement shape function. Its Vandermode matrix must be invertible. The functions must be differentiable so that the strains are uniquely defined over the element. The integral of the strain energy density must have a finite value.

The strain energy approach determines the values of the coefficients that minimize the error in the equilibrium equations in the strain energy norm, under the assumption that the assumed displacements arise.

We refer to an element model as a potential energy element if we can guarantee that when used with like element models, the finite element solution defines a lower bound on the strain energy of the deformed structural system, regardless of the choice of discretization. The simplex element is such a model, as are many of those in Tables 7.2 through 7.5. Potential energy elements produce behavior predictions in which all discretization errors are rectified to yield a lower bound on the response strain energy. These extreme predictions tend to underestimate generalized displacements.

Figure 7.5 illustrates that independently developed finite element models may result in gapping or overlapping along the boundary where two elements interface. If we require that membrane elements that share a common interface before deformation continue to touch all along the boundary during deformation, we guarantee that the strain energy integral will be bounded.

Accordingly, we require this displacement continuity across the shared boundary. This continuity requirement is the requirement that the integral of the strains appearing in the strain energy expression be matched. For a membrane model, u and v displacements must be matched at the boundary. For a plate model, lateral displacements and slopes must be matched.

The simplex triangular model meets all these requirements and hence is a potential energy finite element model. The displacement shape function is differentiable, and the differentials are integrable.

To visualize invertibility of the Vandermode matrix of the simplex membrane element, consider that the membrane lies in the xy plane and its u displacements are plotted normal to the triangle, as Fig. 7.6(A) suggests. The displacement function, Eqs. (a) of Table 7.7, defines a plane in the space. The constant a_0 measures the displacement at the origin. The constants a_1 and a_2 are the two slopes of the plane. Since the same plane is also definable by the u displacements at the nodes, the Vandermode matrix must be invertible for u. The same argument applies to v displacements.

Figure 7.6 shows the geometry displacement space for two adjacent simplex elements. Since the shape function is planar, it defines a straight line along the shared side. This straight line is particularized by the u displacements at nodes $1'$ and $2'$ or $1''$ and $2''$. Therefore, regardless of the value of the u displacements at the four nodes of the substructure, u displacements will match all along the shared boundary. The same argument applies to v displacements.

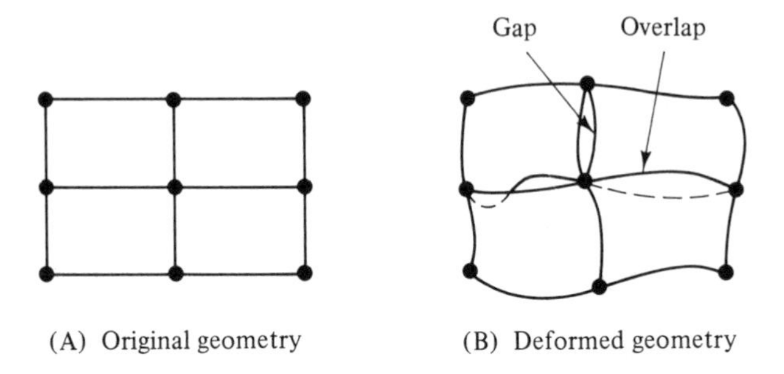

Figure 7.5 Discontinuous element boundary displacements.

In brief, as long as the displacement shape functions define displacements along each side of the membrane element only in terms of displacements at nodes along the side, displacement continuity across element boundaries is guaranteed. With plates, both displacements and slopes must be definable only in terms of the generalized displacement variables along the shared side. With shells, all generalized displacement variables along a shared side must be functions of only generalized displacements at nodes on the side.

If we substitute the simplex element equation for stress, Eqs. (a) through (e) of Table 7.7, in Eqs. (7.14), we can show that the stresses satisfy equilibrium microscopically at every point in the element. The displacement function used in the potential energy approach need not satisfy equilibrium at every point in the element and does not for most element models.

The potential energy method can be used for deriving both field and boundary element macro equations. For the boundary elements, the nonzero stiffness coefficients arise from the strain energy in movement of the support.

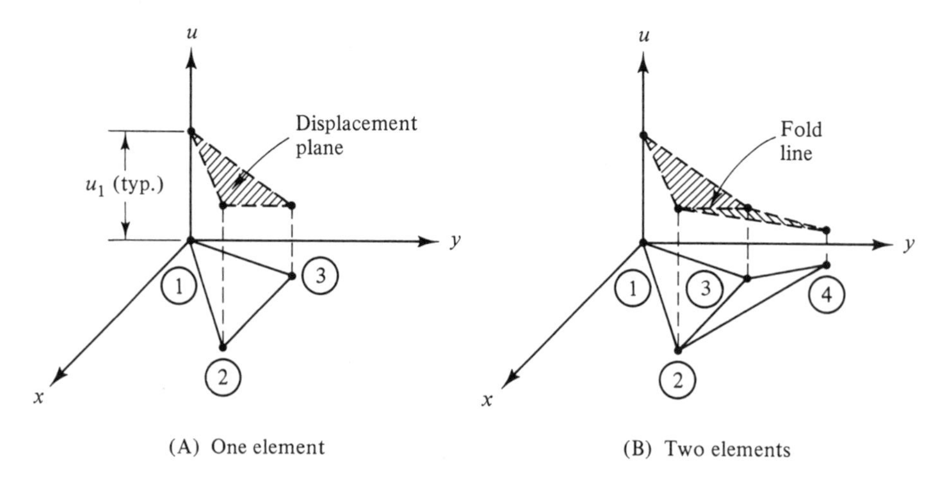

Figure 7.6 Geometry displacement space for a simplex membrane.

7.5 SHAPE AND INTERPOLATION FUNCTIONS

Equations (a) of Table 7.7 define the displacement shape functions for the simplex membrane element. These functions evaluate the displacements everywhere over the element in terms of six arbitrary constants. When we invert the Vandermode matrix to develop expressions for displacements in terms of nodal generalized displacements, we create interpolation functions for the element.

It is useful to recognize equations of the form of Eqs. (c) of Table 7.7 as interpolation functions. Recognition means that we can draw on over a century of development of interpolation technology in mathematics. If we use interpolation functions, we avoid the need to solve the Vandermode equations: The invertibility of the Vandermode matrix is implied in the existence of the interpolation function.

Tables 7.8 through 7.10 present some of the interpolation functions of the mathematical literature. These include functions for univariable,[8] multivariable,[9] and Hermite interpolation.[10] The univariable formulas are useful for line elements and for multivariable interpolation.[9] The multivariable formulas are useful for modeling membranes, plates, shells, and three-dimensional solids. Hermite interpolation formulas include first derivatives of displacements as well as displacements for nodal variables—functions appropriate for developing plate and shell finite element models. Hermite formulas can be extended to "osculating" interpolation functions—functions involving higher than first derivatives of displacements.

Table 7.8 presents both the interpolation formulas and their remainder terms. The remainder measures the error in the interpolation assuming that the function can be represented by a power series with an infinite number of terms.

Unfortunately, we cannot evaluate the remainder without knowing the function we are approximating. For evaluation in the univariable cases, we need to know the $(p + 1)$th derivative of the function over the interpolation interval, where p is the maximum power of the coordinate of the polynomial used to fit the data. We can estimate this derivative, but we cannot rigorously establish its value by using only the data from a finite number of nodes.

The remainder defines the requirements for the interpolations to be exact. For example, the remainder of the three-point univariable interpolation $0.0065h^3\frac{\delta^3 y}{\delta x^3}\Big|_0$ means that the error will vanish as the nodal spacing h approaches zero or when the third derivative of the function is zero everywhere between the endpoints of the interpolation. The first condition is a requirement of mathematical "completeness": the requirement that as the nodal mesh gets finer and finer, we can represent any continuous function as accurately as we like by the given interpolation formula. This condition on the derivative is met when we choose an interpolating polynomial of equal or greater power than a polynomial function we wish to interpolate.

TABLE 7.8 FORMULAS FOR POLYNOMIAL INTERPOLATION IN ONE DIMENSION

$n = 1$, points $= 2$:

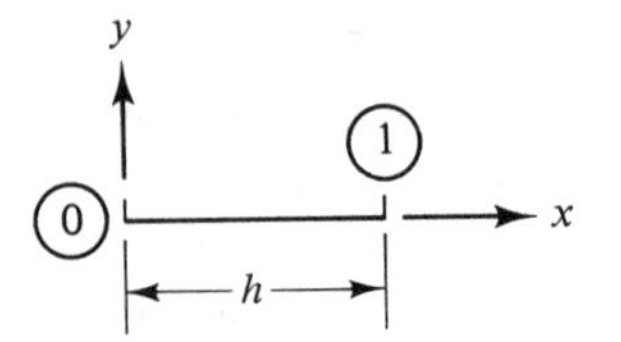

$$y(x) = \frac{(x-h)}{-h} y_0 + \frac{x}{h} y_1$$

$$R = 0.125h^2 \left. \frac{d^2y}{dx^2} \right|$$

$n = 2$, points $= 3$:

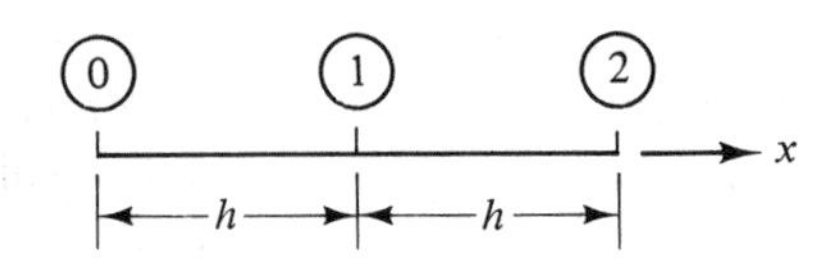

$$y(x) = \frac{(x-h)(x-2h)}{-h(-2h)} y_0 + \frac{x(x-2h)}{h(-h)} y_1 + \frac{x(x-h)}{2h(h)} y_2$$

$$R = 0.065h^3 \left. \frac{d^3y}{dx^3} \right|$$

$n = 3$, points $= 4$:

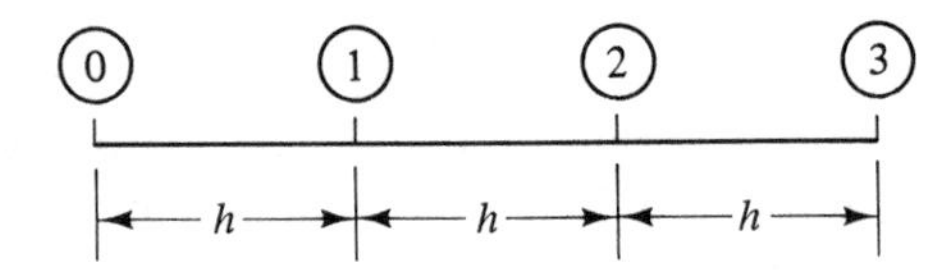

$$y(x) = \frac{(x-h)(x-2h)(x-3h)}{-h(-2h)(-3h)} y_0 + \frac{x(x-2h)(x-3h)}{h(-h)(-2h)} y_1 + \frac{x(x-h)(x-3h)}{2h(h)(-h)} y_2 + \frac{x(x-h)(x-2h)}{3h(2h)(h)} y_3$$

$$R \approx 0.024h^4 \left. \frac{d^4y}{dx^4} \right| \quad h<x<2h; \qquad R = 0.042h^4 \left. \frac{d^4y}{dx^4} \right| \quad x<h, x>2h$$

$n = 4$, points $= 5$:

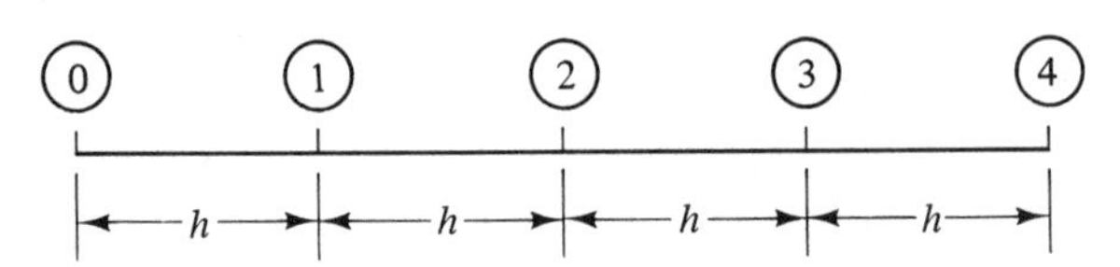

$$y(x) = \frac{(x-h)(x-2h)(x-3h)(x-4h)}{-h(-2h)(-3h)(-4h)} y_0 + \frac{x(x-2h)(x-3h)(x-4h)}{h(-h)(-2h)(-3h)} y_1$$

$$+ \frac{x(x-h)(x-3h)(x-4h)}{2h(h)(-h)(-2h)} y_2 + \frac{x(x-h)(x-2h)(x-4h)}{3h(2h)(h)(-h)} y_3 + \frac{x(x-h)(x-2h)(x-3h)}{4h(3h)(2h)(h)} y_4$$

$$R = 0.012h^5 \left. \frac{d^5y}{dx^5} \right| \quad h < x < 2h; \qquad R = 0.031h^5 \left. \frac{d^5y}{dx^5} \right| \quad x < h, x > 3h$$

TABLE 7.9 MULTIVARIABLE INTERPOLATION FUNCTIONS

Bivariable

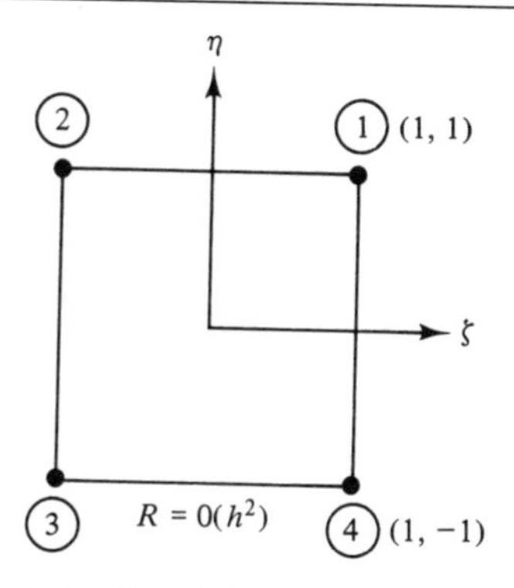

$$4f = (\zeta + 1)(\eta + 1)f_1 - (\zeta - 1)(\eta + 1)f_2 + (\zeta - 1)(\eta - 1)f_3 - (\zeta + 1)(\eta - 1)f_4 \quad \text{(a)}$$

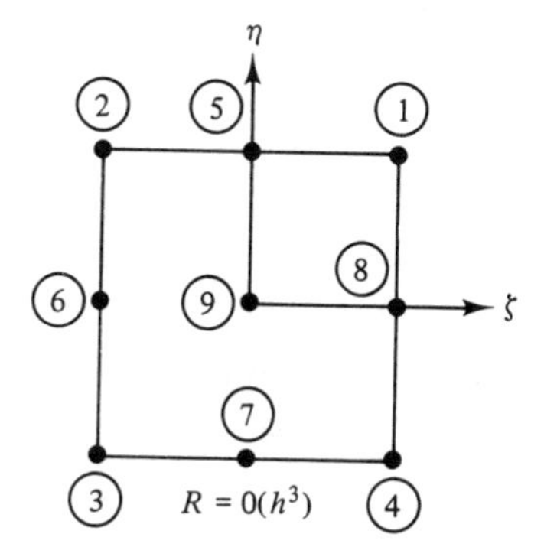

$$\begin{aligned} 4f = {} & \zeta\eta(\zeta + 1)(\eta + 1)f_1 - \zeta\eta(\zeta - 1)(\eta + 1)f_2 + \zeta\eta(\zeta - 1)(\eta - 1)f_3 \\ & - \zeta\eta(\zeta + 1)(\eta - 1)f_4 - 2\eta(\zeta - 1)(\zeta + 1)(\eta + 1)f_5 \\ & + 2\zeta(\zeta - 1)(\eta + 1)(\eta - 1)f_6 - 2\eta(\zeta + 1)(\zeta - 1)(\eta - 1)f_7 \\ & - 2\zeta(\zeta + 1)(\eta + 1)(\eta - 1)f_8 + 4(\zeta + 1)(\zeta - 1)(\eta + 1)(\eta - 1)f_9 \end{aligned} \quad \text{(b)}$$

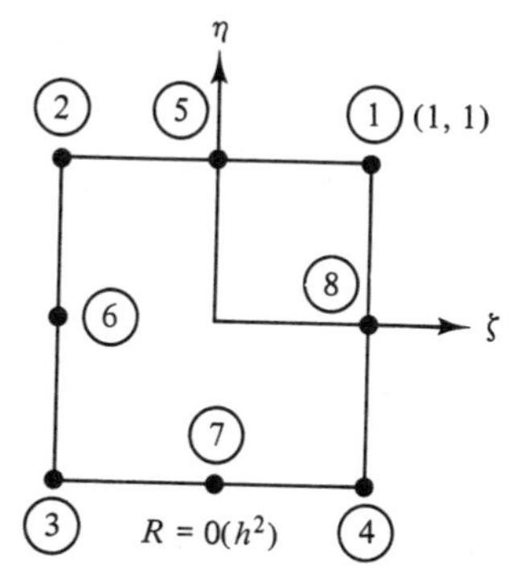

$$\begin{aligned} 4f = {} & -(1 + \zeta)(1 + \eta)(1 - \zeta - \eta)f_1 - (1 - \zeta)(1 + \eta)(1 + \zeta - \eta)f_2 \\ & - (1 - \zeta)(1 - \eta)(1 + \zeta + \eta)f_3 - (1 + \zeta)(1 - \eta)(1 - \zeta + \eta)f_4 \\ & + 2(1 - \zeta^2)(1 + \eta)f_5 + 2(1 - \zeta)(1 - \eta^2)f_6 \\ & + 2(1 - \zeta^2)(1 - \eta)f_7 + 2(1 + \zeta)(1 - \eta^2)f_8 \end{aligned} \quad \text{(c)}$$

TABLE 7.9 (Continued)

Trivariable

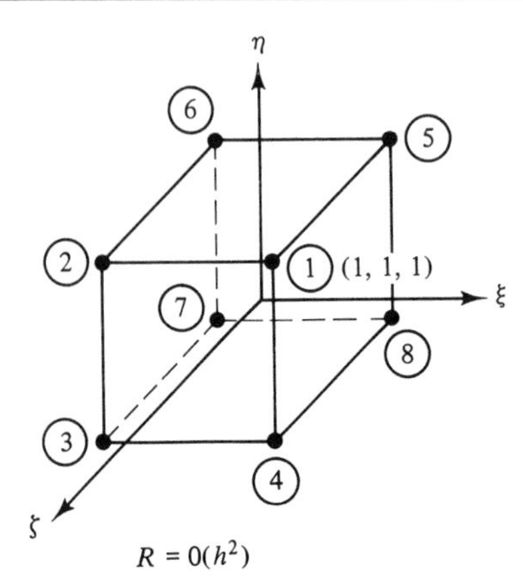

$R = 0(h^2)$

$$\begin{aligned} 8f = &+(1+\zeta)(1+\eta)(1+w)f_1 - (1+\zeta)(1-\eta)(1+w)f_2 \\ &+(1+\zeta)(1-\eta)(1-w)f_3 - (1+\zeta)(1+\eta)(1-w)f_4 \\ &-(1-\zeta)(1+\eta)(1+w)f_5 + (1-\zeta)(1-\eta)(1+w)f_6 \\ &-(1-\zeta)(1-\eta)(1-w)f_7 + (1-\zeta)(1+\eta)(1-w)f_8 \end{aligned} \quad \text{(d)}$$

We may compare interpolation formulas based on the order of their remainders. We define the order as the part of the remainder term that excludes the fraction and the derivative. Therefore, the order of error for the three-point univariable interpolation is h^3.

It is tempting to think that an interpolation function with a smaller remainder order will always be more accurate than one with a higher remainder order. This is not the case, since the value of the derivative is not considered in the order. Accordingly, interpolation of order h^5 may be less accurate than that of order h^3 for some functions.

The remainder measures the degree of completeness of the interpolation with respect to a power series expansion. For the first bivariable case, for example, the lowest derivative in the remainder is a second derivative. Therefore, the interpolation is complete through the first-derivative terms in the power series.

Figure 7.7 shows how a coefficient of a nodal displacement varies over an element for two bivariable cases. Each coefficient is an influence function in the sense that it has a unit value when the coordinates coincide with the nodal variable

TABLE 7.10 INTERPOLATION WITH HERMITE POLYNOMIALS*

For the rectangular figure shown, the osculatory interpolation formula, which is based on the Hermite polynomials, is given for the case when the function parameters are corner zero- and first-order derivatives.†

$$f(x,y) = \sum_{i=1}^{2} \sum_{j=1}^{2} \left\{ H_{0i}(x)H_{\sigma j}(y)f_{ij} + H_{1i}(x)H_{0j}(y)\frac{\partial f_{ij}}{\partial x} + H_{ai}(x)H_{ij}(y)\frac{\partial f_{ij}}{\partial y} \right\}$$

where

$$H_{01}(x) = \frac{1}{a^3}(2x^3 - 3ax^2 + a^3)$$

$$H_{02}(x) = -\frac{1}{a^3}(2x^3 - 3ax^2)$$

$$H_{11}(x) = \frac{1}{a^2}(x^3 - 2ax^2 + a^2x)$$

$$H_{12}(x) = \frac{1}{a^2}(x^3 - ax^2)$$

The formulas for $H(y)$ are the same when y replaces x and b replaces a.

*Hildebrand, F.B., *Introduction to Numerical Analysis*, New York: McGraw-Hill, 1956.

†These formulas can be generated for as many points and function derivatives as desired.

it multiplies and has a zero value when the coordinates coincide with any of the other nodal points.

When displacements corresponding to a rigid translation or rotation are prescribed in an interpolation formula, the interpolation function is consistent with the rigid movement. Accordingly, for any of the univariable interpolation formulas, the coefficients sum to 1 when $f_0, f_1, f_2, \ldots$ are all 1. The first moment of the coefficients are zero about any point for which $f_0, f_1, f_2 \ldots$ define a straight line or plane passing through the point.

Table 7.11 shows a proof of the fact that the second bivariable interpolation

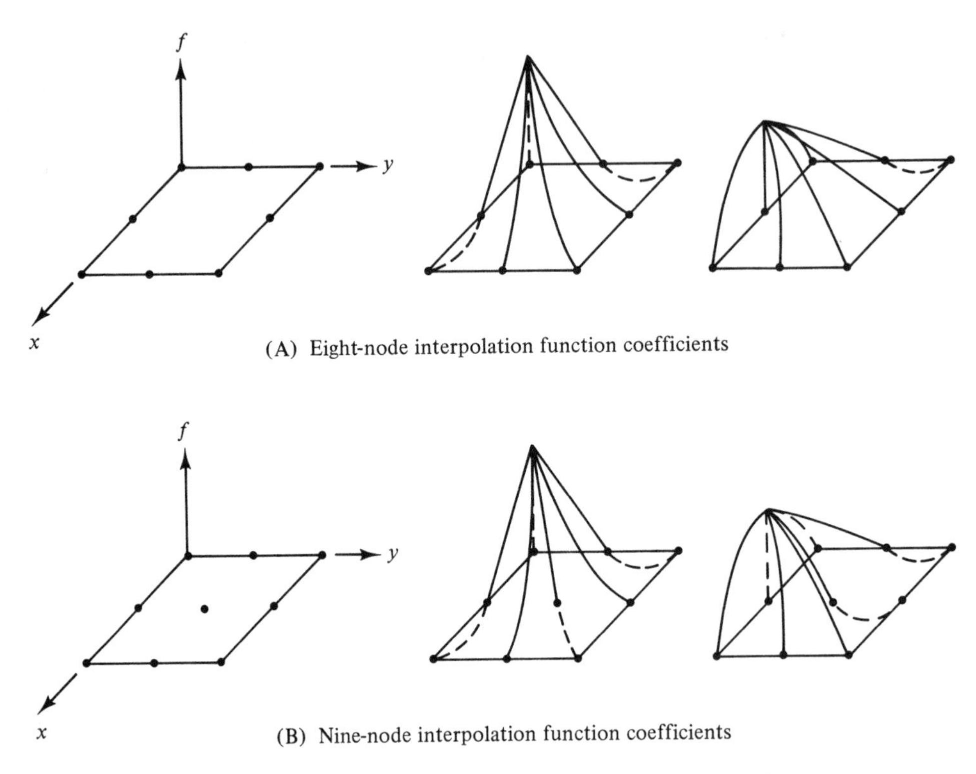

Figure 7.7 Geometry displacement space for bivariable functions.

formula provides a suitable basis for developing a potential energy membrane element. The proof consists of choosing x and y values for one side of the element and checking that the interpolation along the side depends only on displacements at nodes of the side.

It is desirable that the interpolation formulas yield the same interpolated value at a point independent of the choice of coordinate axes used. This is the requirement of coordinate invariance. The consequence of violating coordinate invariance is that quantities that should be uniquely defined, such as the behavioral strain energy and external work, the length of the resultant displacement vectors at a node, and sum of the diagonals of the system stiffness matrix, change with the user's choice of coordinate axes. These consequences are usually avoided by evaluating matrix coefficients with respect to axes embedded in the element, such as principal axes, and transforming the coefficients to relate to the system axes. Then direct consideration of invariance becomes a nonissue.

TABLE 7.11 POTENTIAL ENERGY TEST OF AN INTERPOLATION FUNCTION

Along side $1-8-4, \zeta = 1$. Eq. (b) of Table 7.9 gives

$$4f = 2\eta(\eta+1)f_1 - 2\eta(\eta-1)f_4 - 4(\eta-1)(\eta+1)f_8$$

Along side $2-6-3, \zeta = -1$. Eqs. (b) gives,

$$4f = 2\eta(\eta+1)f_2 - 2\eta(\eta-1)f_3 - 4(\eta+1)(\eta-1)f_6$$

Along side $1-5-2, \eta = 1$. Eq. (b) gives

$$4f = 2\zeta(\zeta+1)f_1 - 2\zeta(\zeta-1)f_2 - 4(\zeta+1)(\zeta-1)f_5$$

Along side $3-7-4, \eta = -1$. Eq. (b) gives

$$4f = 2\zeta(\zeta-1)f_3 + 2\zeta(\zeta+1)f_4 - 4(\zeta+1)(\zeta-1)f_7$$

Since the function f is defined along each side only by nodal valves of f along that side, displacements will be continuous, across element sides, with like elements. The interpolation function satisfies the potential energy continuity requirements.

7.6 GAUSS QUADRATURE

The integration required to evaluate the coefficients of the macro element equations is often performed numerically. Integration is thus replaced by summation.

For evaluating stiffness matrix coefficients using the potential energy approach, for example, the integration is related to the terms of the summation by

$$\mathbf{K} = \iiint (\mathbf{N}^T \mathbf{B}^T \mathbf{E} \mathbf{B} \mathbf{N})dV = \sum w_i(\mathbf{N}^T \mathbf{B}^T \mathbf{E} \mathbf{B} \mathbf{N})_i \qquad (7.21)$$

where

the subscript i identifies the point at x_i, y_i, z_i within the volume of the element,
w_i is the weighting given to the integrand at the point i, and
the summation extends over all samples of the integrand.

Here we have chosen to express the quadrature in terms of a single summation for clarity. In the case of integration over the volume, it could be expressed in terms of a triple summation to imply volume integration.

When the integrand is a polynomial, it is always possible to choose the points where the integrand is evaluated and the weights so that quadrature yields the exact value of the integral. In this case, changing analytic integration for quadrature trades computer calculation penalties in coefficient generation for time for the model creator to develop the coefficients in closed form. Most model creators are willing to accept this penalty.

However, use of quadrature offers the possibility of approximating the integration. These approximations can simultaneously reduce the number of integrand evaluations needed to estimate the equation coefficients and reduce the discretization error that associates with the element.

To clarify these claims, suppose that we want to evaluate the integral

$$A = \int_{-1}^{1} (a_0 + a_1 x + a_2 x^2 + a_3 x^3)\, dx \tag{7.22}$$

where a_j, $j = 1, 2, 3$ are constants. The integration limits have been scaled to 1 by selection of the x variable. Our quadrature problem is to choose points $-1 \leq x_i \leq 1$ and weights w_i so that we develop the exact value of the integral by numerical integration of the right-hand side of Eq. (7.21). Among the quadrature formulas that produce the exact solution, we want one that requires the least number of evaluations of the integrand.

Since the exact integral involves the sum of four independent terms, we require at least four parameters to characterize the integral. We elect two parameters for each evaluation of the integrand: the coordinate x_i and the weight w_i. Therefore, we need at least two integration points to integrate the cubic polynomial; that is, at least two integration points are needed for exact integration to be feasible.

Table 7.12 presents the analysis to develop the equations that must be satisfied by x_i and w_i for integrating the cubic polynomial. These equations require that each of the terms of the cubic will be integrated exactly. The method of undetermined coefficients extracts the necessary relations.

Equations (c) provide four nonlinear relations to be satisfied. Since they are nonlinear, they may have more that one solution. One solution is given by

$$x_1 = \frac{1}{\sqrt{3}} \cdot w_1 = 1; \qquad x_2 = \frac{1}{\sqrt{3}} \cdot w_2 = 1 \tag{7.23}$$

Because these values are real variables, a useful solution has been found.

Table 7.13 lists solutions of similar equations for a range of univariable polynomials.[8] In every case, the quadrature formulas listed produce the exact value of the integral and require the minimum number of evaluations of the integrand. The variables of the integrand are assumed scaled so that the limits of integration are plus and minus 1. We observe that the solutions for x_i often involve irrational numbers. In respect for the input round-off error in evaluating the formulas,

TABLE 7.12 DEVELOPMENT OF A GAUSS QUADRATURE RULE

Given: $f(x) = a_0 + a_1 x + a_2 x^2 + a_3 x^3$

Find: w_i and x_i such that $\int_{-1}^{1} f(x)\,dx = w_1 f(x_1) + w_2 f(x_2)$

1. The exact value of the integral is

$$\int_{-1}^{1} f(x)\,dx = 2a_0 + \frac{2}{3} a_2 \tag{a}$$

2. The value of the integral by the proposed quadrature rule is

$$\begin{aligned}\int_{-1}^{1} f(x)\,dx &= w_1(a_0 + a_1 x_1 + a_2 x_1^2 + a_3 x_3^3) + w_2(a_0 + a_1 x_2 + a_2 x_2^2 + a_3 x_2^3) \\ &= (w_1 + w_2)a_0 + (w_1 x_1 + w_2 x_1)a_1 + (w_1 x_1^2 + w_2 x_2^2)a_2 + (w_1 x_1^3 + w_2 x_2^3)a_3\end{aligned} \tag{b}$$

3. Equating the coefficients of a_i in Eqs. (a) and (b) gives

$$\left.\begin{aligned} w_1 + w_2 &= 2 \\ w_1 x_1 + w_2 x_2 &= 0 \\ w_1 x_1^2 + w_2 x_2^2 &= \frac{2}{3} \\ w_1 x_1^3 + w_2 x_2^3 &= 0 \end{aligned}\right\} \tag{c}$$

Any w_1, w_2, x_1, x_2 set that satisfies Eq. (c) defines a Gauss quadrature rule.

values are given for 15 significant digits. For a computer configuration with a precision of more than 15 digits, even more accurate values should be used.

This type of integration is Gauss integration, so called in honor of its inventor. The Gauss univariable quadrature formula can be applied to simple two- and three-dimensional geometries directly. Figure 7.8 indicates the basis of Gauss quadrature for a square membrane and a solid cube. The Gauss integration points are located by the univariable coordinates; the weights are the products of the one-dimensional weights.

But applying the univariable formulas to two- and three-dimensional models usually does not result in a quadrature rule with the least number of integrand evaluations. This goal is achieved by writing and solving the equations for the multidimensional case by direct extension of the example of Table 7.12. When the order of the polynomial gets large, numerical analysis is used to find solutions of the nonlinear equations.

The integral is "fully integrated" when the quadrature determines the exact value of the integral. The feasibility argument is useful in determining the minimum number of Gauss points needed for full integration. For example, for a rectangular membrane using the first bivariable interpolation formula of Table 7.9,

TABLE 7.13 ABSCISSAS AND WEIGHT FACTORS FOR GAUSSIAN INTEGRATION

$$\int_{-1}^{+1} f(x)dx = \sum_{i=1}^{n} w_i f(x_i)$$

Abscissas = $\pm x_i$; weights = w_j

n	$\pm x_i$	w_i
$n = 2$	0.57735 02691 89626	1.00000 00000 00000
$n = 3$	0.00000 00000 00000	0.88888 88888 88889
	0.77459 66692 41483	0.55555 55555 55556
$n = 4$	0.33998 10435 84856	0.65214 51548 62546
	0.86113 63115 94053	0.34785 48451 37454
$n = 5$	0.00000 00000 00000	0.56888 88888 88889
	0.53846 93101 05683	0.47862 86704 99366
	0.90617 98459 38664	0.23692 68850 56189
$n = 6$	0.23861 91860 83197	0.46791 39345 72691
	0.66120 93864 66265	0.36076 15730 48139
	0.93246 95142 03152	0.17132 44923 79170
$n = 7$	0.00000 00000 00000	0.41795 91836 73469
	0.40584 51513 77397	0.38183 00505 05119
	0.74153 11855 99394	0.27970 53914 89277
	0.94910 79123 42759	0.12948 49661 68870
$n = 8$	0.18343 46424 95650	0.36268 37833 78362
	0.52553 24099 16329	0.31370 66458 77887
	0.79666 64774 13627	0.22238 10344 53374
	0.96028 98564 97536	0.10122 85362 90376
$n = 9$	0.00000 00000 00000	0.33023 93550 01260
	0.32425 34234 03809	0.31234 70770 40003
	0.61337 14327 00590	0.26061 06964 02935
	0.83603 11073 26636	0.18064 81606 94857
	0.96816 02395 07626	0.08127 43883 61574
$n = 10$	0.14887 43389 81631	0.29552 42247 14753
	0.43339 53941 29247	0.26926 67193 09996
	0.67940 95682 99024	0.21908 63625 15982
	0.86506 33666 88985	0.14945 13491 50581
	0.97390 65285 17172	0.06667 13443 08688
$n = 12$	0.12523 34085 11469	0.24914 70458 13403
	0.36783 14989 98180	0.23349 25365 38355
	0.58731 79542 86617	0.20316 74267 23066
	0.76990 26741 94305	0.16007 83285 43346
	0.90411 72563 70475	0.10693 93259 95318
	0.98156 06342 46719	0.04717 53363 86512
$n = 16$	0.09501 25098 37637 440185	0.18945 06104 55068 496285
	0.28160 35507 79258 913230	0.18260 34150 44923 588867
	0.45801 67776 57227 386342	0.16915 65193 95002 538189
	0.61787 62444 02643 748447	0.14959 59888 16576 732081
	0.75540 44083 55003 033895	0.12462 89712 55533 872052
	0.86563 12023 87831 743880	0.09515 85116 82492 784810
	0.94457 50230 73232 576078	0.06225 35239 38647 892863
	0.98940 09349 91649 932596	0.02715 24594 11754 094852
$n = 20$	0.07652 65211 33497 333755	0.15275 33871 30725 850698
	0.22778 58511 41645 078080	0.14917 29864 72603 746788
	0.37370 60887 15419 560673	0.14209 61093 18382 051329
	0.51086 70019 50827 098004	0.13168 86384 49176 626898
	0.63605 36807 26515 025453	0.11819 45319 61518 417312
	0.74633 19064 60150 792614	0.10193 01198 17240 435037
	0.83911 69718 22218 823395	0.08327 67415 76704 748725
	0.91223 44282 51325 905868	0.06267 20483 34109 063570

TABLE 7.13 (Continued)

0.96397 19272 77913 791268	0.04060 14298 00386 941331
0.99312 85991 85094 924786	0.01761 40071 39152 118312
$n = 24$	
0.06405 68928 62605 626085	0.12793 81953 46752 156974
0.19111 88674 73616 309159	0.12583 74563 46828 296121
0.31504 26796 96163 374387	0.12167 04729 27803 391204
0.43379 35076 26045 138487	0.11550 56680 53725 601353
0.54542 14713 88839 535658	0.10744 42701 15965 634783
0.64809 36519 36975 569252	0.09761 86521 04113 888270
0.74012 41915 78554 364244	0.08619 01615 31953 275917
0.82000 19859 73902 921954	0.07334 64814 11080 305734
0.88641 55270 04401 034213	0.05929 85849 15436 780746
0.93827 45520 02732 758524	0.04427 74388 17419 806169
0.97472 85559 71309 498198	0.02853 13886 28933 663181
0.99518 72199 97021 360180	0.01234 12297 99987 199547

**Handbook of Mathematical Functions* edited by Milton Abramowitz and Irene Stegun, Dove Publications, 1965, p. 916.

we determine that we need two Gauss points and hence two evaluations of the integrand for the strain energy by the following feasibility arguments.

1. Since the strains associated with the interpolation are linear in the membrane coordinates x and y, the strain energy integrand depends on up to quadratic terms in x and y.

2. The total number of terms in a bivariable power series expansion through quadratic is six. Therefore, to represent the complete quadratic, we need to assign at least six arbitrary constants.

3. For each Gauss point in bivariable space there will be three arbitrary parameters—two spatial coordinates and the weighting constant. Therefore, we need at least $6/3 = 2$ Gauss points to attain full integration.

When more than the minimum number of Gauss points are used, the integral is "overintegrated." For linearly behaving systems, the only advantage of overintegration is simpler computer program development. In analyses, the quadrature is exact, but the number of calculations is needlessly increased.

When fewer than the minimum number of Gauss points are used, the integral is "underintegrated." Experiments show that underintegration reduces stiffness. Since potential energy models present overstiff representations, reducing stiffness by underintegration usually results in reducing discretization error for these models. Hence the surprising result that reducing the accuracy of quadrature yields more accurate structural behavior predictions with fewer calculations than those of full integration.

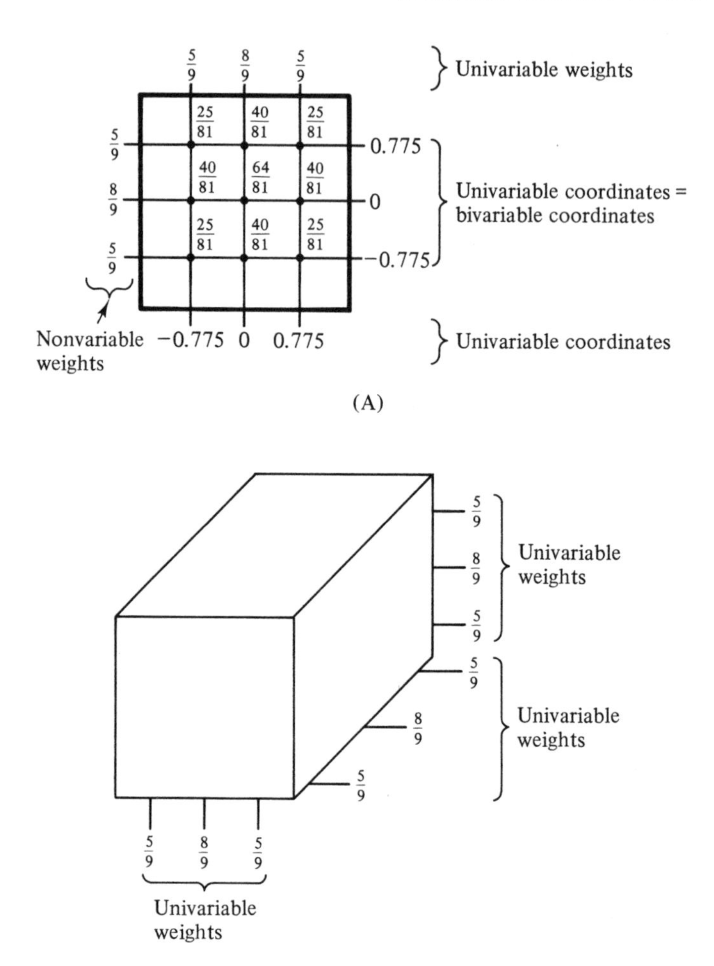

Point Coordinates	Weight*	Point Coordinates	Weight*	Point Coordinates	Weight*
−0.775, −0.775, −0.775	125	0, −0.775, −0.775	200	0.775, −0.775, −0.775	125
−0.775, −0.775, 0	200	0, −0.775, 0	320	0.775, −0.775, 0	200
−0.775, −0.775, 0.775	125	0, −0.775, 0.775	200	0.775, −0.775, 0.775	125
−0.775, 0, −0.775	200	0, 0, −0.775	320	0.775, 0, −0.775	200
−0.775, 0, 0	320	0, 0, 0	512	0.775, 0, 0	320
−0.775, 0, 0.775	200	0, 0, 0.775	320	0.775, 0, 0.775	200
−0.775, 0.775, −0.775	125	0, 0.775, −0.775	200	0.775, 0.775, −0.775	125
−0.775, 0.775, 0	200	0, 0.775, 0	320	0.775, 0.775, 0	200
−0.775, 0.775, 0.775	125	0, 0.775, 0.775	200	0.775, 0.775, 0.775	125

*times 729.

Figure 7.8 Extension of Gauss rules from the univariable case.

7.7 PARAMETRIC MAPPING

Parametric mapping provides a systematic way of changing variables so that integration limits for irregularly shaped element models become constants. It thus facilitates use of simple interpolation formulas, like those of Table 7.9, in developing FEA models for elements with curved sides like those of Figs. 7.2 through 7.4.

Suppose, for example, we let

$$\begin{aligned} 4x &= (\zeta + 1)(\eta + 1)x_1 - (\zeta - 1)(\eta + 1)x_2 + (\zeta - 1)(\eta - 1)x_3 \\ &\quad - (\zeta + 1)(\eta - 1)x_4 \\ 4y &= (\zeta + 1)(\eta + 1)y_1 - (\zeta - 1)(\eta + 1)x_2 + (\zeta - 1)(\eta - 1)y_3 \\ &\quad - (\zeta + 1)(\eta - 1)y_4 \end{aligned} \tag{7.24}$$

where

x and y are the coordinates of points of a trapezoid in real space, and
ζ and η are the coordinates in the mapped space.

Equations (7.24) define a parametric mapping. It maps points in the x, y space uniquely into points in the ζ, η space. It maps the line 1–2 into the line 1′–2′. Similarly, it maps the other three edges into straight edges in the ζ, η space. It maps points inside the trapezoid into points inside the two-unit square. If we use this mapping to change any function of x and y integrated over the surface area, the limits of integration will change from being linear functions of x and y to constants in the mapped space.

According to calculus, the differential geometry in the mapped space is related to the differential geometry in the real space by

$$dx\,dy = |\mathbf{J}|\,d\zeta\,d\eta \tag{7.25}$$

where

$\mathbf{J}$ is the Jacobian of the mapping and
$|\mathbf{J}|$ is the determinant of the matrix that relates the differentials of a function in the real and mapped spaces.

This relation involves the chain rule of differentiation:

$$\begin{pmatrix} dS/d\zeta \\ dS/d\eta \end{pmatrix} = \begin{bmatrix} dx/d\zeta & dy/d\zeta \\ dx/d\eta & dy/d\eta \end{bmatrix} \begin{pmatrix} dS/dx \\ dS/dy \end{pmatrix} = \mathbf{J} \begin{pmatrix} dS/dx \\ dS/dy \end{pmatrix} \tag{7.26}$$

Using a polynomial interpolation function for the mapping, the coefficients of $\mathbf{J}$ are polynomials. Evaluation of the determinant is a tedious analytical task in finite element model development. Therefore, it is assigned to the computer.

Suppose we use isoparametric mapping: We describe the displacements in the mapped space by the same interpolation function used to describe the original geometry. Then, for the sample rectangular membrane, corresponding to Eqs. (7.24), we choose the interpolation functions,

$$\begin{aligned}4u &= (\zeta + 1)(\eta + 1)u_1 - (\zeta - 1)(\eta + 1)u_2 + (\zeta - 1)(\eta - 1)u_3 \\ &\quad - (\zeta + 1)(\eta - 1)u_4 \\ 4v &= (\zeta + 1)(\eta + 1)v_1 - (\zeta - 1)(\eta + 1)v_2 + (\zeta - 1)(\eta - 1)v_3 \\ &\quad - (\zeta + 1)(\eta - 1)v_4\end{aligned} \tag{7.27}$$

Then there is only one Jacobian determinant to evaluate. (We could choose to map the original geometry with a lower-order polynomial than the displacements for a "subparametric" mapping or, conversely, for a "superparametric" model, but this is rarely done for finite element models.)

Another advantage of isoparametric mapping is that it preserves the continuity of displacements between like elements. As long as the mapped space interpolation function along a side is a function of only the generalized nodal variables on the side, the stiffness coefficients for the real space will reflect this continuity. Therefore, we can create potential energy analysis models using isoparametric models and exact quadrature.

To relate the mapping to element development, we recognize that Eqs. (7.27) express the displacements in the mapped space in the form

$$\begin{pmatrix} u \\ v \end{pmatrix} = \begin{bmatrix} 0.25 & \mathbf{N}(\zeta, \eta) & 0 \\ & 0 & \mathbf{N}(\zeta, \eta) \end{bmatrix} \begin{pmatrix} \mathbf{u}_i \\ \mathbf{v}_i \end{pmatrix} = \mathbf{N}\,\mathbf{g}_e \tag{7.28}$$

resulting in membrane strains

$$\begin{pmatrix} e_{xx} \\ e_{yy} \\ e_{xy} \end{pmatrix} = \begin{bmatrix} \delta/\delta x & 0 \\ 0 & \delta/\delta y \\ \delta/\delta y & \delta/\delta x \end{bmatrix} \begin{pmatrix} u \\ v \end{pmatrix} = \mathbf{B}\,\mathbf{u} = \mathbf{B}\,\mathbf{N}\,\mathbf{g}_e \tag{7.29}$$

Choosing the function $S = N$ in Eqs. (7.26) and inverting the relationship yields

$$\begin{pmatrix} \delta N/\delta x \\ \delta N/\delta y \end{pmatrix} = \mathbf{J}^{-1} \begin{pmatrix} \delta N/\delta \zeta \\ \delta N/\delta \eta \end{pmatrix} \tag{7.30}$$

Equations (7.25), (7.26), and (7.30) define all the terms needed for evaluating the potential energy in Eq. (7.18) and hence the coefficients of the element macro equations.

The parametric mapping approach, detailed here for the membrane case, is applicable to plates, shells, and three-dimensional solid finite element models as well. By introducing this mapping into finite element model development, Irons provided the means for more than doubling the library of elements from that which preceded his paper.[11]

Note that parametric mapping results in approximate quadrature when irregular geometries are involved. In general the mapping converts the integrand to a rational fraction. This fraction can only be integrated exactly by Gauss quadrature when $|\mathbf{J}|$ is constant. For the membrane example, $|\mathbf{J}|$ is constant only when the quadrilateral is a rectangle.

7.8 STIFFNESS MODELS: MINIMUM REQUIREMENTS

The minimum requirement of field element stiffness models is that they represent strain-free and constant strain states as the element size approaches zero. This requirement implies that each element model must have the capacity to represent the constants and first-order derivative terms in a Taylor series expansion of the exact solution of the elasticity or structural equations being addressed.[11]

From the structural engineering viewpoint, this requirement means that when nodal generalized displacements are consistent with rigid translations and rotations of the element, the stiffness matrix will reflect zero strain energy. In terms of generalized nodal forces, this requirement enforces the condition that forces of each column of the stiffness matrix be in macroscopic equilibrium.

The Taylor series requirement also means that when nodal displacements are consistent with a constant-strain state in the element, the strain energy, any undefined nodal displacements, and the nodal forces must be consistent with the constant-strain state. This requirement must be met for any linearly independent rigid movement and any linearly independent constant strain state.

The table below defines the number of linearly independent rigid and constant-strain states for the principal subclasses of element models. It notes that in the element-embedded axes, the axes lying in the midplane of the flat membrane, three rigid modes are possible: translation in two directions and rotation about the surface normal. When the membrane is transformed for general use with three-dimensional nodal variables (u, v, w), three additional linearly independent rigid movements can be described by the nodal variables, though the number of constant strain states still includes only linear combinations of uniaxial extensions and shear in the midplane. Similar interpretations apply for the plate, shell, and solid elements.

Displacement States for Continuum Models

ELEMENT MODEL	RIGID STATES LOCAL	x, y, z	CONSTANT-STRAIN STATES
Membrane	3	6	3
Plate	3	6	3
Shell	6	6	6
3D solid	6	6	6

Since the stiffness matrix is independent of element size for any membrane model, we must represent the strain-free and constant-strain states of this model without regard to element size. For generality, we extend this characteristic into a requirement for all element models. Accordingly, the minimum requirement of a field element model is that it represent all independent rigid movements as strain-free and independent constant-strain states exactly.

We also require that the rank of the element stiffness matrix equal its order minus the number of independent rigid movements. The addition of this requirement, though not a mathematical necessity, ensures that the constrained system stiffness matrix can only be singular when the model represents kinematic instability, just as for lattice models.

The final minimum requirement of a finite element field continuum stiffness model is that its stiffness matrix be positive semidefinite whenever the material properties matrix is positive definite. By adding this requirement, we guarantee that the system stiffness matrix will be positive definite when the material model is not degenerate. We provide for associating indefinite and negative definite system stiffness matrices with material and/or geometric buckling.

7.9 EQUIVALENT LOADING: MINIMUM REQUIREMENTS

Clough defines the minimum requirement for equivalent loading models: The nodal forces that replace the distributed loading must preserve the total force in each direction and for each side of the element.[7]

The principle behind this requirement is that as the grid becomes finer, the equivalent loads will be prescribed over shorter internodal distances. Therefore, the distributed forces will be described more accurately. By appropriate selection of the number of nodes between any two points, the analyst can ensure that the distributed loading is defined as accurately as desired.

In deriving the macro equations using the potential energy approach, the equivalent loading model is defined by the external work integral. The equivalent nodal forces that arise by differentiating the integral with respect to the generalized nodal displacements must preserve the total force and the first moment of the force. This is true because the displacement states include rigid movements as can be shown by considering the external work for displacements corresponding to rigid movements.

The minimum requirements for equivalent loading models tolerates a variety of models. Table 7.14 gives data for assessing four equivalent loading models for the rectangular membrane model. Since all the loading models satisfy the minimum requirements, the comparison is based on the decline in error in representing the first and second moment of the loading.

The external work model assumes the bilinear shape functions, Eqs. (7.27). To calculate the lumped loads, the total force and first moment of the equivalent loading are required to match those of the loading itself. The external work and

TABLE 7.14 Validation of Equivalent Loading on Edge*

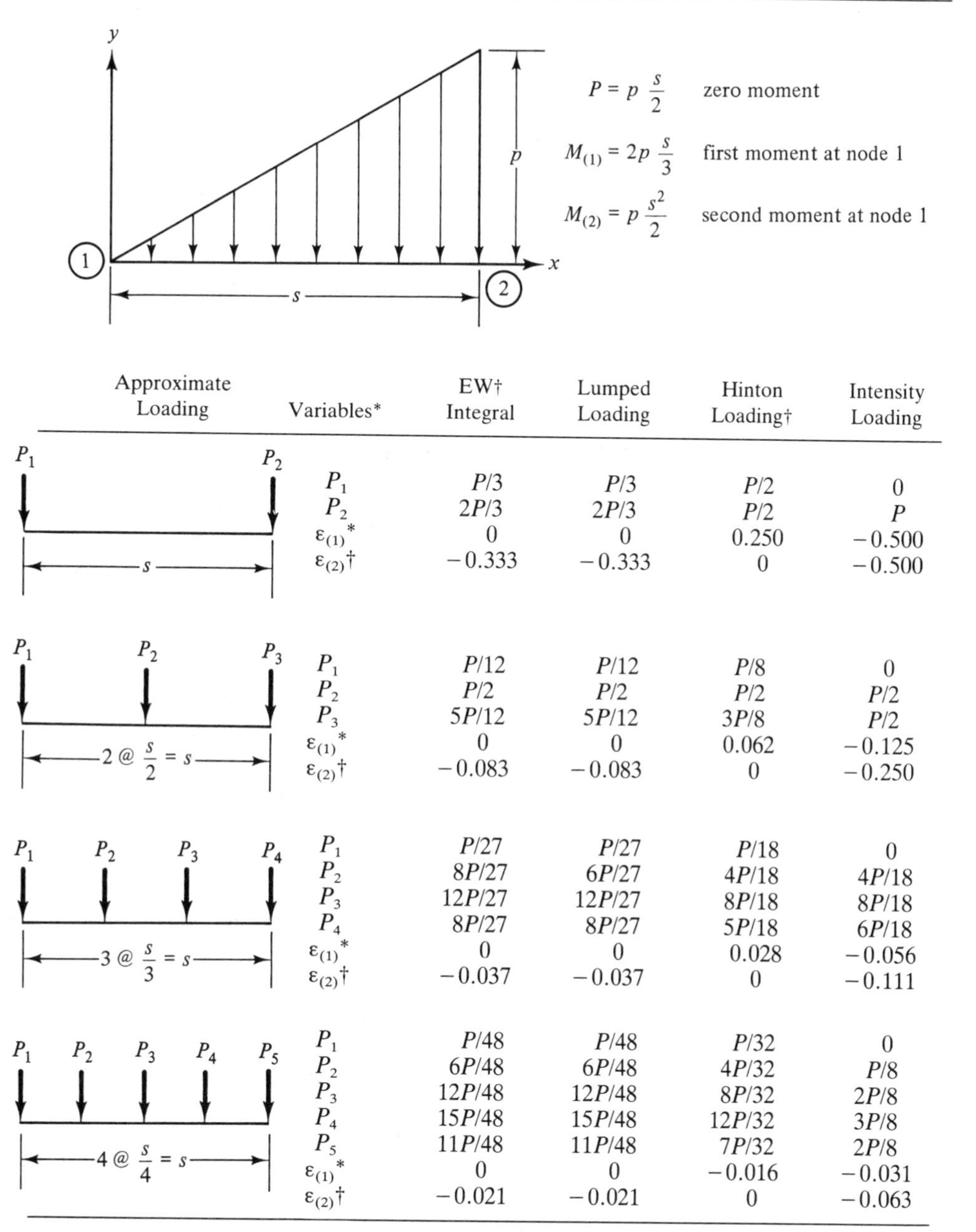

Approximate Loading	Variables*	EW† Integral	Lumped Loading	Hinton Loading†	Intensity Loading
1 @ s	P_1	$P/3$	$P/3$	$P/2$	0
	P_2	$2P/3$	$2P/3$	$P/2$	P
	$\varepsilon_{(1)}$*	0	0	0.250	−0.500
	$\varepsilon_{(2)}$†	−0.333	−0.333	0	−0.500
$2 @ \frac{s}{2} = s$	P_1	$P/12$	$P/12$	$P/8$	0
	P_2	$P/2$	$P/2$	$P/2$	$P/2$
	P_3	$5P/12$	$5P/12$	$3P/8$	$P/2$
	$\varepsilon_{(1)}$*	0	0	0.062	−0.125
	$\varepsilon_{(2)}$†	−0.083	−0.083	0	−0.250
$3 @ \frac{s}{3} = s$	P_1	$P/27$	$P/27$	$P/18$	0
	P_2	$8P/27$	$6P/27$	$4P/18$	$4P/18$
	P_3	$12P/27$	$12P/27$	$8P/18$	$8P/18$
	P_4	$8P/27$	$8P/27$	$5P/18$	$6P/18$
	$\varepsilon_{(1)}$*	0	0	0.028	−0.056
	$\varepsilon_{(2)}$†	−0.037	−0.037	0	−0.111
$4 @ \frac{s}{4} = s$	P_1	$P/48$	$P/48$	$P/32$	0
	P_2	$6P/48$	$6P/48$	$4P/32$	$P/8$
	P_3	$12P/48$	$12P/48$	$8P/32$	$2P/8$
	P_4	$15P/48$	$15P/48$	$12P/32$	$3P/8$
	P_5	$11P/48$	$11P/48$	$7P/32$	$2P/8$
	$\varepsilon_{(1)}$*	0	0	−0.016	−0.031
	$\varepsilon_{(2)}$†	−0.021	−0.021	0	−0.063

*$\varepsilon_{(1)}$ is the relative error in the value of the first moment of the edge loading.

†$\varepsilon_{(2)}$ is the relative error in the value of the second moment of the edge loading.

lumped loading models are the same for this example because the element interpolation function is linear along a side. The Hinton model defines the loading forces to be proportional to the stiffness matrix diagonals.[12] The force vector is then scaled so that the total force is preserved. The intensity model defines the loading forces to be proportional to the intensity of loading at each node.[13] Again, the force vector is scaled so that the total force is preserved.

The data shows the convergence of all four models toward zero error in the first and second moments. Neither the Hinton nor the intensity model involves as little error as the external work and lumped loading models.

A column of Table 7.14 lists the kind of data that can be developed to validate the equivalent loading model of a computer configuration. Producing this data in numerical form using the particular computer will show directly that the equivalent loading model meets its minimum requirement.

7.10 MICRO MODELS: MINIMUM REQUIREMENTS

The minimum requirement for a microscopic model is that it produce values of stresses at a point that, as the size of the element model approaches zero, satisfies the differential equations of equilibrium. As with stiffness and equivalent loading models, the minimum requirement tolerates a variety of models, and the model can be selected independently of the choice of models for the other components of the element model.

When the element development is based on displacement interpolation functions, as it is for 96 percent of the elements in the library,[2] the displacement functions are differentiated at the point, to evaluate strains, and the pointwise stress-strain equations used to evaluate stresses. Except for the simplex membrane and tetrahedron, this process results in approximate values of point stresses that do not satisfy the microscopic equilibrium equations.

Nevertheless, most finite element computer configurations deal with these stress estimates as if they satisfy microscopic equilibrium. They direct calculation of principal stresses under the assumption that stress estimates approximate equilibrium with negligible error. They report these values as stresses. Interpretation of these "stresses" as the value of nodal forces averaged over the element cross section is much more in keeping with their approximate nature. This interpretation cautions the analyst that errors in stress estimates are sensitive to the magnitude of discretization error as well as the choice of the microscopic element model.

Consistent with this interpretation, we validate the micro element model implementation by requiring that the stresses at interior points be consistent with the nodal forces on the element. Validation consists of testing that for all linearly independent nodal displacement states for the element and for all geometries, the total force across a section, represented by integrating the stresses over the cross section, equals the sum of the nodal forces on the section.

7.11 METHODS OF DERIVING MODELS

The general mathematical approach to derive a continuum element model is to choose shape functions that represent the independent variable as linear functions of arbitrary constants and then evaluate the constants so that the function satisfies the governing equations of the element in some best sense. The shape functions may be displacement, stress, or displacement and stress functions for the FEA of structures. The governing equations may be the equations of elasticity or equations of the structural theory. The function may be fit to satisfy the equations at a number of points over the element, to control the integral of the error in satisfying the equations, or by weighted point summations or integrations.

The role of minimum requirements for continuum models is to restrict the choice of shape functions to those with the prospect of associating with zero discretization error as the number of nodal variables approaches infinity. On this basis, shape functions can be rejected by the model developer, or element models can be shown to be valid by the analyst.

The choice of shape function is critical in establishing whether stiffnesses will tend to be overestimated or underestimated. Displacement shape functions are necessary for ensuring low estimates of strain energy by the potential energy approach and usually associate with low estimates even when the discretization error is not controlled in the energy norm. Similarly, equilibrium-satisfying stress states usually associate with high estimates of energy even though the complementary energy approach is not used as the basis of error control. When shape functions include both displacement and stress functions, stiffness may be over- or underestimated.

When the strain energy estimates are low, generalized displacements tend to be low, and vice versa. This is only a tendency—some nodal displacement variables may be higher than the zero discretization error solution, but since the strain energy is lower than the exact, in this averaging sense, displacements must be smaller.

Regardless of whether the strain energy estimate is high or low, stress estimates will cluster around the zero discretization error solution. We cannot identify the tendency to have high or low strain energy with the tendency for high or low estimates of stresses.

Pointwise error control (collocation) is used in classical finite difference model derivations and in the integral equation method of analysis—the "boundary integral" method.[14] Finite difference derivations replace the derivatives in the governing equations with difference approximations. This approach was used in deriving the plate stiffness matrix of Table 7.4. The approach is unpopular, probably due to the high sensitivity of solutions to the choice of location of the nodes.

Boundary integral elements are founded on superimposing exact solutions of the governing equations to satisfy boundary conditions at a finite number of points on the boundary of the element. This approach seems to be gaining popularity, probably because its need for only nodes on and definition of the boundary of a region suggests that it may require fewer nodal variables than a comparably accurate finite element model. Its need for exact solutions within the element restricts its scope of application to isotropic materials and linear analysis, or forces compromises in the approach.

Weighted integral methods of derivation are popular. These, like the simplex derivation, couch the element behavior equations in the form of a variational calculus statement. The potential energy approach is by far the most popular approach to element model derivation, though element models based on stationary principles[2] and complementary energy[15] are available.

Based on the methods of solving elasticity structural problems in the classical literature, there are at least 768 ways of deriving an element model for a particular geometry, set of nodes, and nodal displacement variables.[16] We can hope that researchers sort out the relation between the choice of method and the efficiency of the associated element model, but it has not happened since the problem was identified in 1971.

The number of ways of deriving element equations and the number of geometries of interest mean that we can expect a continual flow of new models into the literature and into FEA computer codes. These numbers also suggest the need for element model testing that is insensitive to the method of derivation of the macro and micro models.

7.12 SELECTION OF ELEMENT MODELS

Selection of element models for the analysis of a structural system depends on the purpose of the analysis and the models available to the analyst. Both the capability and the efficiency (accuracy per unit of computer resources needed) are of concern.

Existing computer programs support model selection for three types of analysis: conventional, potential energy, and self-qualified. In conventional analysis, any of the models in the program's library may be used with any others. In potential energy analysis, selection is made of a subset of elements that can be used together to perform an analysis consistent with potential energy theory. In self-qualified analysis, only elements that can generate accurate assessments of discretization error and are guaranteed to ensure zero discretization error as the mesh interval approaches zero are used.

In conventional analysis, element models need only satisfy the minimum requirements. The element types to use in a particular analysis are selected based on the analyst's experience with elements on relevant problems. In the automobile industry, for example, cumulative experience in applying the finite element to automobile structures furnishes the basis for modeling new car structure.[17]

Conventional analysis tends to be the most efficient analysis, since it provides for the broadest selection of models. It is well suited to production analyses where modeling experience can accumulate. It can play an important role in design analyses at the stages of design when high-accuracy response predictions can be traded for efficient approximations.

In potential energy analyses, continuous displacements must be assumed in the interior of each element, and continuity across element boundaries must be maintained to ensure integrability and boundedness of cross-boundary integrals. Besides meeting minimum continuum element requirements, elements must be tested to ensure that they deform without interelement gapping or overlapping.

Potential energy analyses are generally less efficient than conventional ones. They are necessary when the guarantee of a low estimate of stiffness ensures a safe structure, such as a structure for which imposed displacements define the critical loading condition. Traditionally, analysts use potential energy models in assessing structural integrity against dynamic loadings.

Self-qualified analyses are least efficient because they usually embrace two or more analyses. They are desirable when other independent checks of analysis results would be too expensive or unreliable. They are justifiable when the structural behavior predictions must be very accurate, say, with error less than 1 percent.

Both element model and structural system tests are useful for analyses for continua. Continuum element model tests, like the tests of two-node elements, involve validating element performance independent of any particular structural system. System tests involve verification of element model performance in problems where the exact solution may be unknown. The next two chapters address continuum model validation by element tests and verification by system analyses that include discretization error estimates.

7.13 HOMEWORK

7:1. Two-dimensional stress-strain equations.

Given: The three-dimensional isotropic stress-strain equations, Eqs. (7.8).

Derive: The two-dimensional stress-strain equations.

(a) Assuming plane strain.

(b) Assuming plane stress.

7:2. Singularity of the material properties matrix.

Given: The plane stress equations, Eqs. (7.9).

Find: The values of Poisson's ratio for which **E** is singular.

***7:3.** Definiteness of the material properties matrix.

Given: The plane strain equations, Eqs. (7.10).

Determine: The conditions on Poisson's ratio that associate with a nonpositive definite material properties matrix.

7:4. Rank of triangular membrane stiffness matrix.

Given: The stiffness matrix of Table 7.3.

Prove: That the rank of this matrix is less than 4.

***7:5.** Derivation of a boundary element matrix.

Given: The displacement function for the first boundary element of Table 7.6 is $u = u_1 + (u_2 - u_1)x/L$.

Derive: The boundary element stiffness matrix.

7:6. Potential energy approach to stiffness derivation.

Given: The straight beam of the figure, modeled as follows:

(a) Nodes are only at the ends and involve v deflection and t rotation.
(b) Movement occurs only in the xy plane.
(c) The Bernouilli-Euler beam assumptions apply.
(d) The displacement shape function is

$$v = a_0 + a_1x + a_2x^2 + a_3x^3$$

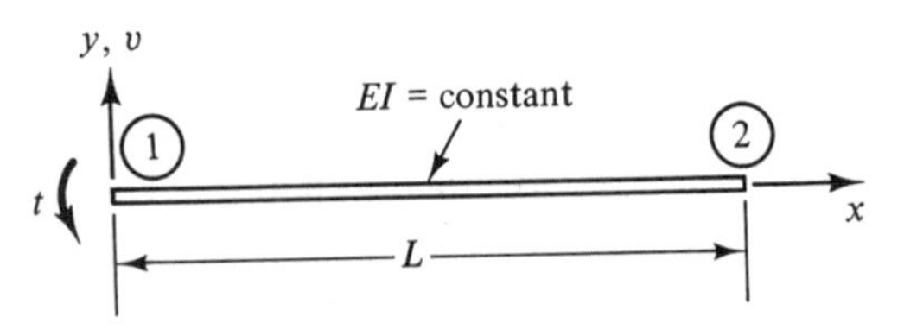

Homework Problem 7:6

Derive: The stiffness matrix for the element.

Method: Use the potential energy approach.

***7:7.** Vandermode matrix.

Given: The displacement function $u = a_0 + a_1x$ for a rod of length L. The origin is at the first node.

Find: The Vandermode matrix and the interpolation function.

7:8. Vandermode matrix requirement.

Given: The displacement function $u = a_0 + a_1x^2$ for a rod of length L. The origin is at the first node.

Find: The Vandermode matrix and the interpolation function.

7:9. Stress equilibrium.

Given: The bilinear displacement function of Table 7.2.

Find: If there are any points in the rectangle where the microscopic equations of equilibrium will be satisfied when the material is isotropic.

7:10. Remainder term in interpolation.

Given: The function $x = \sin\theta$ when $\theta = 0, \pi/2$, and π.

Find: The maximum error of interpolation in the interval $0 < \theta < \pi$ when quadratic interpolation is used.

Method: Use the remainder.

***7:11.** Interpolation error estimation.

Given: The function $x = \sin\theta$ when $\theta = 0, \pi/2$, and π.

Calculate:

(a) The estimate of x at $\theta = \pi/4$ using linear interpolation.

(b) The estimate of x at $\theta = \pi/4$ using quadratic interpolation between data at $\theta = 0$ and $\theta = \pi$.
(c) The estimate of the error using the results of items (a) and (b).
(d) The error estimate for quadratic interpolation at $\pi/4$ using the remainder formula.
(e) The actual error of the quadratic fit.

7:12. Potential energy shape function test.

Given: The second bivariable interpolation function of Table 7.9.

Determine: If this function will result in a potential energy analysis.

Method: Test the displacements along the side $x = 1$.

7:13. Gauss integration equations.

Given: A quadratic polynomial,

$$x = a_0 + a_1x + a_2x^2 + a_3x^3 + a_4x^4$$

Prove: That the weights of a three-point Gauss integration formula are 5/9, 8/9, and 5/9 for the points at $x = \sqrt{3}/\sqrt{5}$, $x = 0$, and $x = -\sqrt{3}/\sqrt{5}$.

7:14. Gauss integration of matrix coefficients.

Given: The following matrix of coefficients is required to be integrated exactly over the interval 0 to 1 by a single Gauss rule.

$$\int_{-1}^{1} \begin{bmatrix} 1 & 2x^2 \\ 2x^2 & 3x^3 \end{bmatrix} dx$$

Find: The integral of the matrix using Gauss quadrature.

7:15. Bivariant Gauss integration.

Given: The claim that the Gauss points and weights given for the two-unit square exactly integrate the polynomial

$$u = a_0 + a_1x + a_2y + a_3x^2 + a_4xy + a_5y^2$$

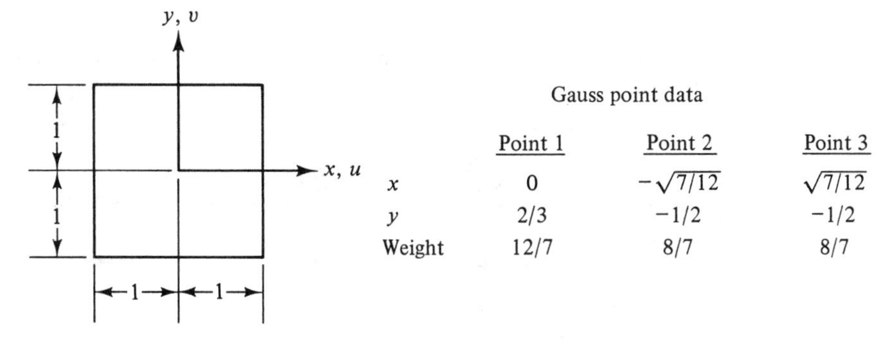

Gauss point data

	Point 1	Point 2	Point 3
x	0	$-\sqrt{7/12}$	$\sqrt{7/12}$
y	2/3	−1/2	−1/2
Weight	12/7	8/7	8/7

Homework Problem 7:15

Verify or discredit: The claim for the xy term.

Method: Compare exact and Gauss quadrature results.

***7:16.** Selection of Gauss rules.

Given: We need to represent a rectangular membrane using the first bivariable interpolation formula of Table 7.9. The thickness of the membrane varies as $t = a_0 + a_1x + a_2y + a_3xy$ over the midsurface.

Determine: The minimum number of Gauss points for full integration.

7:17. Parametric mapping.

Given: The parametric mapping of Eq. (7.24).

Show: That the point at $(\zeta, \eta) = (0.5, 0.25)$ is mapped from a point in the upper right-hand quadrant of the unmapped geometry.

7:18. Isoparametric mapping and potential energy elements.

Given: The parametric mapping of Eq. (7.24) and the fact that the displacements of Eqs. (7.27) maintain displacement continuity across each side of the rectangular element.

Show: That the displacements along a side of the unmapped element will vary linearly and that the element is a potential energy element model.

***7:19.** Evaluation of the Jacobian.

Given: The first bivariable interpolation function of Table 7.9.

Evaluate: The coefficients of the Jacobian of the mapping for x and y at the point (1, 1).

7:20. Validation of equivalent loading.

Given: The loading along a straight side of a planar element is $p = 1.0 + 0.2x^4$, and the interpolation function along the side is the quadratic LaGrange interpolation function.

Show: That the error in the first moment of the loading decreases in every step as the number of uniformly spaced nodes increases along the side.

7.14 COMPUTERWORK

7-1. Membrane program validation test.

Given: The narrow membrane modeled in plane stress with three membrane elements of the figure.

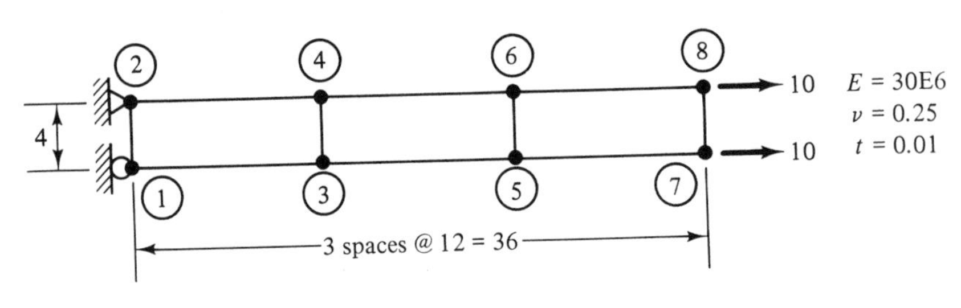

Computerwork Problem 7-1

Show: That the deflection at node 4 in the direction of the applied loads equals PL/AE where P is the total of the two end forces.

***7-2.** Gauss integration rules.

Given: The structure of Prob. 7-1 and a membrane analysis program that allows the user to select from a variety of Gauss quadrature rules.

Show:

(a) That the stiffness matrix gives the same responses when the quadrature rule is exact or results in overintegration.

(b) What happens when the element is underintegrated.

7-3. Rigid movement stiffness matrix validation.

Given: A single rectangular membrane element of the figure.

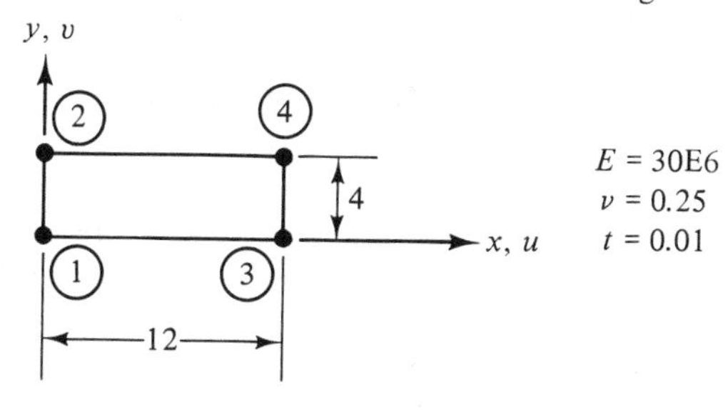

Computerwork Problem 7-3

Show:

(a) That the program can predict displacements when the element is fully integrated and the system has at least determinate supports.

(b) That the program cannot predict displacements when there are fewer external support constraints than those of a determinate support, even when the element is overintegrated.

7-4. Constant-strain validation.

Given: The membrane of Prob. 7-3 subjected to two linearly independent loadings, each of which will cause a constant-strain state.

Do the following:

(a) Define the two independent loading conditions.

(b) Show that they are linearly independent.

(c) Show that your computer program predicts displacements of a node exactly using a one-element model.

***7-5.** Partial validation of a stress model.

Given: A rectangular membrane element with a width-to-length ratio of the analyst's choice.

Determine:

(a) Whether the micro model predicts zero stresses in the interior for displacements of the nodes corresponding to a rigid translation in the x direction.

(b) Whether the micro model predicts zero stresses for a rigid rotation about the normal to the membrane.

(c) Whether the micro model predicts constant stresses for displacements of a uniform expansion in the x and y directions.

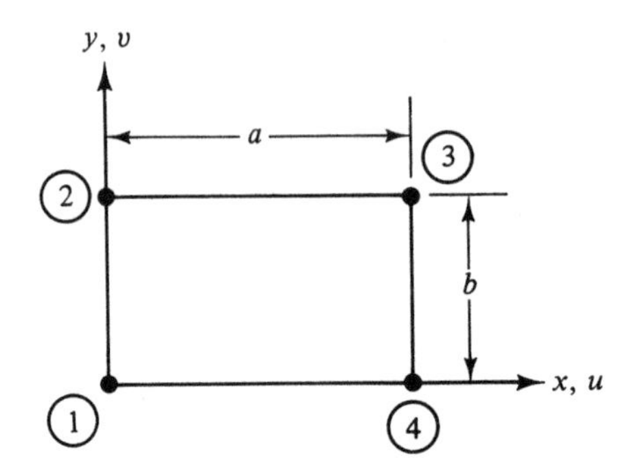

Computerwork Problem 7-5

7-6. Membrane analysis problem.

Given: The structural configuration of the figure and the exact values (from the elasticity solution) of some of the nodal displacements.

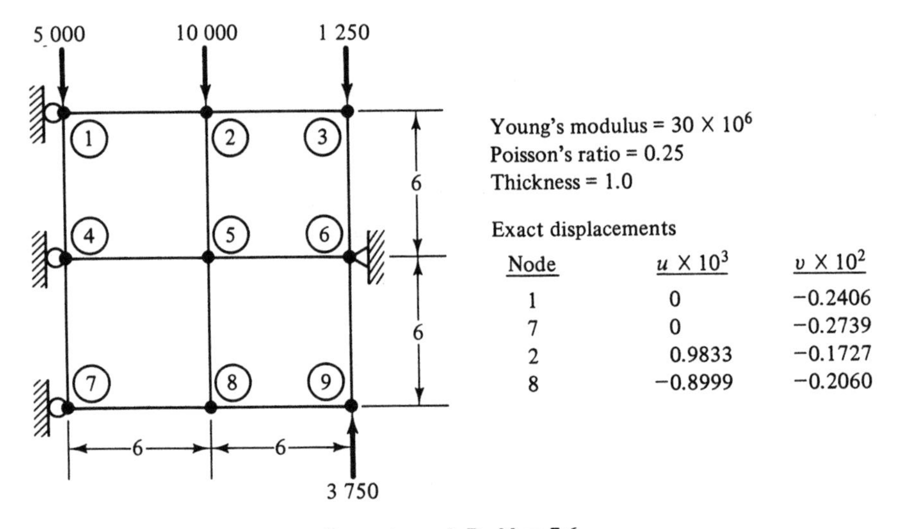

Node	$u \times 10^3$	$v \times 10^2$
1	0	−0.2406
7	0	−0.2739
2	0.9833	−0.1727
8	−0.8999	−0.2060

Computerwork Problem 7-6

Do the following:

(a) Find the FEA solution for the mesh shown.

(b) Find the percentage by which the displacements are underestimated.

7.15 REFERENCES

1. *MSC/NASTRAN User's Manual*, Version 64, MacNeal-Schwendler Corp., July 1984.
2. Fredriksson, B.; Mackerle, J., "Finite Element Review," Report AEC-1003, Advanced Engineering Corporation, Linköping, Sweden, 1984.

3. Turner, M.J.; Clough, R.W.; Martin, H.C.; Topp, L.J., "Stiffness and Deflection Analysis of Complex Structures," *Journal of Aerospace Sciences*, vol. 23, no. 9, Sept. 1956, pp. 805–823.
4. Melosh, R.J., "Basis for Derivation of Matrices for the Direct Stiffness Method," *AIAA Journal*, vol. 1, no. 7, July 1963, pp. 1631–1637.
5. Melosh, R.J., "A Flat Triangular Shell Element Stiffness Matrix," AFFDL-TR-66-80, Wright-Patterson AFB report, Dayton, Ohio, Dec. 1965, pp. 445–455.
6. Melosh, R.J., "A Stiffness Matrix for the Analysis of Thin Plates in Bending," *Journal Aerospace Sciences*, vol. 28, no. 1, Jan. 1961, pp. 34–43.
7. Melosh, R.J., "Structural Analysis of Solids," *J. Struct. Div.* ASCE, vol. 93, no. ST4, Aug. 1968, pp. 205–223.
8. Abramowitz, M.; Stegun, I., eds., *Handbook of Mathematical Functions*, New York: Dover, 1964, chap. 25.
9. Steffensen, J.F., *Interpolation*, Baltimore: Williams & Wilkins, 1927, pp. 203–214.
10. Hildebrand, F.B., *Introduction to Numerical Analysis*, New York: McGraw-Hill, 1956.
11. Bazeley, G.P.; Cheung, Y.K.; Irons, B.M.; Zienkiewicz, O.C., "Triangular Elements in Plate Bending-Conforming and Nonconforming Solutions," *Proc. First Conf. on Matrix Methods in Struct. Mech.*, Wright-Patterson AFB, Ohio, 1965.
12. Clough, R.W., "Analysis of Structural Vibration and Response," in *Recent Advances in Matrix Methods of Structural Analysis and Design*, ed. R.H. Gallagher, Y. Yamada, J.T. Oden, Tuscaloosa; U. of Alabama Press, 1971, pp. 25–46.
13. Hinton, E.; Rock, T.; Zienkiewicz, O.C., "A Note on Mass Lumping and Related Processes in the Finite Element Method," *Earthquake Eng. and Struct. Dynamics*, vol. 4, 1976, pp. 245–249.
14. *Boundary-Integral Equation Method: Computational Applications in Applied Mechanics*, New York: Applied Mechanics Division, ASME, 1975, vol. 11.
15. Fraeijs de Veubeke, F., "Upper and Lower Bounds in Matrix Structural Analysis," in *Matrix Methods in Structural Analysis*, ed. B. Fraeijs de Veubeuke, New York: Macmillan, 1964, pp. 166–201.
16. Melosh, R.J., "The Optimum Approach to Analysis of Elastic Continua," *Computers and Structures J.*, vol. 1, 1971, pp. 241–263.
17. Barone, M.R.; Chang, D.C., "Finite Element Modeling of Automotive Structures," *Modern Automotive Structural Analysis*, ed. M.M. Kamal, J.A. Wolf, Jr., New York: Van Nostrand Reinhold, 1982, chap. 4.

Chapter 8

Preassessing Performance of Finite Element Models

The choice of element model to represent parts of a structure is the principal decision affecting FEA accuracy and efficiency. The choice affects accuracy by its influence on discretization and round-off error. Alternative element models consume different amounts of computer resources in element coefficient evaluation and equation-solving data processing.

With respect to the approximations of discretization, we need to know that a given element model will produce behavior predictions that approach those of the theory as the number of nodal variables approaches infinity. We will want models that guarantee too stiff or, alternately, too soft a representation, as measured by strain energy of the solution, and guarantee monotonically improving behavior predictions as the number of nodal variables increases. We desire element performance data that indicates how accurate the response predictions of a coarse grid analysis will be and how rapidly the accuracy will improve as the mesh is refined.

With respect to round-off error, we will be biased in element selection and use by performance data indicating the values of the of the element shape, local geometry parameters, and material properties when the model becomes numerically unstable. We will be influenced by limitations imposed on use of each element by the precision of computer calculations.

To a large extent, we can assess element model performance by numerical tests involving few elements. Numerical tests are necessary because analytical assessments are available for only a handful of the elements of the library of element models.

The chief drawback of assessments using numerical experiments is the disadvantage of the empirical approach. Since we can never perform the infinity of tests required for mathematical rigor, these tests can only lead to worst-case assessments, tests for which conclusions are based on no-exception hypotheses, or tests whose conclusions are encumbered by statistical qualifications.

However, numerical testing offers important advantages. It can be applied to element models created by the computer configuration regardless of their derivation basis. Testing can yield performance data for elements embodied in procedural specifications; elements generated by ad hoc or faulty logic; elements whose basis is incompletely or poorly documented; models based on displacement, stress, or both; potential energy, complementary energy, or mixed models; and elements using collocation or weighted integral accuracy control. Testing can encompass the full range of model parameters, thereby catering to the broadest use of a particular model.

This chapter presents two-node and multinode element tests. The eigenvalue and extended lattice tests address numerical stability and convergence for two-node elements.[1] The eigendata and grid refinement tests furnish data for assessing discretization and round-off performance for multinode elements. The direct comparison test provides a rigorous basis for comparing the computational efficiency of two elements possessing the same geometry, nodes, and nodal variables.

8.1 TWO-NODE ELEMENT TESTS

Two-node elements permit much simpler testing than multinode, and testing results are more readily interpreted.

For efficient analyses, two-node element models should always reduce to lattice elements for some values of the element parameters. Consequently, two-node models invite self-consistency testing for determining the ability of these models to provide an exact solution of the appropriate differential equations in the limit.

We need not test two-node elements for all possible loading conditions. We simply require that equivalent loading be developed in the same way as that of lattice element equivalent loading.

Almost all two-node element models are expressed in algebraic form. By inspection, we can determine the appropriate nondimensional parameters of each model. The required rank of each model is known *ab initio*. The characteristic elastic states of behavior are known from the theory.

The two-dimensional beam element model illustrates these points. The stiffness matrix for this element is recited in Table 8.1. This element is a lattice element when the beam has a straight axis, is prismatic, undergoes plane stress, and is

TABLE 8.1 CONVENTIONAL BEAM STIFFNESS MATRIX

$AE(x) = AE$

$EI(x) = EI$

$$\mathbf{K}_e\,\mathbf{g}_e = \mathbf{p}_e$$

$$\begin{bmatrix} AE/L & & & & & \text{Sym.} \\ 0 & 12EI/L^3 & & & & \\ 0 & 6EI/L^2 & 4EI/L & & & \\ -AE/L & 0 & 0 & AE/L & & \\ 0 & -12EI/L^3 & -6EI/L^2 & 0 & 12EI/L^3 & \\ 0 & 6EI/L^2 & 2EI/L & 0 & -6EI/L^2 & 4EI/L \end{bmatrix} \begin{pmatrix} u_1 \\ v_1 \\ t_1 \\ u_2 \\ v_2 \\ t_2 \end{pmatrix} = \begin{pmatrix} U_1 \\ V_1 \\ T_1 \\ U_2 \\ V_2 \\ T_2 \end{pmatrix}$$

composed of an isotropic and Hookean material. Its rank is 3. The elastic states are constant extensional strain, constant curvature, and bending with linearly varying curvature.

Assume that this element is to be used to represent a tapered beam with a curved axis as Fig. 8.1 suggests. In this context, the element is a continuum element model. To be satisfactory, we require that as the number of elements in the finite element model of this structure approaches infinity, the solution approaches that of the beam differential equation as closely as we desire.

We intend to approximate the actual geometry of the tapered beam as a stepped beam with a folded neutral axis as suggested by Fig. 8.1(C). The equivalent loading is therefore the equivalent loading for each segment of the beam model of Table 8.1. Testing can establish the strategy to use to ensure the acceptability of the element model.

In summary, we need three types of two-node element tests: one to establish

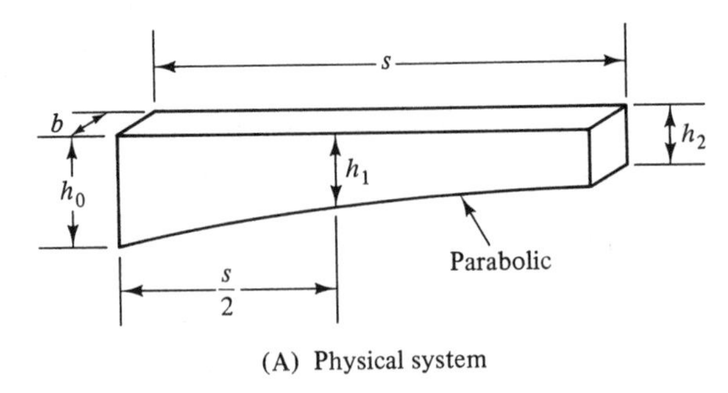

(A) Physical system

Figure 8.1 Tapered beam models.

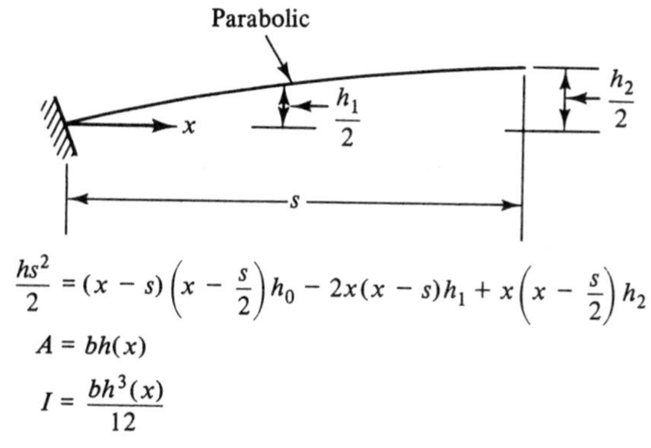

Parabolic
x
$\frac{h_1}{2}$
$\frac{h_2}{2}$
s
$\frac{hs^2}{2} = (x - s)\left(x - \frac{s}{2}\right)h_0 - 2x(x - s)h_1 + x\left(x - \frac{s}{2}\right)h_2$
$A = bh(x)$
$I = \frac{bh^3(x)}{12}$

Plane stress, E = constant

(B) Mathematical model

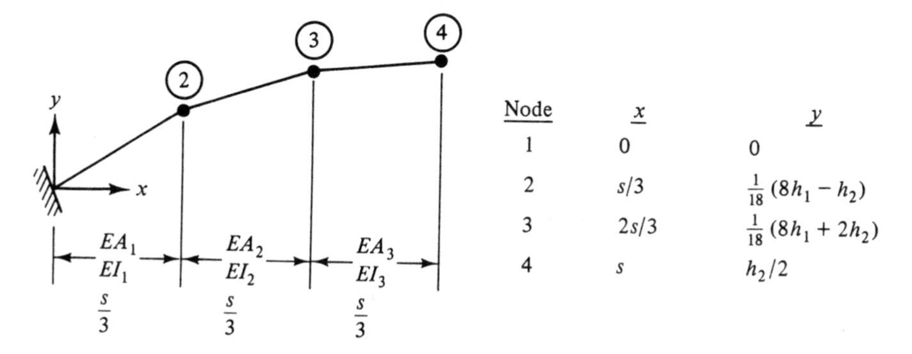

Node	x	y
1	0	0
2	$s/3$	$\frac{1}{18}(8h_1 - h_2)$
3	$2s/3$	$\frac{1}{18}(8h_1 + 2h_2)$
4	s	$h_2/2$

(C) Typical finite element model

Figure 8.1 (Continued)

that the element model reduces to a lattice stiffness model, one to ensure that the strategy for using the element as a continuum element is adequate, and one to determine numerical stability as a function of modeling parameters.

8.2 EIGENVALUE TEST

This test uses the eigenvalues of the element stiffness matrix for assessing numerical stability of the model.

Consider the equation

$$(\mathbf{K}_e - \lambda\mathbf{I})\mathbf{g} = 0 \tag{8.1}$$

where

λ is a scalar and

$\mathbf{I}$ is an identity matrix.

Any λ_i and associated $\mathbf{g}_i$ that satisfy Eqs. (8.1) form an "eigenpair," that is, an eigenvalue and its modal vector. There may be as many distinct eigenpairs as the order of $\mathbf{K}$.

The eigenvalues are roots of the characteristic equation. Observing that Eqs. (8.1) can only have a non-null $\mathbf{g}$ when the determinant of the parenthesized matrix is zero, we see that the determinant is a polynomial that expresses the characteristic equation in the form,

$$\lambda^N + a_{N-1}\,\lambda^{N-1} + a_{N-2}\,\lambda^{N-2} + \ldots + a_1\lambda + a_0 = 0 \tag{8.2}$$

where the a_i are constants. We can factor Eq. (8.2) to

$$(\lambda - \lambda_1)(\lambda - \lambda_2)(\lambda - \lambda_3)\ldots(\lambda - \lambda_N) = 0 \tag{8.3}$$

where $\lambda_1, \lambda_2, \lambda_3, \ldots, \lambda_N$ are the eigenvalues.

We can establish the definiteness of the element stiffness matrix by inspecting the eigenvalues using the following table.

Eigenvalue Rules for Definiteness

EIGENVALUES	INTERPRETATION
$\lambda_i > 0$, all i	$\mathbf{K}_e$ is positive definite
$\lambda_i > 0$; $\lambda_i = 0$ some i	$\mathbf{K}_e$ is positive semidefinite
$\lambda_i < 0$; $\lambda_i = 0$ some i	$\mathbf{K}_e$ is negative semidefinite
Otherwise	$\mathbf{K}_e$ is indefinite

To justify these statements, set $\lambda = \lambda_i$ in Eq. (8.1) and premultiply by $0.5\mathbf{g}_i^T$ to get

$$0.5\,\mathbf{g}_i^T\,\mathbf{K}_e\,\mathbf{g}_i = 0.5\lambda_i\,\mathbf{g}_i^T\,\mathbf{g}_i \tag{8.4}$$

noting that if $\mathbf{g}_i$ is a eigenvector of Eq. (8.1), so is a scalar multiple of $\mathbf{g}_i$. Accordingly, we normalize the modal vector by setting

$$\mathbf{g}_i = \sqrt{(\mathbf{g}_i^T\,\mathbf{g}_i)}\;\mathbf{m}_i$$

where $\mathbf{m}_i$ is normalized so that its Euclidean norm is 1. Then Eqs. (8.4) reduces to

$$0.5\mathbf{m}_i^T\,\mathbf{K}_e\,\mathbf{m}_i = 0.5\lambda_i \tag{8.5}$$

We thereby establish the relation between definiteness and the eigenvalues and justify the eigenvalue rules of definiteness.

Knowing the eigenvalues, the degeneracy of $\mathbf{K}_e$ is given by the number of zero eigenvalues. For all two-node elements, we require that the degeneracy never exceed the number of rigid body motions of the element, just as we do for lattice elements. Accordingly, the degeneracy of a uniaxial model must not exceed 1; of a biaxial model, 3; and of a triaxial model, 6.

Rigid motion degeneracy is the primary degeneracy of the element model. Element models will add secondary degeneracies for extreme values of the model parameters. For the beam of Table 8.1, for example, degeneracy will equal the order when Young's modulus is zero. It will be 2 greater than the number of rigid modes when EI is zero or span becomes infinite. It will increase by 1 when AE or span becomes zero.

When secondary degeneracy depends on differences of product terms, the values of element model parameters that induce secondary degeneracy will depend on the precision of calculations. To deal with this complication, we define the condition number of the element stiffness matrix as the maximum eigenvalue divided by the smallest nonzero eigenvalue. This definition implies that for any element, rigid motions must be constrained by other elements or by external supports, thereby avoiding a singular stiffness matrix.

We characterize the secondary degeneracy of the element model by

$$D_L = \left| \log_{10}\left(2^{-t} \frac{\lambda_{\max}}{\lambda_{\min}} \right) \right| \tag{8.6}$$

where

$\lambda_{\max}$ is the absolute value of the maximum eigenvalue and

$\lambda_{\min}$ is the absolute value of the minimum nonzero eigenvalue.[1]

The eigenvalues establish the definiteness and primary and secondary degeneracies of the element stiffness matrix. They thus measure the stability and numerical stability of the element model as a function of element parameters.

8.3 EXTENDED LATTICE TEST

This test uses data from undivided- and subdivided-grid finite element analyses for assessing convergence properties of an element model. The test is an extension of the lattice property test of chapter 4 to embrace two-node continuum elements.

Reconsider the lattice property test. Selective Gauss elimination reduces the equations of the subdivided element to the order and variables of the undivided. Subtracting the two sets of equations yields

$$\mathbf{K}_u \, \mathbf{g}_u - \mathbf{K}_r \, \mathbf{g}_r = 0 \tag{8.7}$$

where

$\mathbf{K}_u$ is the stiffness matrix of the undivided element,

$\mathbf{K}_r$ is the reduced two-element stiffness matrix, and

$\mathbf{g}_u$ and $\mathbf{g}_r$ are the corresponding displacement vectors.

The right-hand side of Eqs. (8.7) is zero because the loading on the undivided element is the equivalent loading of the subdivided. As in the lattice property test, the reduced stiffness matrix is evaluated by decoupling the variables of the interior node from those of the model end nodes.

Letting the reduced displacements be

$$\mathbf{g}_r = \mathbf{g}_u + \mathbf{g} \tag{8.8}$$

simplifying yields

$$\mathbf{g}_u^T \mathbf{K}_r \mathbf{g} = \mathbf{g}_u^T \mathbf{K}_u \mathbf{g}_u - \mathbf{g}_u^T \mathbf{K}_r \mathbf{g}_u \tag{8.9}$$

Evaluating the strain energy associated with the subdivided element using Eq. (8.8) directly furnishes

$$SE = 0.5\,(\mathbf{g}_u^T \mathbf{K}_r \mathbf{g}_u + 2\mathbf{g}_u^T \mathbf{K}_r \mathbf{g} + \mathbf{g}^T \mathbf{K}_r \mathbf{g}) \tag{8.10}$$

Comparing Eqs. (8.9) and (8.10), we conclude that when $\mathbf{K}_u - \mathbf{K}_r = 0$, the element is a lattice model and the discretization error–free solution associates with both the undivided and subdivided element solutions. By comparison with this conclusion, we observe the following:

If $\mathbf{K}_u - \mathbf{K}_r$ is positive definite, successive grids associate with decreasing stiffness.

If $\mathbf{K}_u - \mathbf{K}_r$ is negative definite, successive grids associate with increasing stiffness.

Otherwise, we cannot guarantee that stiffness is increasing or decreasing by this criterion.

The ratio

$$c_r = 2\,\frac{\|\mathbf{K}_u - \mathbf{K}_r\|}{\|\mathbf{K}_u + \mathbf{K}_r\|} \tag{8.11}$$

where c_r is the convergence rate. The variable c_r measures the rate of convergence toward the exact solution. The norms of $\mathbf{K}_u$ and $\mathbf{K}_r$ lead to measures of the discretization error for a particular grid.

8.4 PERFORMANCE OF THE BEAM ELEMENT MODEL

Data from eigenvalue testing of the model of Table 8.1 is the basis for Fig. 8.2. This figure describes the zones of numerical stability for the Bernouilli-Euler beam model as a function of the nondimensional parameters of the element beam stiffness matrix. The central cross-hatched zone covers values of the parameters for which the number of digits losable is less than 6.92, the number of digits of single precision of the IEEE specification. The hatched zone and the cross-hatched zone identify parameter values where IEEE double precision is adequate.

The zones are bounded by folded lines because the modal vectors that define

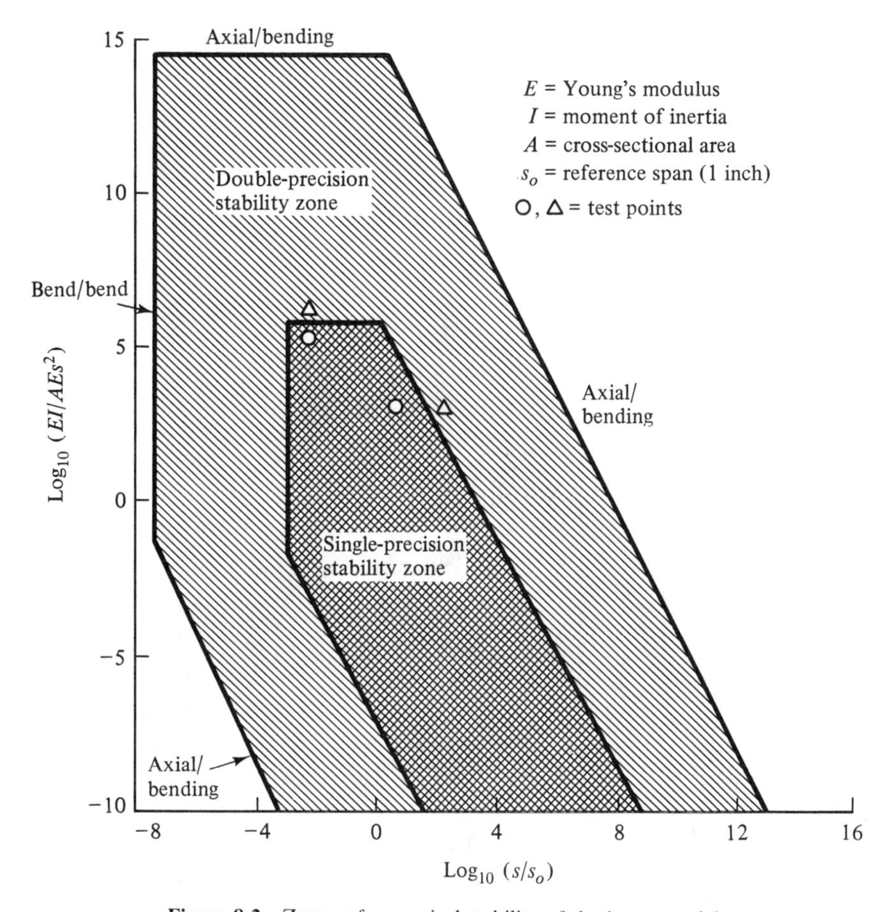

Figure 8.2 Zones of numerical stability of the beam model.

the maximum and minimum nonzero eigenvalues change. The upper, horizontal bound is sensitive only to the ratio of axial to bending stiffness. The vertical left bound depends only on the span of the beam.

The vertical bounds reflect the fact that the eigenvalues of the beam matrix are dependent on the choice of the length unit because the nodal variables include dimensional deflection and nondimensional angular rotation. Therefore, it is important to observe that the reference length of Fig. 8.2 is in inches.

The data of Fig. 8.2 addresses the secondary degeneracies of single-element representations of the beam model. They are applicable to assessment of intra-element round-off numerical stability and define some of the limiting requirements to ensure that interelement round-off stability is maintained. For example, the four data points on Fig. 8.2 confirm loss of all retained digits in crossing the single-precision boundary for analyses of cantilevered beams.

Consider the three-dimensional line model formed by adding stiffnesses of

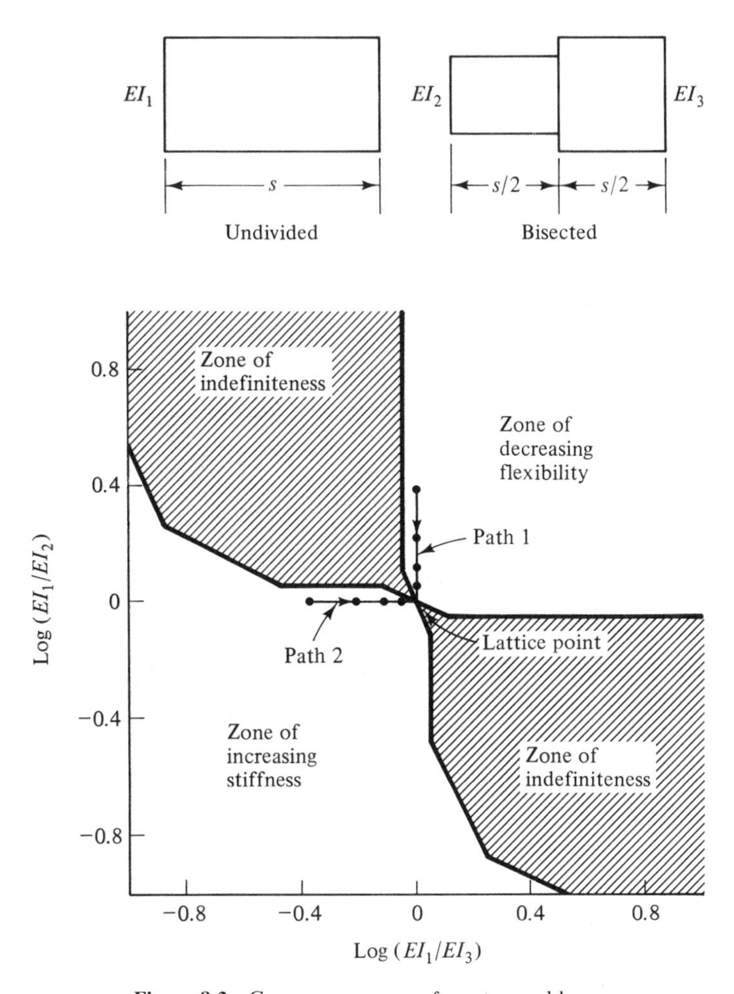

Figure 8.3 Convergency zones for a tapered beam.

bending about the axes perpendicular to the line, twisting of the line, and stretching along the line. The limits on values of the relative stiffnesses (EI/s^2AE, EI/s^2GJ, EI/EI, and AE/GJ) limit both the relative geometries of an individual element and the relative geometries of elements of the structure being analyzed.

The extended lattice test provides the data of Fig. 8.3. This data characterizes the convergence of the beam model as a function of relative bending stiffnesses for the tapered beam, which has a straight axis.

The graph shows that convergence characteristics depend on the strategy for assigning EI for elements of the model. The element model is a lattice model at the point where the four zones of the figure touch. This occurs only when the beam is prismatic. If the discretization strategy is such that all element EI ratios fall in the upper right-hand zone of the figure, convergence will be monotonically stiffening. If the strategy places all elements in the lower left-hand zone, convergence will be monotonic from the soft side. Otherwise, monotonicity is not guaranteed by the test.

The values of the convergence ratio increase as the radial distance from the lattice point increases in this figure. Consequently, when the discretization strategy ensures monotonic convergence, the rate of convergence decreases monotonically. In Fig. 8.3, the distance between a point and the lattice point is a measure of the discretization error for a particular section.

Figure 8.4 illustrates use of particular discretization strategies for a particular tapered beam. The stiffening strategy uses the smallest EI in the section for the

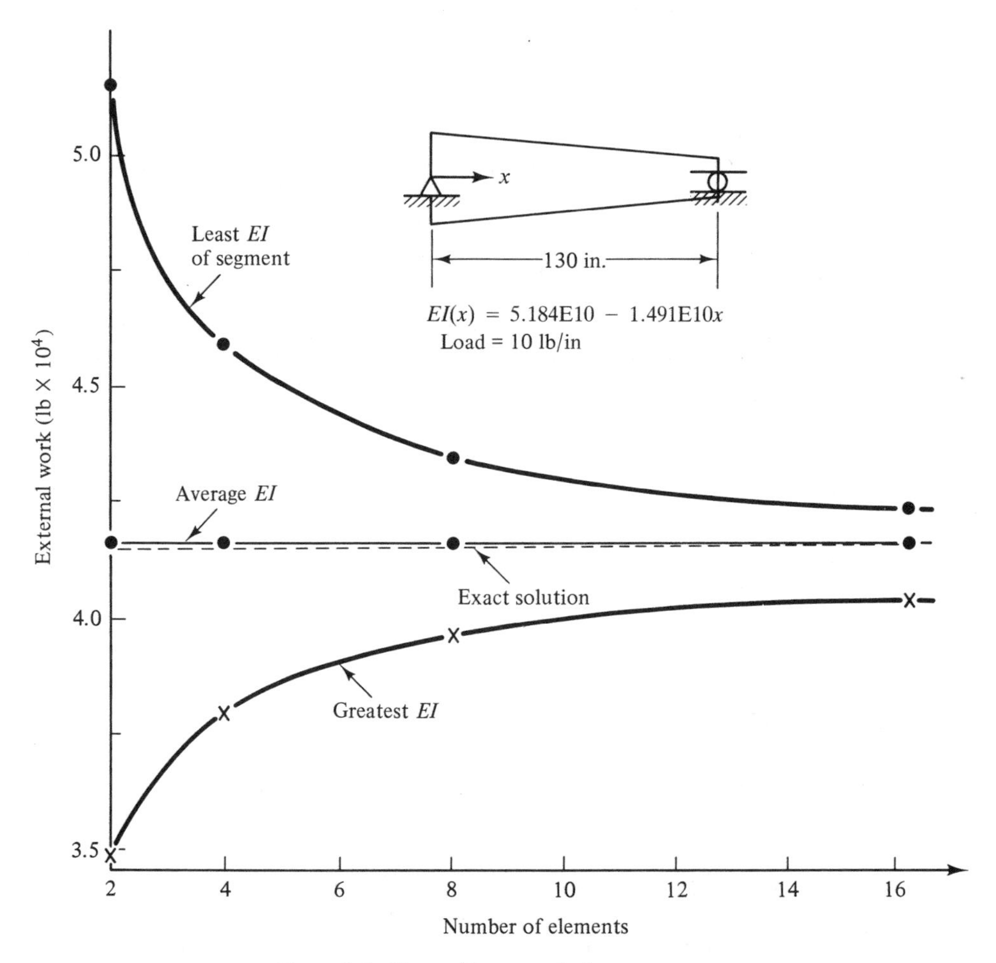

Figure 8.4 Tapered beam analysis convergence.

element; the softening strategy the largest EI. These strategies associate with the lower and upper curve of the figure, respectively.

The figure also shows the results of discretization using the average EI of the section for the element. Although this strategy is not guaranteed to produce a monotonically convergent sequence in general, it does so for this case of linear taper.

For the straight beam, the axial and bending stiffnesses are uncoupled. Furthermore, the axial stiffness involves only one parameter. Therefore, it is always possible to choose for the rod element a representative stiffness factor that results in a lattice model for the rod component of the stiffness matrix, regardless of the variation of the cross-sectional area along the rod.

Finding a strategy for discretizing a curved-axis beam by a series of straight-axis beam elements so that convergence is guaranteed to be monotonic is more difficult. It cannot be guaranteed, by the test, for a discretization strategy in which the element model of Table 8.1 is used for the folded axis defined by points along the curved axis as suggested by Fig. 8.1.

The following strategy is guaranteed by the extended lattice test to produce a monotonically decreasing convergent sequence of estimates of behavior for a curved-axis tapered beam:

1. Assign the values of AE and EI to be the minimum AE and EI of the section.

2. Assign the length L of each segment to be a quadratic function of the rise-to-span ratio of the segment of the beam being modeled. Thus we choose the length to be

$$L = \left[1 + a\left(\frac{\text{rise}}{\text{span}}\right)^2\right] L_0$$

where

a is a nondimensional constant and

L_0 is the original straight length of the rod segment.

8.5 MULTIPLE-NODE ELEMENT TESTS

Multiple-node tests encompass all elements with three or more nodes. These element models are necessarily continuum elements. We consider here tests for field elements models only.

Testing multiple-node macro models is more complex than testing two-node models. There is a greater variety of modeling bases; procedures for evaluating coefficients are more intricate, abstruse, and often ad hoc; and errors in program logic are much more likely to occur and be more difficult to pinpoint.

In addition, multiple-node models require much more exhaustive testing. Tests must not only span all possible loadings but must also circumscribe many more values of geometry and material parameters.

For multinode models, tests need to assess both the stiffness matrix and the equivalent loading basis. The stiffness matrix is critical to both convergence and numerical stability. Selection of an incommensurate loading basis cannot only destroy the monotonicity of convergence but can change the direction of convergence as well.[2]

In summary, we need four types of multiple-node element assessments: one to establish that the macro equations contain the ingredients to represent the structural or elasticity equations as the number of nodal variables approaches infinity (the admissibility requirement), another to ensure that the finite element process will select the exact solution in the limit (the preferentiality requirement), a third to identify the character of primary degeneracies, and a fourth to assay secondary degeneracies. The eigendata and subdivisibility tests embody these four assessments.

8.6 EIGENDATA TEST

Calculations of this test use the eigenvalues and eigenvectors of the stiffness matrix to assess matrix admissibility, maximum discretization error, primary degeneracies, and secondary degeneracies.[3] The eigendata test, like the Irons Patch Test,[4] determines whether an element's behavior, as measured by nodal displacements and strain energy, faithfully reflects behavior in constant-strain states.

The eigendata test consists of the following steps:

1. Find the eigenvalues and modal vectors of the equations:

$$\mathbf{K}_e\,\mathbf{m}_i = \lambda \mathbf{m}_i \tag{8.12}$$

2. Solve the linear system of equations

$$\mathbf{A}^T\,\mathbf{A}\,\mathbf{Y} = \mathbf{A}^T\,\mathbf{M} \tag{8.13}$$

where

A is a rectangular matrix whose columns define the nodal displacements of each of the rigid and constant strain modes of the element

Y is a matrix of unknown multipliers and

M is a matrix of the modal vectors of step 1. The solution of Eqs. (8.13) provides the least squares error solution of the equations $\mathbf{A}\,\mathbf{Y} = \mathbf{M}$.

3. Evaluate the residuals of the least squares fit; that is, find **R** in

$$\mathbf{R} = \mathbf{A}\,\mathbf{Y} - \mathbf{M} \tag{8.14}$$

where the coefficients of **R** quantify the residuals of the equations $\mathbf{A}\,\mathbf{Y} = \mathbf{M}$.

4. Find the exact value of the strain energy in the modal vectors that involve only constant-strain participation. This calculation is facilitated by the fact that the strains are constant. Accordingly, it is given by

$$SE(\text{exact}) = 0.5 \iiint \mathbf{e}_0^T\,\mathbf{E}\,\mathbf{e}_0\,dV \tag{8.15}$$

where

SE(exact) is precisely the strain energy associated with the modal vector and

$\mathbf{e}_0$ is the vector of element constant strains.

8.7 SUBDIVISIBILITY TEST

This test incurs FEA calculations of a single and a subdivided patch of the structure to assess convergence, monotonicity, and direction of convergence.

The subdivisibility test evokes the following calculations:

1. Consider a patch of the structure represented by a number of finite elements of the type of interest. Generate a random set of nodal forces for nodes on the patch boundary, assuming that the patch has statically determinate supports. This step evaluates the vector $\mathbf{p}_s$. [If the element is a flat quadrilateral membrane, the patch could be the patch of Fig. 8.5(A).

2. Calculate the diplacements of all the nodes of the patch using the finite

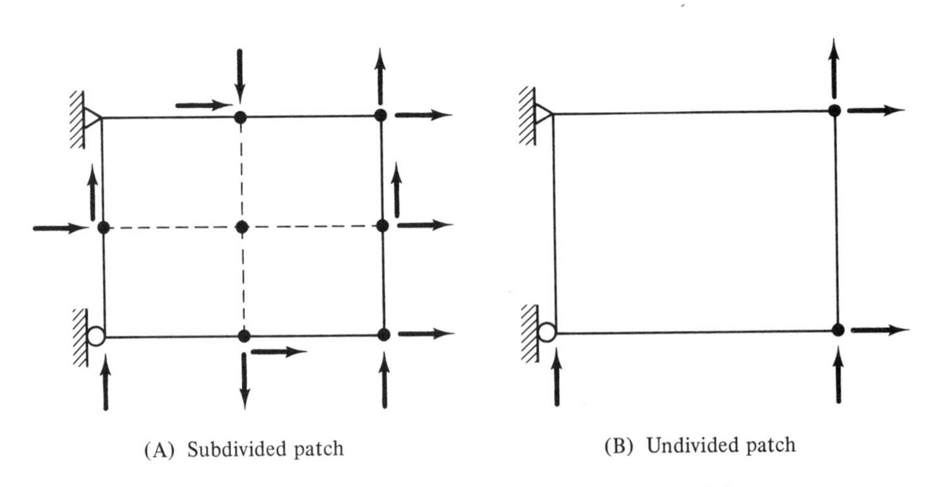

Figure 8.5 Subdivisibility patch: Divided and undivided.

element analysis process. This step produces the value of the components of the displacement vector for the subdivided system $\mathbf{g}_s$.

3. Determine the total strain energy in the elements of the subdivided patch

$$SE_s = 0.5\ \mathbf{g}_s^T\ \mathbf{K}_c\ \mathbf{g}_s \tag{8.16}$$

where

SE_s designates the strain energy of the subdivided patch.

4. Calculate the constant-strain strain energy of the elements of the subdivided patch

$$SE_s' = 0.5\ \mathbf{g}_s^T\ \mathbf{K}_c'\ \mathbf{g}_s \tag{8.17}$$

where

SE_0 designates the value of the strain energy consumed by the constant-strain states and

$\mathbf{K}_c'$ is the constant-strain stiffness matrix of the subdivided structure.

The constant-strain stiffness matrix is the conventional stiffness matrix in which all the contributions due to nonconstant strains are discarded.

5. Find the nodal forces on the boundary nodes of the patch when the patch is represented by a single finite element model of the type of interest

$$\mathbf{p}_u = \mathbf{Q}\ \mathbf{p}_s \tag{8.18}$$

where

$\mathbf{p}_u$ is the loads for the undivided patch and

$\mathbf{Q}$ is the equivalent load transformation matrix.

6. Use the finite element process to evaluate the displacements of the nodes of the undivided patch $\mathbf{g}_u$.

7. Determine the total strain energy in the element of the undivided patch

$$SE_u = 0.5\ \mathbf{g}_u^T\ \mathbf{K}_c\ \mathbf{g}_u \tag{8.19}$$

where SE_u is the total strain energy in the undivided model.

8. Calculate the strain energy that occurs in the constant-strain states of the undivided patch

$$SE_u' = 0.5\ \mathbf{g}_u^T\ \mathbf{K}_c'\ \mathbf{g}_u \tag{8.20}$$

where SE_u' is the undivided-patch constant-strain energy.

9. Repeat steps 1–8 as many times as deemed necessary to represent all possible loading conditions.

TABLE 8.2 DERIVATION OF A UNIFORM STRAIN STIFFNESS MATRIX

Given: The four-node rectangular membrane of Table 7.2.

Find: The uniform-strain stiffness matrix.

Steps:

1. Find the expressions for strains from the definition of strains, Eq. (7.1), and the interpolation function.

$$
\begin{aligned}
e_{xx} &= \frac{\delta u}{\delta x} = \frac{1}{4ab}\{(y-b)u_1 - (y+b)u_2 + (y+b)u_3 - (y-b)u_4\} \\
e_{yy} &= \frac{\delta v}{\delta y} = \frac{1}{4ab}\{(x-a)v_1 - (x-a)v_2 + (x+a)v_3 - (x+a)v_4\} \\
e_{xy} &= \frac{\delta u}{\delta y} + \frac{\delta v}{\delta x} = \frac{1}{4ab}\{(x-a)u_1 - (x-a)u_2 + (x+a)u_3 - (x+a)u_4 \\
&\qquad + (y-b)v_1 - (y+b)v_2 + (y+b)v_3 - (y-b)v_4
\end{aligned} \tag{a}
$$

2. Rewrite Eqs. (a), omitting all terms in x and y:

$$
\begin{aligned}
e_{xx} &= \frac{1}{4a}\{-u_1 - u_2 + u_3 + u_4\} \\
e_{yy} &= \frac{1}{4b}\{-v_1 + v_2 + v_3 - v_4\} \\
e_{xy} &= \frac{1}{4}\left\{\frac{1}{b}(-u_1 + u_2 + u_3 - u_4) + \frac{1}{a}(-v_1 - v_2 + v_3 + v_4)\right\}
\end{aligned} \tag{b}
$$

3. Evaluate $\hat{\mathbf{K}} = \int_{-t/2}^{t/2}\int_{-b}^{b}\int_{-a}^{a} \mathbf{N}^T\mathbf{B}^T\mathbf{E}\mathbf{B}\mathbf{N}\, d\,Vo = 4abt\, \mathbf{N}^T\mathbf{B}^T\mathbf{E}\mathbf{B}\mathbf{N}$, or

$$
\hat{\mathbf{K}}_e = \frac{abtE}{4(1-\nu^2)}
\begin{bmatrix}
-1/a & 0 & -1/b \\
-1/a & 0 & 1/b \\
1/a & 0 & 1/b \\
1/a & 0 & -1/b \\
0 & -1/b & -1/a \\
0 & 1/b & -1/a \\
0 & 1/b & 1/a \\
0 & -1/b & 1/a
\end{bmatrix}
\begin{bmatrix}
1 & \nu & 0 \\
\nu & 1 & 0 \\
0 & 0 & \dfrac{(1-\nu)}{2}
\end{bmatrix}
\begin{bmatrix}
-1/a & -1/a & 1/a & 1/a & 0 & 0 & 0 & 0 \\
0 & 0 & 0 & 0 & -1/b & 1/b & 1/b & -1/b \\
-1/b & 1/b & 1/b & -1/b & -1/a & -1/a & 1/a & 1/a
\end{bmatrix} \tag{c}
$$

10. Repeat steps 1–9 for all the element geometries of interest.

11. Repeat steps 1–10 for all material models of interest.

The new analysis feature of the subdivisibility test is use of the constant-strain stiffness matrix for the finite element model being used. Table 8.2 illustrates how such a matrix is derived from a given interpolation function. The key idea is to discard all terms of the interpolation function that reflect nonconstant strains. In the case of the four node membrane elements, these are all terms that are functions of the coordinates of points in the element.

Direct use of the constant-strain eigenvectors, discovered in the eigendata test, provides an alternate approach to calculating the constant-strain strain energy.[5] In this approach, the response displacements are synthesized by the eigenvectors. Discarding the participation of the nonconstant-strain vectors leaves the basis for calculating the constant-strain strain energy.

8.8 PERFORMANCE OF A MEMBRANE ELEMENT MODEL

Table 8.2 defines the stiffness matrix and interpolation function for a rectangular membrane in a state of either plane stress or plane strain. The element is constructed using hyperbolic displacement shape functions. We will present the results of performance tests of this element to clarify the interpretation of eigendata and subdivisibility test information.[5]

The eigendata test results confirm that the element stiffness model meets admissibility needs. Figure 8.6 shows the modal vectors of a square element as an example. Regardless of the choice of element width-to-length ratio, regardless of whether the choice is plane stress or plane strain, and regardless of the values of Young's modulus and Poisson's ratio, three of the modal vectors are linear combinations of only constant-strain deformations, and the associated eigenvalue yields the correct value of the strain energy.

For the hyperbolic-based element, since we know the shape function, it is easy to show analytically that the constant-strain states are included. Therefore, for this case the admissibility calculations become a necessary check of the element stiffness matrix generation logic.

The eigenvalues of the hyperbolic membrane element, produced for a range of element aspect ratios, and Eq. (8.6) are the basis for the curve of Fig. 8.7. This plot emphasizes that width-to-length ratios of this rectangular membrane element may exceed 100 while losing only three digits to secondary degeneracy. The optimum width-to-length ratio with respect to round-off is 1.

Figure 8.8 exhibits the growth in round-off error bound, as measured by the condition number of the element stiffness matrix as a function of Poisson's ratio. This data indicates not only the secondary degeneracy when Poisson's ratio is 0.5 but also the need to obtain accurate estimates of the ratio when it exceeds 0.4 because of solution sensitivity.

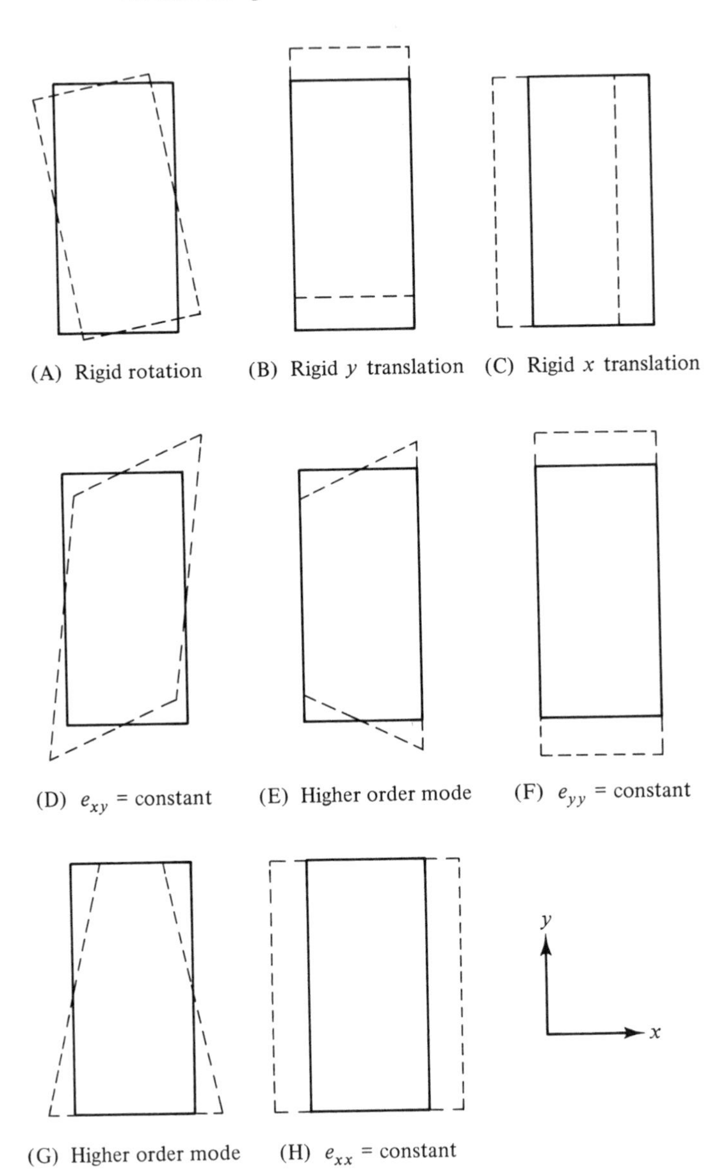

Figure 8.6 Eigenmodes for a rectangular element.

Figure 8.9 indicates the maximum relative discretization error of the hyperbolic shape membrane as a function of the width-to-length ratio of the element. This curve shows that the least error associates with square elements and that the error increases monotonically as the width-to-length ratio departs from 1.

The relative maximum discretization error is based on the least nonzero

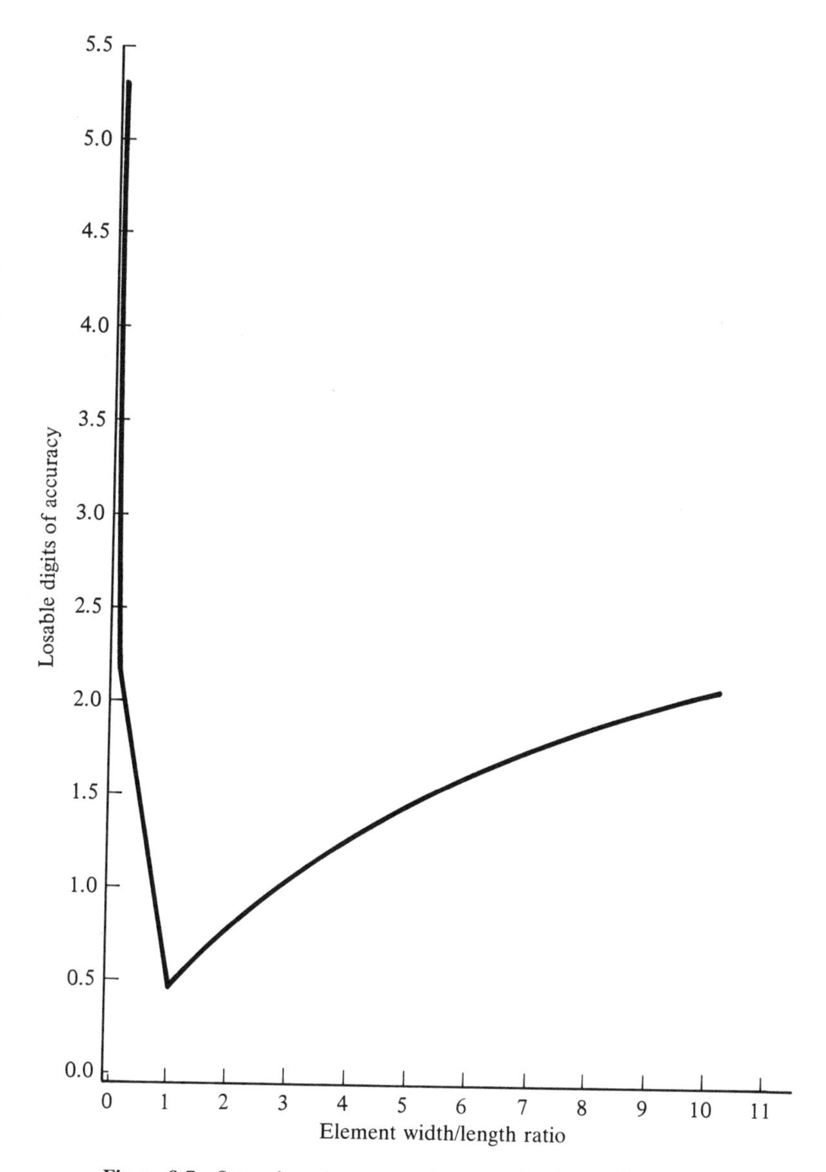

Figure 8.7 Secondary degeneracy (geometry): Hyperbolic rectangle.

eigenvalue of the element matrix. Assuming that the loading activates the eigenvector associated with the least nonzero eigenvalue and that the loading vector is normalized to 1, the eigenvalue determines the maximum strain energy excitable in the element by a loading vector of unit norm.[5]

Figure 8.10 shows the preferentiality data for the hyperbolic model in repre-

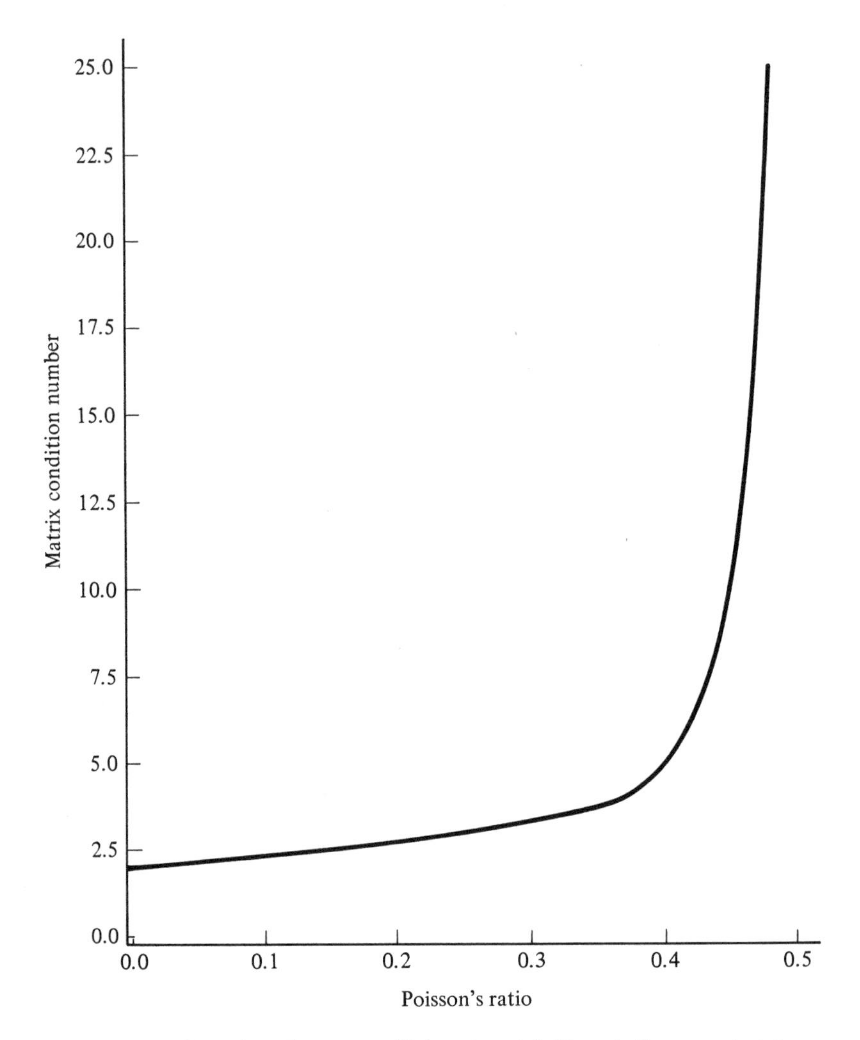

Figure 8.8 Secondary degeneracy (Poisson's ratio): Hyperbolic rectangle (plane strain).

senting a square steel membrane. The figure displays the results for 1000 random loadings.

The abscissa of Fig. 8.10 measures the total strain energy of the undivided patch divided by the total strain energy of the subdivided patch, SE_u/SE_s. It thus provides a measure of how close the finite element model is to obtaining the exact solution. To ensure monotonic convergence in the strain energy norm, each test point must have an abscissa less than or equal to 1.

The ordinate defines the ratio of that part of the strain energy in constant

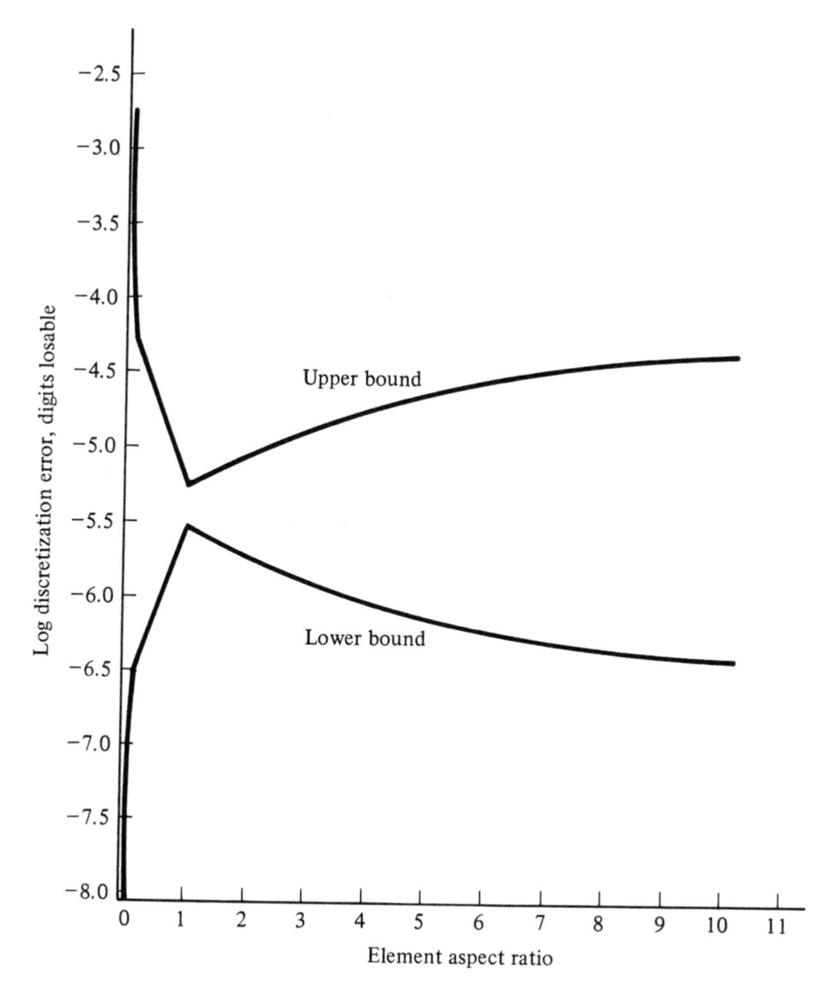

Figure 8.9 Relative discretization error bounds: Hyperbolic rectangle.

straining of the undivided patch divided by that part in the subdivided patch, SE'_u/SE'_s. The ordinate measures whether participation of the constant-strain states in the response energy is increasing or decreasing with refinement. Each test point that has an ordinate less than or equal to 1 represents monotonic convergence in the constant-strain energy norm.

Since all of the 1000 cases depicted in Fig. 8.10 fall in the unit square, we conclude that the square hyperbolic element model will lead to solutions with zero discretization error as the grid size approaches zero for the potential energy equivalent loading. (If the analyst is not satisfied with 1000 load cases, the testing can be continued to accumulate more convincing evidence.)

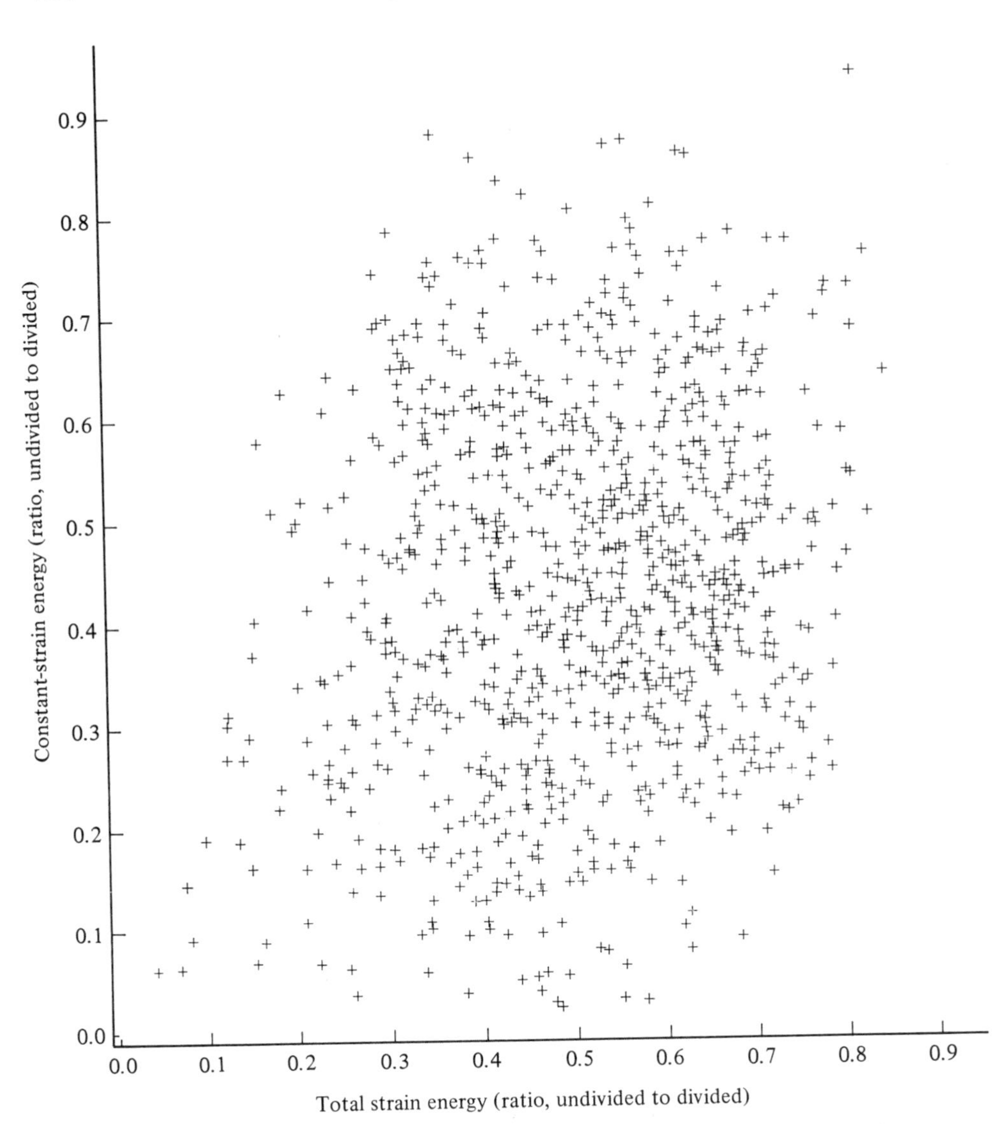

Figure 8.10 Preferentiality behavior: Hyperbolic rectangle (admissible load model).

Figure 8.11 displays the preferentiality of the hyperbolic element when the equivalent loading is evaluated by selective Gauss elimination of the subdivided patch. This is the same process as that used in quantifying equivalent loads for lattice models. The data signals conditional convergence; the condition is that convergence is guaranteed as long as $SE_u/SE_x > 0.4$ for all elements of the membrane structure.

The data of Figs. 8.7 and 8.8 pertains to a square element with a Poisson's ratio of 0.3. To complete the picture of element performance, preferentiality data

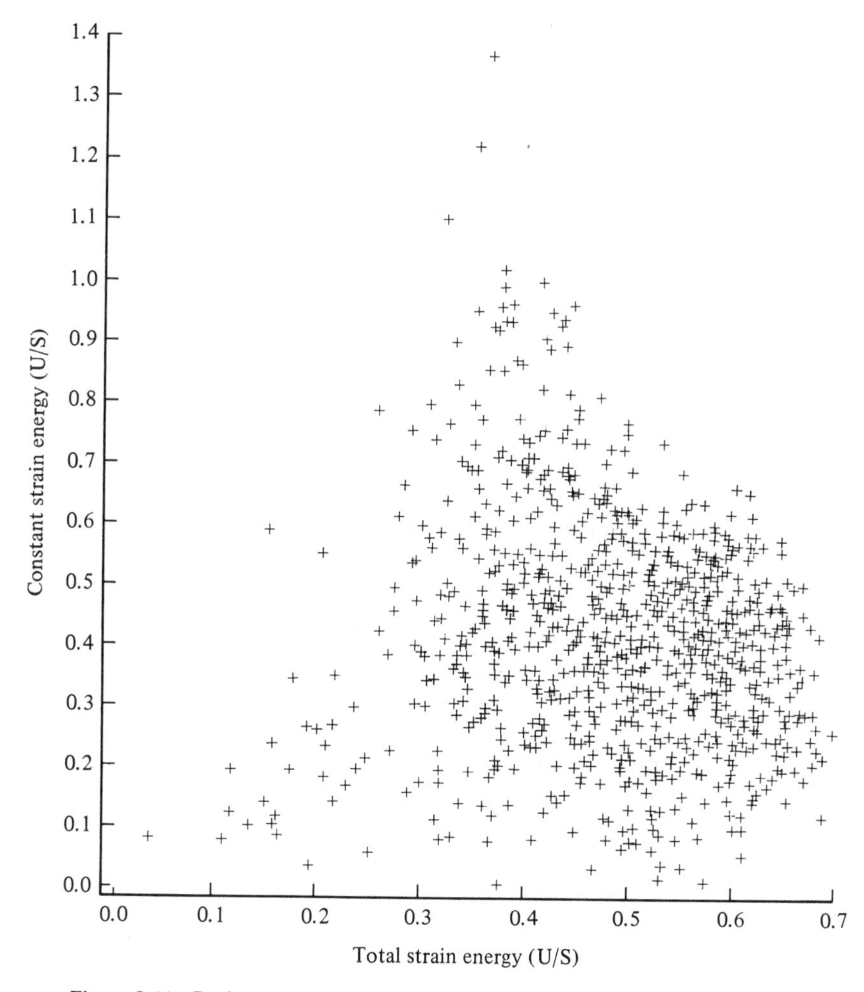

Figure 8.11 Preferentiality behavior: Hyperbolic rectangle (inadmissible model).

would be added covering other width-to-length and Poisson's ratios. If anisotropic materials are involved, tests would be needed to circumscribe these material models.

8.9 DIRECT COMPARISON OF ELEMENT MODELS

Comparisons of performance data from the eigendata and subdivisibility tests provide a basis for indirect comparison of element models. The Khanna and Hooley test is useful for directly comparing two models with respect to total strain energy.[6]

The Khanna and Hooley test facilitates comparison of two-element models that involve the same nodes, nodal degrees of freedom, and equivalent loading.

Then the difference of the stiffness equations of the two models is

$$\mathbf{K}_1 \mathbf{g}_1 - \mathbf{K}_2 \mathbf{g}_2 = 0 \tag{8.21}$$

where the subscripts 1 and 2 reference the element models.

Equations (8.21) are of the same form as Eqs. (8.7). Reuse of the arguments leading to Eqs. (8.8) and (8.9) leads to the rules of the following table.

Rules in Comparing Stiffness Matrices

$\mathbf{K}_1 - \mathbf{K}_2$	INTERPRETATION
Positive definite	Matrix $\mathbf{K}_1$ will involve less strain energy than $\mathbf{K}_2$ in the analysis for any particular loading.
Positive semidefinite	Matrix $\mathbf{K}_1$ will involve the same or less strain energy than $\mathbf{K}_2$.
Negative semidefinite	Matrix $\mathbf{K}_1$ will involve the same or more strain energy than $\mathbf{K}_2$.
Negative definite	Matrix $\mathbf{K}_1$ will involve more strain energy than $\mathbf{K}_2$.
Otherwise	One-sidedness of the energy cannot be guaranteed by this test.

One important use of the comparison test involves preassessment of the relative efficiency of two potential energy models. Based on the fact that the more efficient model has less stiffness, the test data provides a basis for identifying the better element model. For example, this test distinguishes the fact that use of the hyperbolic membrane to represent a square must result in better estimates of system strain energy than representation of the square by two simplex triangular membranes.

8.10 PREASSESSMENT TESTS

Much can be learned from data generated by numerical testing of element models. The following table summarizes information available from tests discussed in this chapter.

Information from Preassessment Tests

TEST	INFORMATION DEDUCIBLE FROM TEST DATA
Eigendata	Definiteness of $\mathbf{K}_e$, rank, number of independent constant-strain states represented, matrix condition number, maximum discretization error

Subdivisibility	Convergence preferentiality, solution too stiff or too soft, monotonicity of convergence
Comparison	Relative strain energy of solution for alternate stiffness matrices

The tests in Chapter 7 establish that the macro and micro models are capable of representing elasticity and structural problems accurately. The preassessment performance tests in this chapter lead to convergence guarantees and characterization.

8.11 HOMEWORK

8:1. Eigenvalues of a rod element matrix.

Given: The stiffness matrix for a rod and its eigendata.

STIFFNESS MATRIX

$$\mathbf{K}_e = AE/L \begin{bmatrix} 1 & -1 \\ -1 & 1 \end{bmatrix}$$

MODE	λ	g_1	g_2
1	0.00	1.00	1.00
2	$0.5AE/L$	1.00	-1.00

Do the following:

(a) Prove that the vectors and scalars given are the eigendata.

(b) Define the matrix definiteness.

(c) Normalize the vectors to obtain m_1 and m_2.

8:2. Losable digits in beam analysis.

Given: The eigendata listed for the first beam element model of Table 4.2.

MODE	λ	g_1	g_2	g_3	g_4
1	0.00	1.00	0.00	1.00	0.00
2	0.00	$-L/2$	1.00	$L/2$	1.00
3	$2EI/L$	0.00	1.00	0.00	-1.00
4	λ_4	1.00	$.5L$	-1.00	$.5L$

$$\lambda_4 = 2EI\left(\frac{12 + 3L^2}{L^3}\right)$$

Find:

(a) The maximum number of digits that can be lost when $L = 10$.

(b) The value of L at which the maximum number of digits lost would equal the IEEE standard of single-precision arithmetic.

8:3. Beam model fourth-modal vector.

Given: The fourth-modal vector listed for Prob. 8:2.

Show: That the modal vector is consistent with the assumption that the moment varies linearly along the span.

***8:4.** Membrane constant-strain states.

Given: Membrane strain displacement equations (7.1), a rectangular membrane with side lengths $2a$ and $2b$, and a rectangular membrane finite element model that has nodes at the corners only, u and v nodal variables, and the shape function given in Table 7.2.

Find: Nodal displacement vectors when

(a) e_{xx} is a nonzero constant, $e_{yy} = 0$, and $e_{xy} = 0$.
(b) e_{yy} is a nonzero constant, $e_{xx} = 0$, and $e_{xy} = 0$.
(c) e_{xy} is a nonzero constant, $e_{xx} = 0$, and $e_{xy} = 0$.

8:5. Plate constant-strain states.

Given: Plate strain displacement equations (7.12), a flat rectangular plate with sides $2a$ and $2b$, and a rectangular plate finite element model that has nodes at the corners only and w, r, and s nodal variables.

Find: Nodal displacement vectors when

(a) $\delta^2 w/\delta x^2$ is a nonzero constant, $\delta^2 w/\delta y^2 = 0$, and $\delta^2 w/(\delta x\ \delta y) = 0$.
(b) $\delta^2 w/\delta y^2$ is a nonzero constant, $\delta^2 w/\delta x^2 = 0$, and $\delta^2 w/(\delta x\ \delta y) = 0$.
(c) $\delta^2 w/(\delta x\ \delta y)$ is a nonzero constant, $\delta^2 w/\delta x^2 = 0$, and $\delta^2 w/\delta y^2 = 0$.

8:6. Three-dimensional solid constant-strain state.

Given: The interpolation function for the prism of Table 7.9 and the three-dimensional strain displacement equations (7.1).

Find: A set of nodal displacements that involves only constant strains.

***8:7.** Least squares equation solution.

Given: The following equations.

$$\begin{bmatrix} 1 & -1 \\ -1 & 2 \\ -2 & 3 \end{bmatrix} \begin{pmatrix} u_1 \\ u_2 \end{pmatrix} = \begin{pmatrix} -3 \\ 5 \\ 8 \end{pmatrix}$$

Do the following:

(a) Find the least squares solution of the equations.
(b) Find the residuals of the least squares solution.

8:8. Least squares solution residuals.

Given: The following equations.

$$\begin{bmatrix} 2 & -1 \\ -1 & 1 \\ 0 & 2 \end{bmatrix} \begin{pmatrix} u_1 \\ u_2 \end{pmatrix} = \begin{pmatrix} 4 \\ -4 \\ -2 \end{pmatrix}$$

Do the following:

(a) Find the least squares solution.
(b) Find the residuals of the least squares solution.

8:9. Least squares equations.

Given: The equations $\mathbf{A}\ \mathbf{x} = \mathbf{b}$, where $\mathbf{A}$ has more rows than columns.

Prove: The solution $\mathbf{x} = (\mathbf{A}^T\mathbf{A})^{-1}\mathbf{A}^T\mathbf{b}$ is a least squares solution.

8:10. Uniform-strain stiffness matrix.

Given: The interpolation function of the second bivariable case of Table 7.9.

Find:

(a) The interpolation function that neglects the nonconstant-strain terms.

(b) A Gauss quadrature rule that will fully integrate the strain energy associated with the uniform-strain interpolation function and involves the fewest Gauss points for a rectangular membrane.

***8:11.** Comparison of torsion models.

Given: The stiffness matrix for a right triangular cross-sectional segment of a shaft in torsion, as shown.

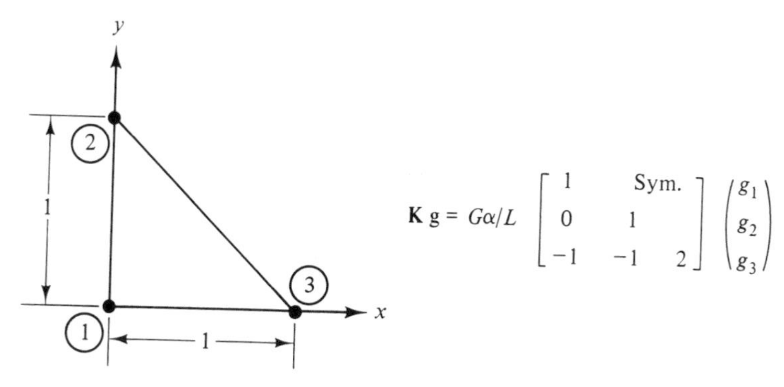

$$\mathbf{K}\,\mathrm{g} = G\alpha/L \begin{bmatrix} 1 & & \text{Sym.} \\ 0 & 1 & \\ -1 & -1 & 2 \end{bmatrix} \begin{pmatrix} g_1 \\ g_2 \\ g_3 \end{pmatrix}$$

Homework Problem 8:11

Note that coefficients of **K** are independent of the angle of rotation about the z axis.

Determine:

(a) The substructure stiffness matrix for a square when the diagonal of the square runs from the upper left corner to the lower right corner.

(b) The substructure stiffness matrix for a square when the diagonal of the square runs from the lower left corner to the upper right corner.

(c) If the two substructure models are capable of yielding the same finite element solution for some loading.

8:12. Comparison of fully and reduced integration elements.

Given: The stiffness matrix for the fully integrated square membrane based on the first interpolation function of Table 7.9 and the stiffness matrix using the three-point Gauss quadrature rule based on the first interpolation function of Table 7.9. The rows and columns relate to the same nodal variables as for the fully integrated model.

Determine: What, if anything, can be said about the relative strain energies of solutions using the two models.

$$\mathbf{K}\ \text{(fully integrated)} = \begin{bmatrix} 843\,750 & & & \text{Sym.} & \\ 93\,750 & 843\,750 & & & \\ 656\,250 & -93\,750 & 843\,750 & & \\ 93\,750 & -656\,250 & -93\,750 & 843\,750 & \\ -843\,750 & -93\,750 & -656\,250 & -93\,750 & 843\,750 \end{bmatrix}$$

$$
\mathbf{K} \; \text{(three-point Gauss)} = \begin{bmatrix} 1\,125\,000 & & & & \text{Sym.} \\ 93\,750 & 1\,125\,000 & & & \\ 375\,000 & -93\,750 & 1\,125\,000 & & \\ 93\,750 & -937\,500 & -93\,750 & 1\,125\,000 & \\ -562\,500 & -93\,750 & -937\,500 & -93\,750 & 1\,125\,000 \end{bmatrix}
$$

Constraints of determinate supports have been applied to the element matrix to reduce calculations for this problem.

8.12 COMPUTERWORK

8-1. Secondary degeneracy in a frame analysis.

Given: The A frame shown here.

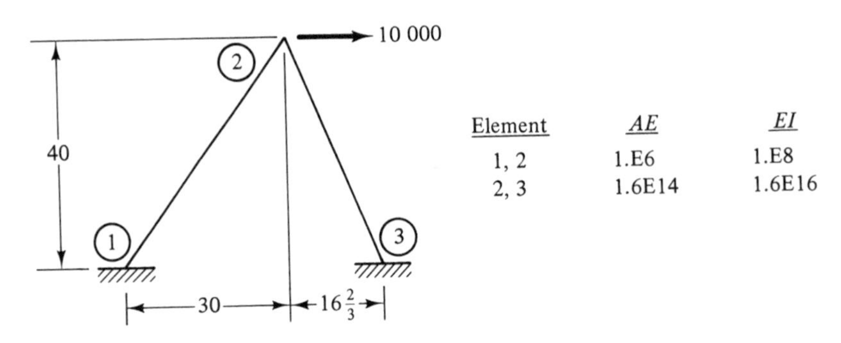

Element	AE	EI
1, 2	1.E6	1.E8
2, 3	1.6E14	1.6E16

Computerwork Problem 8-1

Determine: Whether the data of Fig. 8.2 would forecast degeneracy of the stiffness matrix.

8-2. Finding an eigenvalue.

Given: The third-modal vector for the beam from the table in Homework Prob. 8:2.

Find: The associated eigenvalue.

Method: By treating the modal vector as a static loading.

8-3. Tapered beam analysis.

Given: The tapered beam of the figure.

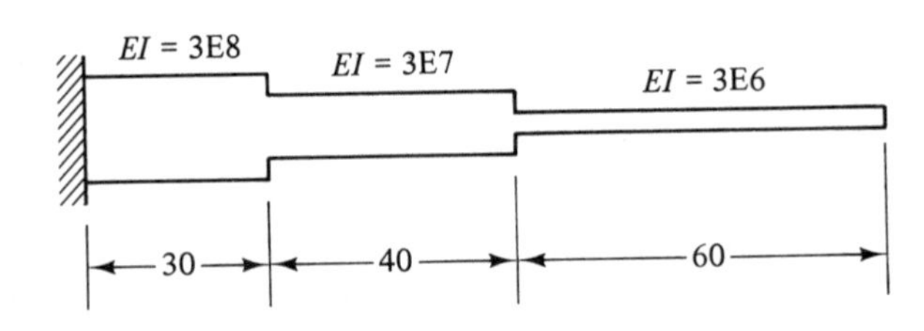

Computerwork Problem 8-3

Find:

(a) The exact solution by FEA for tip deflection.

(b) A solution that gives a nonzero lower-bound estimate of the exact solution.

(c) A solution that gives a finite upper-bound estimate of the exact solution.

***8-4.** Membrane constant-strain state check.

Given: The rectangular membrane element in plane stress with imposed nodal displacements (settlements) of the figure.

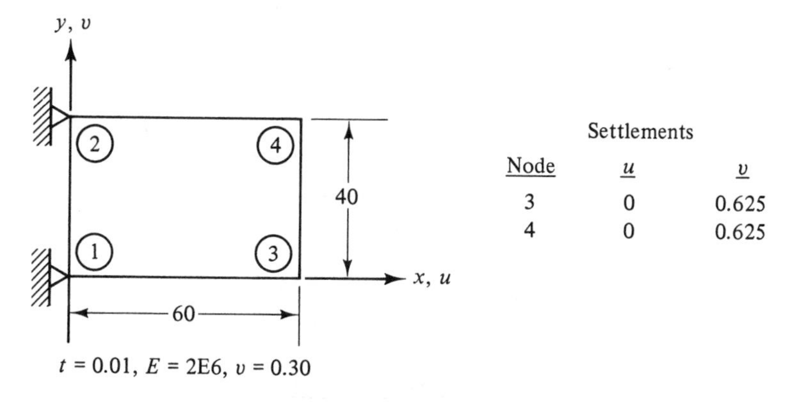

Computerwork Problem 8-4

Show: That the settlements impose a state of constant strain.

Method: Use the stress predictions for the membrane model based on four-Gauss-point integration.

8-5. Membrane constant-stress state check.

Given: The rectangular membrane loaded as in the figure.

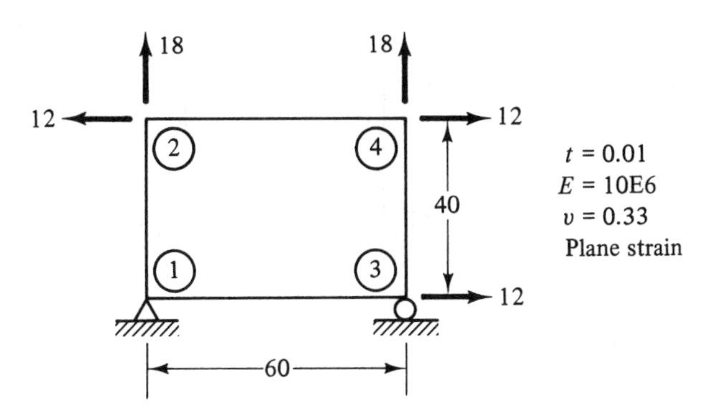

Computerwork Problem 8-5

Show: That this loading excites a constant-stress state.

8-6. Calculation of constant-strain strain energy.

Given: The one-element membrane model of the figure of Computerwork Prob. 8-5.

Find: The constant-strain strain energy.

Method:

(a) By using the formulas of Table 8.2 to find the strains and integrating over the element volume.

(b) By imposing displacements on an element integrated by a single Gauss point.

***8-7.** Increasing constant-strain strain energy.

Given: The membrane of figure part (A) modeled in plane stress by one and four elements as in parts (B) and (C) of the figure.

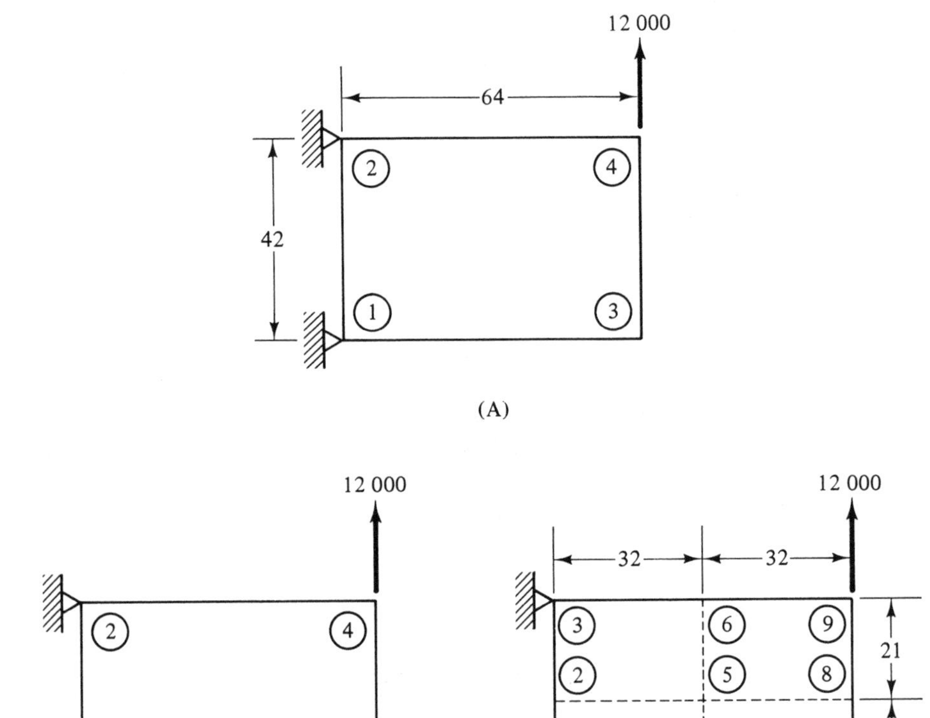

Computerwork Problem 8-7

Show: That the total constant-strain strain energy increases as the grid is refined as required by the data exhibited in Figs. 8.9 and 8.10.

Method: Get the constant-strain strain energy by

(a) Using data produced by the FEA code for the loading of part (A) on the models of parts (B) and (C).

(b) Using the fact that one-point Gauss models only represent the constant-strain strain energy and the model of part (B).

8.13 REFERENCES

1. Melosh, R.J., "Accuracy Control of Finite Element Analysis of Articulated Structures," *Proc. Ninth Electronic Computation Conf.*, University of Alabama, Feb. 1986, pp. 1–13.
2. Bolkir, A.B., "Extrapolation in Finite Element Analysis," Ph.D. thesis, Duke University, 1984.
3. Verma, A.; Melosh, R.J., "Numerical Tests for Assessing Finite Element Model Convergence," *Int. J. Num. Meth. in Engineering*, vol. 24, 1987, pp. 843–857.
4. Bazeley, G.P.; Cheung, Y.K.; Irons, B.M.; Zienkiewicz, O.C., "Triangular Elements in Plate Bending-Conforming and Nonconforming Solutions," AFFDL-TR-66-80, *Proc. Conf. on Matrix Meth. in Struct. Mech.*, Dayton, Ohio, 1965, pp. 547–576.
5. Verma, A., "Numerical Finite Element Model Performance Tests," Ph.D. dissertation, Duke University, 1986.
6. Khanna, J.; Hooley, R.F., "Comparison and Evaluation of Stiffness Matrices," *AIAA Journal*, vol. 4, no. 12, Dec. 1966, pp. 2105–2110.

Chapter 9

Self-qualifying Finite Element Analyses

An FEA that produces a measure of the solution error along with predictions of behavior of an engineering system is a qualified FEA. It is self-qualifying when the error measures are created by the FEA computer program.

Self-qualifying FEA results means defining the magnitudes of round-off, discretization, and other analysis inaccuracies. In this chapter, we focus on self-qualification of discretization errors.

The characteristics of discretization errors are different from those of round-off error. Discretization errors associate with the structural modeling rather than the computer number representation. Discretization errors can be guaranteed to decrease as the number of nodal variables increase. Since we assume that we do not know the exact solution to the structure we are modeling with finite elements, our measures of discretization error must be indirect.

Self-qualified analysis capabilities are a valuable enhancement of conventional capabilities. They can establish the limitations of modeling accuracy for a given production analysis, determine how many additional nodal variables are needed and where they should be placed to increase modeling accuracy to the required level by reanalysis, and improve response predictions of a sequence of analyses by extrapolation.

To achieve these capabilities, it is generally conceded that we must select the element models and discretization with special care. We will not be satisfied with element models that meet only the minimum requirements of element models.

We require that models satisfy preferentiality requirements and that they ensure monotonic convergence. We require that each system model use element models that work together to define either a lower-bound (potential energy) solution or an upper-bound solution. We demand that successive meshes be designed so that monotonic convergence guarantees remain intact.

This chapter centers on a "typical" finite element analysis: prediction of the torsional rigidity of a square shaft. It examines the relevant elasticity theory, available finite element models, and the analysis concepts that associate with discretization error qualification. It follows the analysis steps for attacking the typical problem with digressions to elaborate on the concepts for nontorsion analyses.

9.1 MATHEMATICAL BASIS FOR TORSION ANALYSIS

The theory of the torsion of solids is structural. The equations are derived by St. Venant's semi-inverse method. In this method, we reduce the equations of elasticity by assuming some of the displacements and solving a differential equation to find the rest of the displacements. For FEA modeling, we usually replace the differential equation with a minimum potential energy statement of the problem.

Following St. Venant, we begin by assuming that the solid, as Fig. 9.1 suggests, has a straight axis about which cross sections rotate. The shaft is cylindrical; the outer boundary is generated by a line parallel to the axis. Thus all cross sections have the same geometry, though their outer boundary need not be a circle.

To describe the deformed geometry, we place the z axis along the shaft axis. When a twist is applied to the end of the shaft, we assume that the displacements in the coordinate system of Fig. 9.1 are

$$[u \quad v \quad w] = \alpha[-yz \quad xz \quad \phi(x, y)] \tag{9.1}$$

where

α is the angle of twist and

ϕ defines the displacements normal to the plane of the cross section (the "warping" displacements).

Substituting Eq. (9.1) in the strain displacement equations (7.1) yields the strains in terms of the warping function:

$$\begin{pmatrix} e_{xx} \\ e_{yy} \\ e_{zz} \\ e_{xy} \\ e_{xz} \\ e_{yz} \end{pmatrix} = \alpha \begin{pmatrix} 0 \\ 0 \\ 0 \\ 0 \\ \delta\phi/\delta x - y \\ \delta\phi/\delta y + x \end{pmatrix} \tag{9.2}$$

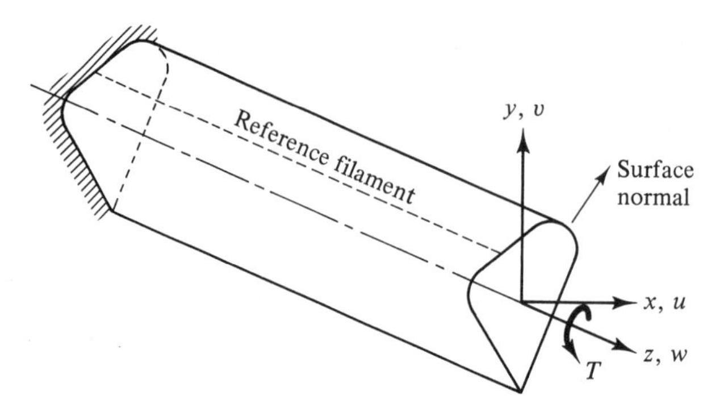

(A) Undeformed geometry

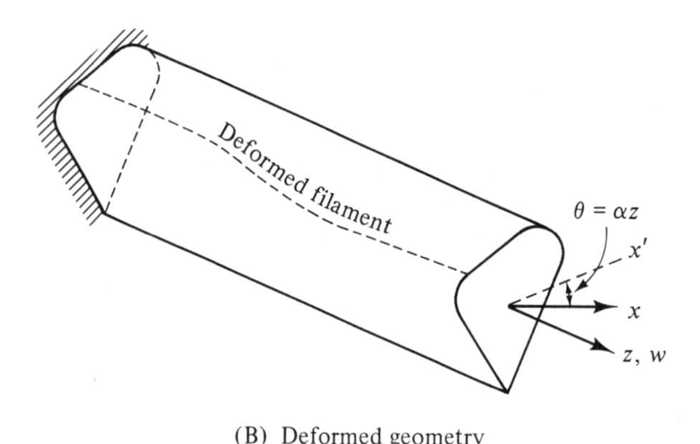

(B) Deformed geometry

Figure 9.1 Twisting of a cylindrical shaft.

Substituting these strains in the isotropic stress-strain equations (7.8) reveals that

$$\begin{pmatrix} \sigma_{xx} \\ \sigma_{yy} \\ \sigma_{zz} \\ \sigma_{xy} \\ \sigma_{xz} \\ \sigma_{yz} \end{pmatrix} = \frac{\alpha E}{2(1 + \nu)} \begin{pmatrix} 0 \\ 0 \\ 0 \\ 0 \\ \delta\phi/\delta x - y \\ \delta\phi/\delta y + x \end{pmatrix} \tag{9.3}$$

Substituting these stresses in the stress equilibrium equations (7.14) gives

$$\frac{\delta^2\phi}{\delta x^2} + \frac{\delta^2\phi}{\delta y^2} = 0 \tag{9.4}$$

which is Laplace's equation.

It is useful to recognize that the torsion problem involves Laplace's equation. Other important engineering problems, such as steady-state heat conduction, seepage, irrotational flow of fluids, electromagnetic and magnetostatic field potential, lubrication, and beam-bending problems require stiffness matrices that must, in the limit, imply satisfaction of Laplace's equation. Exact solutions of Laplace's equation have been known for some time and are available for generating a variety of element models.

The assumption of a displacement shape function and use of the isotropic material properties has led to a single equation whose solution implies satisfaction of the equilibrium and stress-strain and strain-displacement equations microscopically.

The boundary conditions for the equation solution are that the boundaries of the solid be stress-free. Thus on the lateral surface of the cylinder,

$$\begin{aligned} p_x &= \sigma_{xx} \cos(x, n) + \sigma_{xy} \cos(y, n) = 0 \\ p_y &= \sigma_{yx} \cos(x, n) + \sigma_{yy} \cos(y, n) = 0 \\ p_z &= \sigma_{zx} \cos(x, n) + \sigma_{zy} \cos(y, n) = 0 \end{aligned} \tag{9.5}$$

where

$\cos(x, n)$ is the cosine of the angle between a vector parallel to the x axis and the vector, which is normal to the lateral surface of the cylinder, and

$\mathbf{p}$ denotes the surface traction vector.

Since $\sigma_{xx} = \sigma_{yy} = \sigma_{xy} = 0$, from Eq. (9.3), only the third of Eqs. (9.5) is nontrivial. Rewriting this equation in terms of the warping function of Eq. (9.3) gives

$$\left(\frac{\delta\phi}{\delta x} - y\right) \cos(x, n) + \left(\frac{\delta\phi}{\delta y} + x\right) \cos(y, n) = 0 \tag{9.6}$$

But from the geometry at the boundary we know that

$$\frac{\delta\phi}{\delta n} = \frac{\delta\phi}{\delta x} \cos(x, n) + \frac{\delta\phi}{\delta y} \cos(y, n) \tag{9.7}$$

So we can simplify Eq. (9.6) to

$$\frac{\delta\phi}{\delta n} = y \cos(x, n) - x \cos(y, n) \tag{9.8}$$

Equation (9.8) defines the boundary condition to particularize the solution of Laplace's equation.

Summing the moments induced by the shear stresses about the axis furnishes the resultant torque on the cross section of the cylindrical solid:

$$T = \iint (x\sigma_{zy} - y\sigma_{zx})\, dx\, dy \tag{9.9}$$

where T is the resultant torque and integration is over the differential cross-sectional area $dx\, dy$. We use the stresses of Eqs. (9.3) to express Eq. (9.9) in terms of the warping function, finding that

$$T = \alpha G \iint \left[(x^2 + y^2) + \left(\frac{x\delta\phi}{\delta y} - \frac{y\delta\phi}{\delta x} \right) \right] dx\, dy \tag{9.10}$$

Now we define

$$\begin{aligned} J &= \iint (x^2 + y^2)\, dx\, dy + \iint \left(\frac{x\delta\phi}{\delta y} - \frac{y\delta\phi}{\delta x} \right) dx\, dy \\ &= I_p + \iint \left(\frac{x\delta\phi}{\delta y} - \frac{y\delta\phi}{\delta x} \right) dx\, dy \end{aligned} \tag{9.11}$$

and

$$I_p = \iint (x^2 + y^2)\, dx\, dy \tag{9.12}$$

where

J is the torsional stiffness and
I_p is the polar moment of inertia.

The torsional stiffness, when the warping is zero, is the polar moment of inertia. Since the polar moment of inertia must be greater than zero, and since permitting warping must reduce the torsional rigidity, the integral in Eq. (9.12) must be negative or zero. Given the particular solution of Laplace's equation, evaluation of the right-hand side of Eq. (9.11) supplies the torsional stiffness.

The torsion problem differs from the general problem of elasticity in two ways: The governing equation in torsion involves only displacements normal to the cross section, and the boundary condition involves the first derivative of the variable of interest rather than the simple variable.

Rather than dealing with Laplace's equation, we can express the torsion problem in energy terms. In the strain energy expression,

$$SE = 0.5 \iiint \boldsymbol{\sigma}^T \mathbf{e}\, dx\, dy\, dz \tag{9.13}$$

with integration over the differential volume $dx\, dy\, dz$. We substitute Eqs. (9.3) and (9.4) to get

$$SE = 0.5\alpha G \iint \left[\left(\frac{\delta\phi}{\delta x} \right)^2 + \left(\frac{\delta\phi}{\delta y} \right)^2 + \frac{x\delta\phi}{\delta x} - \frac{y\delta\phi}{\delta y} \right] dx\, dy \tag{9.14}$$

where the term in $x^2 + y^2$ in the integrand is dropped because it will vanish when the strain energy is minimized.

Equations (9.14) and (9.8) express the energy form of the torsion equation. Our goal is to find the warping function that minimizes the strain energy and satisfies the derivative boundary condition.

We have arrived at two bases for constructing finite element models for determining the torsional rigidity of a prismatic cylindrical shaft: a differential equation and an energy basis. Comparing Eq. (9.4) and Eq. (9.14), we note a fundamental difference in the equations. If we use the differential equations directly, our shape functions must approximate the warping function through second derivatives, whereas the energy form requires only approximating first derivatives. The energy approach both simplifies element model development and intrinsically involves an integral measure of solution error.

9.2 FINITE ELEMENTS FOR ANALYSIS OF TORSION

Use of the potential energy method of deriving stiffness matrices for the torsion problem is essentially the same as for other continuum models. The steps are the same, but the strain energy form is that of Eq. (9.14). Minimization of the strain energy yields both the stiffness matrix and a vector of equivalent loads. Introducing the interpolation function in the boundary condition, Eq. (9.8) states this condition as a general linear function of nodal variables on, and next to, the cross section boundary.

Table 9.1 provides an illustrative torsion stiffness matrix. The shape function for warping is linear in the x and y coordinates, so the element is another simplex model. The nodal variable is the warping displacement. Since this represents movement normal to the plane, the matrix does not change when the reference axes are rotated about the z axis. Accordingly, the warping function can be regarded as a potential and the warping amplitudes as scalars.

The stiffness matrix has a degeneracy of 1. The degeneracy associates with axial translation of the solid whose cross section is triangular and whose span is the length of the shaft along the z axis. This type of degeneracy is characteristic of all torsion element models.

TABLE 9.1 SIMPLEX TORSION ELEMENT MODEL

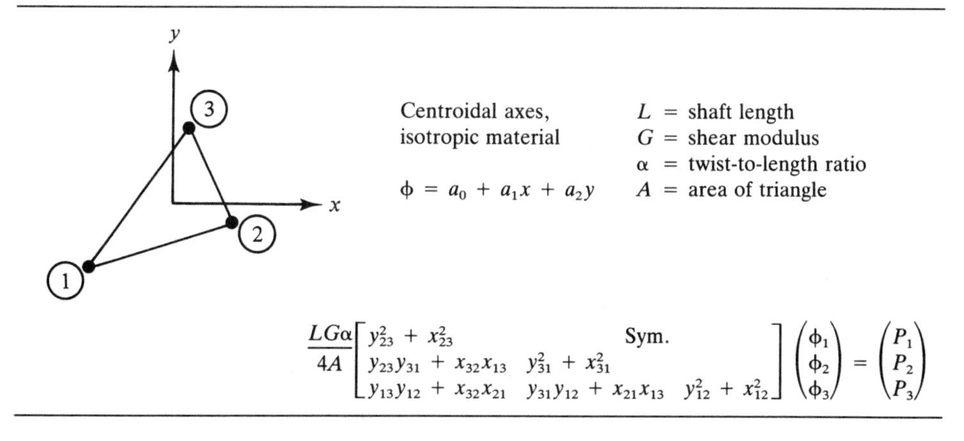

$$\frac{LG\alpha}{4A}\begin{bmatrix} y_{23}^2 + x_{23}^2 & & \text{Sym.} \\ y_{23}y_{31} + x_{32}x_{13} & y_{31}^2 + x_{31}^2 & \\ y_{13}y_{12} + x_{32}x_{21} & y_{31}y_{12} + x_{21}x_{13} & y_{12}^2 + x_{12}^2 \end{bmatrix}\begin{pmatrix}\phi_1\\ \phi_2\\ \phi_3\end{pmatrix} = \begin{pmatrix}P_1\\ P_2\\ P_3\end{pmatrix}$$

Available torsion models have cross sections like the planform of available membrane models. They are based on the bivariable interpolation functions of Table 7.9, but warping interpolation only is used. They use Gauss quadrature rules for intergration over the cross section. They incorporate parametric mapping to improve the element's ability to represent irregular boundary geometries.

Figure 9.2 illustrates the torsion models that are used in developing data for studying discretization error control in this chapter. These models represent four different elements because the superquadratic element includes both a fully integrated Gauss rule (3 × 3 Gauss point array) and a reduced integration array

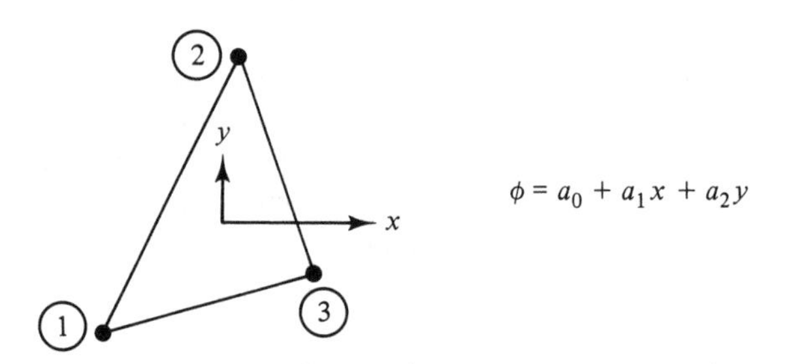

(A) Simplex torsion element

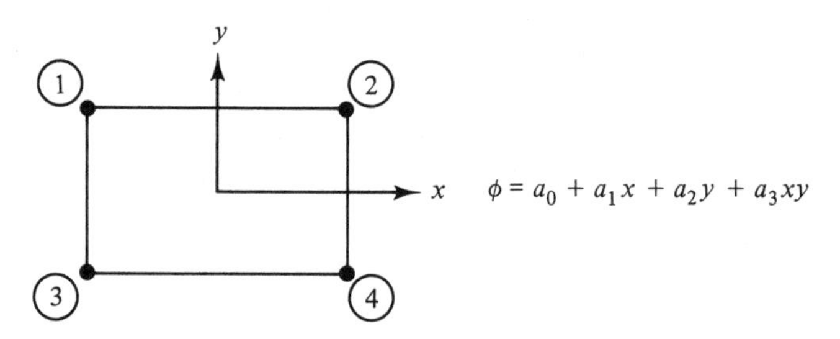

(B) Hyperbolic shape function torsion model

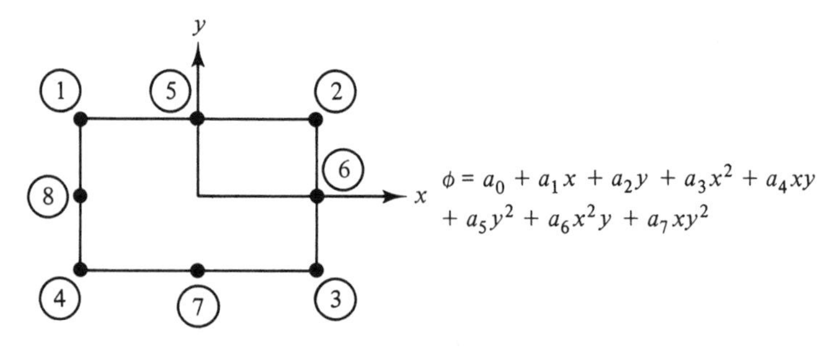

(C) Superquadratic shape function

Figure 9.2 Torsion element model cross sections.

(2 × 2). All these element models use displacement shape functions that satisfy the potential energy continuity requirement across the element boundaries, that is, continuity of warping through the zeroth derivative. The simplex and hyperbolic shape function models are known to guarantee monotonic convergence.[1]

In many production codes, heat transfer element models are provided instead of torsion elements. Since these elements model the Laplace equation, they can be used for torsion analysis. Membrane stiffness models can also be used by letting u displacements play the role of warping displacements, setting all v displacements to zero, and using the computer program's capability of modeling orthotropic materials to nullify the nonshear energy terms.

9.3 TORSIONAL STIFFNESS OF A SQUARE SHAFT

In this section we define a torsion analysis problem. The problem will serve to focus the discussion of discretization error control. We choose a problem with a known solution so that we have a direct check on the effectiveness of control methods.

Figure 9.3 is a sketch of the problem. The structure is a straight prismatic shaft that has a square cross section. The shaft is fixed at the origin of the z axis and unsupported at its other end. It is loaded with a torque at its unsupported end. The shaft is made from an isotropic material. It is assumed to be free to warp. The goal of analysis is to determine the torsional stiffness of the shaft with less than 1 percent discretization error in the FEA.

The exact solution for the warping and the torsional stiffness is expressible

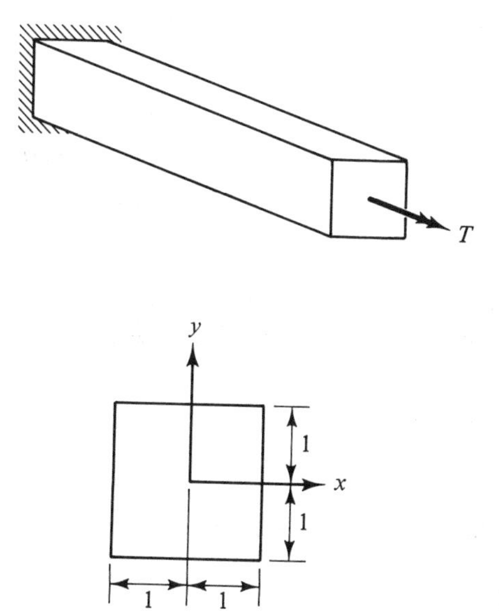

Figure 9.3 Square shaft in torsion.

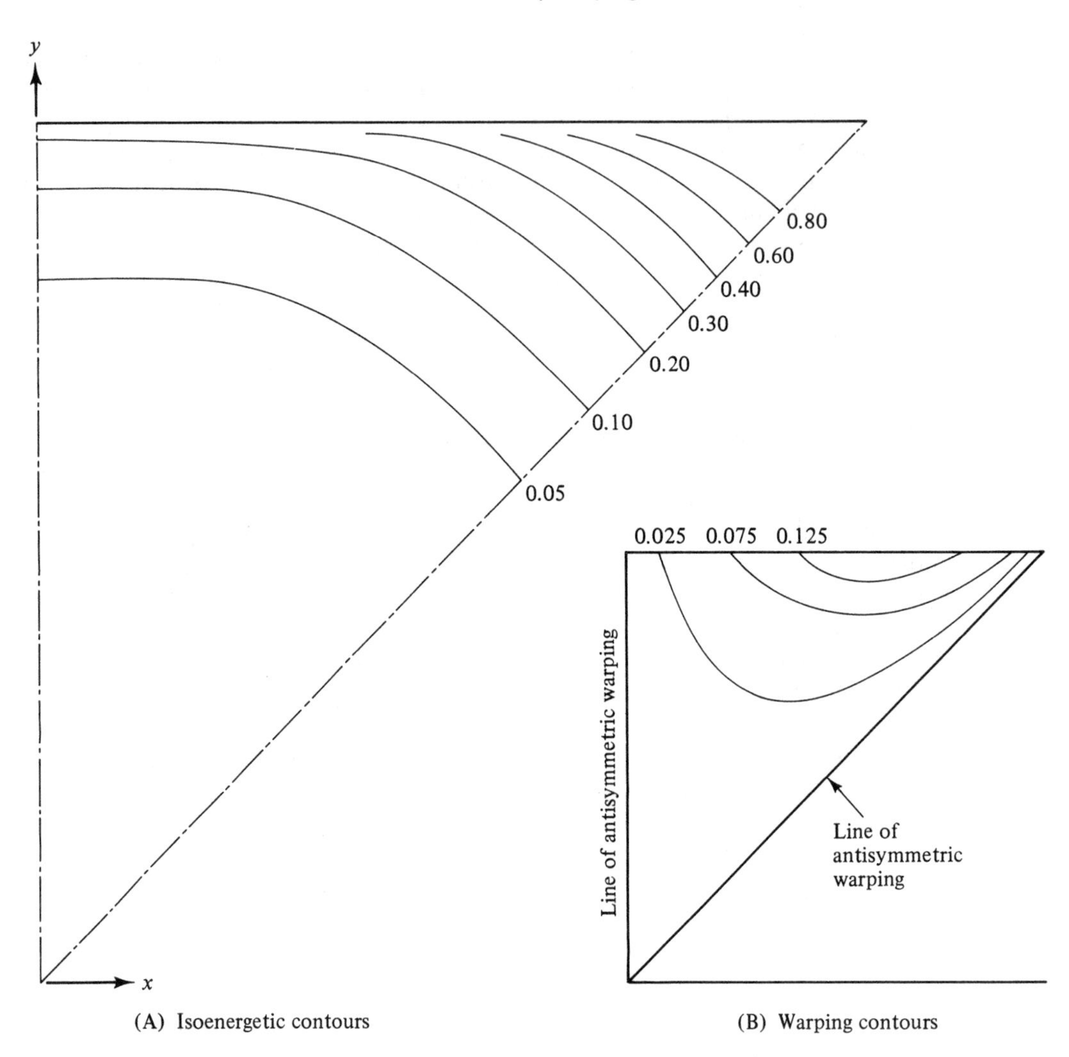

Figure 9.4 Square shaft: Isoenergetic and warping contours.

in terms of an infinite series.[2] Figure 9.4 shows the contours of the relative strain energy and warping amplitudes of the exact solution for a quadrant of the cross section. The torsional stiffness is

$$J = 0.140577s^4 = 2.249232 \tag{9.15}$$

where

J is the torsional stiffness of the cross section and
s is the length of a side of the square (2 units).

We seek to solve this torsion problem by FEA using a minimum of computer resources. As a first step toward this objective, we observe that the warping function is antisymmetric across the coordinate axes and across the diagonals of the square. Therefore, we will deal with an octant of the square, as suggested by the finite element model of Fig. 9.5. Because of the response antisymmetries, the same number and relative location of nodal lines will be used along the x and y axes of the square to imply a symmetric grid across the diagonal. Warping displacements must be zero for all nodes falling on the x and y axes or along the diagonal of the square to imply antisymmetric response across these boundaries.

We choose a rectangular shape as the basic finite element module and make the module up from two right triangles for the simplex representation. Since the module is rectangular, the octant will be represented by a stepped edge along the diagonal. To deal with this, we require that the nodal displacements at an overlapped node, such as node 8 in Fig. 9.5, be loaded to induce equal and opposite nodal displacements of its mirror-image counterpart, node 3.

Since we will analyze only one-eighth of the cross section, the exact torsional stiffness will be one-eighth that of Eq. (9.15). Furthermore, the FEA determines only the reduction in torsional stiffness from the polar moment of inertia. Therefore, the exact value of the warping reduction for the 2-unit square shaft is

$$J_w = 0.052179 \tag{9.16}$$

where J_w is the warping stiffness.

Figure 9.4(A) indicates that the highest gradients of strain energy occur near the free edge of the square. Therefore, to maximize the model's efficiency, more elements should be allocated to this region than to the region near the center of the square.

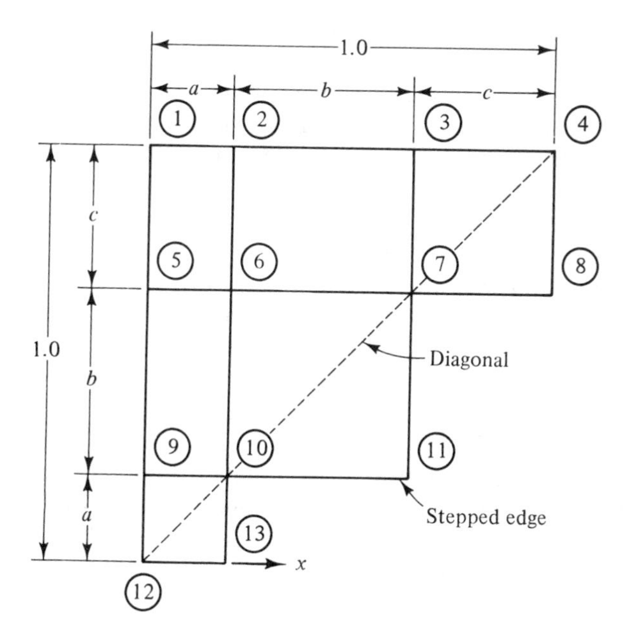

Figure 9.5 Typical finite element model.

9.4 OPTIMUM GRIDS

An optimum finite element grid is one for which the position of the nodes minimizes the discretization error. When potential energy modeling is used, the optimum grid defines a maximum of the potential with respect to the coordinates of the nodes and a minimum with respect to the generalized nodal displacements.

The grid design problem is an illustration of a classical nonlinear optimization problem. To solve it we seek values of the nodal coordinates that will minimize discretization error subject to inequality constraints that require that the nodes must remain on the surface or within the structural material. The constraints are nonlinear when the bounding surfaces are nonlinear functions of the system coordinates. The cost function, the discretization error, is nonlinear in the nodal coordinates because stiffnesses usually vary as a nonlinear function of element length and size.

Use of optimization algorithms to find the best finite element grid is impractical.[3] The computer resources needed to solve the optimization problem are at least a factor of 10 times that of the FEA of a given grid because there will be about twice as many design variables as nodal variables, the optimization will involve at least as many constraints as unknown nodal coordinates, and the solution process is necessarily interative. These resources must be amortized for a single analysis because the optimum grid design will change with the structural geometry, material composition, loading, and displacement boundary conditions.

Attached to this lining is a silver cloud. The concept of an optimum grid is important and useful. It establishes ideal grids that can serve as standards of comparison for all other grids. It leads to an understanding of the characteristics of the most computationally efficient grids for a particular application.

Figure 9.6 illustrates these points. It shows the relation between optimum grids and the number of active nodal variables and accuracy of stiffness estimates for the square shaft in torsion. The number of active variables is the number of simultaneous equations solved to get the stiffness. Thus the abscissa is a rough measure of the computer time to solve the equations. The ordinate measures the number of digits of accuracy of the FEA prediction.

A curve represents each of the element models of Fig. 9.2 and an underintegrated element model based on the superquadratic shape function. The curves demonstrate that the optimum grid varies with element selection. Note that for any given number of active variables, the hyperbolic shape function element produces more accurate stiffness estimates than the simplex. This must be so; it can be proven by the comparison test. Similarly, the reduced-integration superquadratic shape function element appears to be more efficient than its fully integrated counterpart. This cannot be proved by the comparison test.

Study of Fig. 9.7 adds further insights into the relation between element selection and grid design. This figure shows the envelope of solutions for all possible grids for two of the element types. Each envelope is constructed by taking the optimum grid curve for an element and adding a worst-grids line: a horizontal

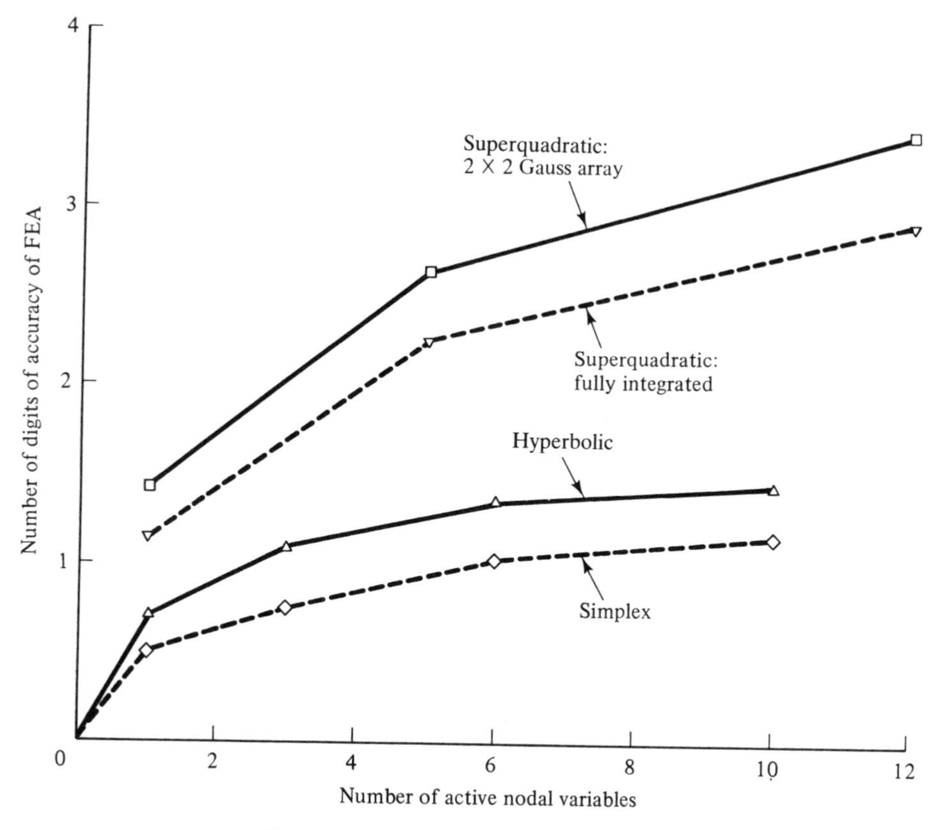

Figure 9.6 Solutions with an optimum grid.

line that represents no improvement in solution accuracy from the coarsest grid. The no-improvement case is realized when nodes added to the coarsest grid coalesce with existing nodes. Each envelope encloses an infinite area because the horizontal line extends to infinity and the curve of optimum grid solutions must be distinct as long as the element models will develop the exact solution when the number of nodal variables approaches infinity.

Consider a single envelope of Fig. 9.7. Assume that only potential energy element models are used in the analyses and that monotonic convergence of strain energy estimates is guaranteed with the element. Nevertheless, depending on the strategy for grid refinement, it is possible to get a sequence of estimates that converges from the too-high stiffness side, a sequence that appears to converge from the too-low stiffness side, and a sequence in which successive estimates ostensibly diverge from a too-high stiffness estimate.

These seeming contradictions to the guarantee of monotonic convergence are readily explained. The analytical and numerical proofs of monotonicity require a sequence of hereditary grids. Figure 9.8 illustrates a set of such grids. Any two successive grids of the sequence has a parent-child relationship. The child has all

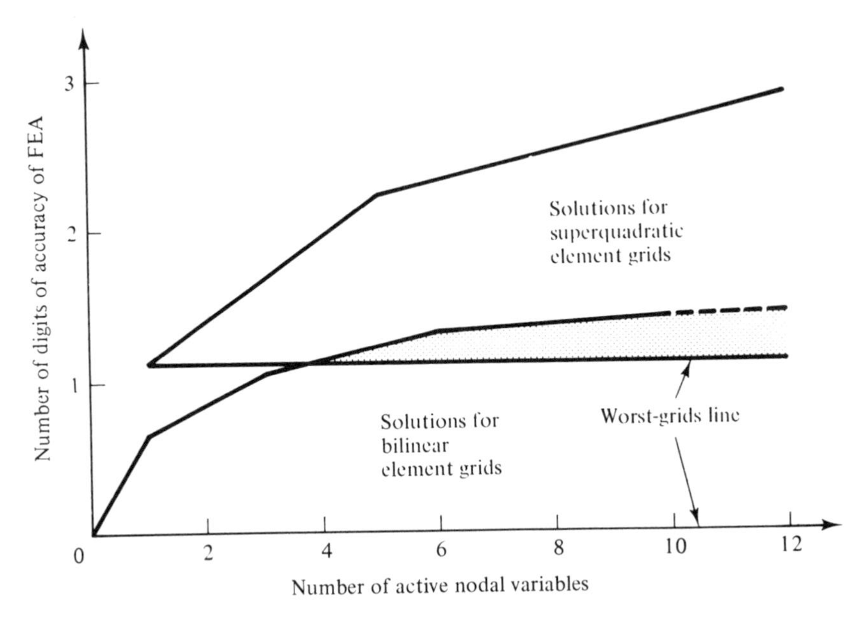

Figure 9.7 Torsion element grid envelopes.

the nodes and node lines of its parent. It involves the same element models covering the same geometry. It has more grid lines and nodes than its parent. Pejoratively, the child has all the faults of its parent and may have some vices of its own. Regardless, the sequence of energy estimates from two hereditarily related grids is guaranteed to be monotonic when the element guarantees monotonic convergence.

It can also be proved that a sequence of optimum grids can be guaranteed to produce a monotonically convergent sequence of energy estimates.[4] This fact is useful as a check on optimum grids and establishes the understanding that monotonic convergence may occur with some nonhereditary grids.

Considering the collection of envelopes of Fig. 9.7 evokes additional conclusions about element selection as a function of grid design:

1. Two different elements may be capable of producing identical energy estimates; that is, the envelopes of two element models may intersect.

2. An element with a higher error term may produce better or worse energy estimates than one with a lower error term; that is, the envelope of the hyperbolic shape element intersects with the envelope of the superquadratic element.

3. The choice of the best element to use depends on the accuracy required of the solution as well as the grids to be used; that is, since the intersection of two elements may be less than the union, each element may have a region of grid designs in which more efficient analyses lie.

4. For a sequence of grids, the most efficient designs involve a strategy in

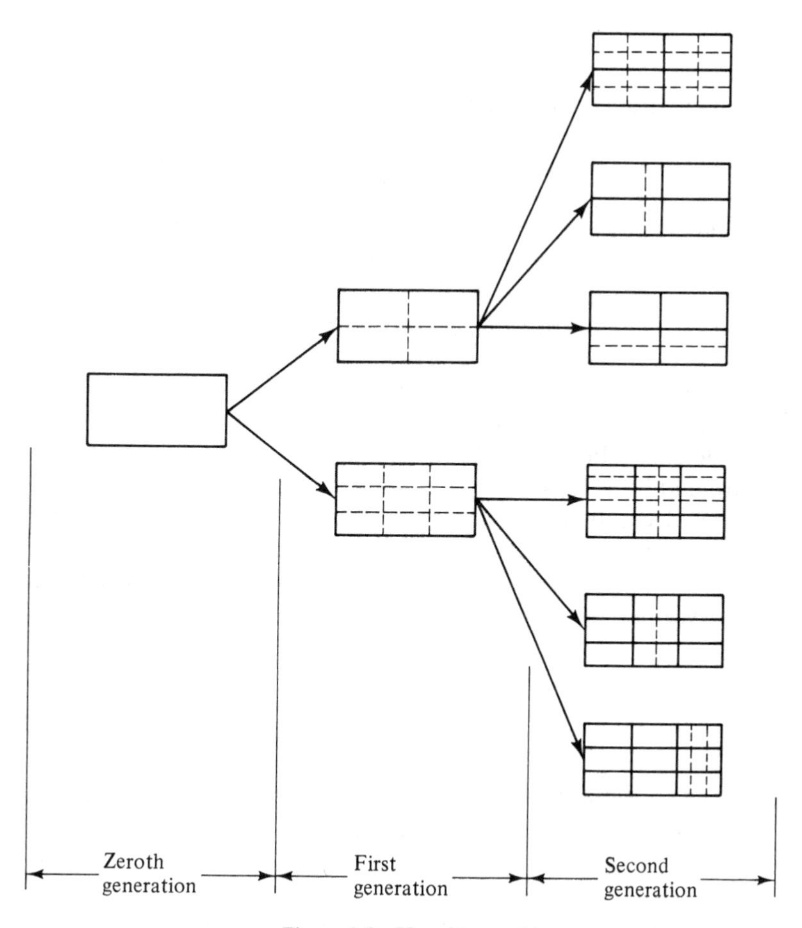

Figure 9.8 Hereditary grids.

which both grid refinement and element model changes are acceptable; that is, the leftmost folded curve of optimum meshes recommends changing the nodes and element models for maximum accuracy per active generalized displacement variable.

These conclusions relate to computer time as well as to the number of active nodal variables.

9.5 DISCRETIZATION ACCURACY MEASURES

In controlling discretization accuracy, the analyst needs data to answer three questions. Are the most recent analysis results acceptable with respect to discretization accuracy? If not, should the next grid involve a local or global refinement of the current grid? If local, where should the local refinement occur? These questions suggest the need for three accuracy measures.

A number of accuracy measures have been reported in the engineering literature.[4] Measures have been based on forward differencing, stress imbalances, constant-strain contributions, bounds on the norm of the residual of the differential equation being addressed, gradients measured by perturbing nodal coordinates, the relative intensity of strain energy in each element, and the relation of node lines to isoenergetic contours. None of the measures directly deduces the magnitude or distribution of discretization accuracy. Their reliability depends on the accuracy of the analysis on which they are based and necessarily improves as the solution accuracy improves. Their ability to reflect error magnitude is often poor when predicted errors are greater than 10 percent.

We can compare accuracy measures on the basis of the number of calculations required to evaluate them. On this basis, most of the measures require a small fraction of the calculations needed for evaluating stresses given generalized displacements. Forward differencing and gradient measures require in addition forward and back substitutions in the triangularized stiffness matrix. Until evidence is provided exhibiting superior accuracy of these measures compared with others, forward differencing and gradient methods should be avoided.

For this chapter, we choose strain energy accuracy measures.

We define the resolution accuracy measure by

$$A_r = \frac{SE'}{SE} \tag{9.17}$$

where

A_r is the discretization resolution accuracy ratio and

SE' is the part of the strain energy involved in the constant-strain modes of behavior.

The resolution measure senses the accuracy of the strain energy estimate of the structural system. It has a value between zero and 1. The resolution ratio will be 1 only for the perfect mesh—the mesh in which each element is undergoing only constant strain, though higher-order strains are permitted by the modeling. The resolution sensor provides optimistic measures of analysis accuracy when the stimulation is a potential energy analysis, because the denominator of Eq. (9.17) is always underestimated. Reference 5 presents the resolution measure and illustrates its effectiveness in analysis of membranes, plates, and shells.

We define the optimality ratio measure by

$$a_{oi} = \frac{SE_i'}{SE_i} \qquad \text{for } i = 1, 2, \ldots, n_e \tag{9.18}$$

where

a_{oi} is the discretization optimality ratio for element i and

n_e is the number of finite elements in the model.

When mesh design is an unconstrained optimization problem, all a_{oi} will have the value a/n_e. Constraints on the redesign of meshes can be expected to inhibit these ratios from taking on the value of $1/n_e$. We measure the optimality of the mesh by the variance of the optimality ratios:

$$A_o = \left[\sum_i (a_{oi} - a_o)^2\right]^{0.5}, \qquad i = 1, 2, \ldots, n_e \tag{9.19}$$

where

A_o is the optimality variance,
a_o is the mean optimality ratio, and

$$a_{oi} = SE_i - SE_i' \qquad \text{for } i = 1, 2, \ldots, n_e \tag{9.20}$$

a_{oi} is the locality residual for element i. The value of a_{oi} must be positive and, ideally, is zero everywhere for an exact analysis. Use of this measure in mesh redesign implies that the elements undergoing the maximum nonconstant-strain strain energy for the pertinent loading are the most likely to benefit from subdivision.

An advantage of the measures of Eqs. (9.17) through (9.20) is that they relate to finite element performance criteria. The resolution ratio is based on the fact that as the solution approaches the exact solution, the constant-strain states must be preferred. The optimality ratio is based on the fact that the element strain energy contributions must be balanced throughout the mesh, when the mesh is optimum. The locality residuals are based on the fact that the element with the highest nonconstant-strain strain energy makes the largest approximation to the current estimate of system strain energy.

A philosophical advantage of these discretization error measures is that they are in energy terms. They involve the error norm, which is being directly minimized by the finite element process. They are scalar measures of accuracy rather than vector measures.

Evaluation of the measures adds relatively few calculations to the FEA. Total and constant-strain strain energy calculations require at most the calculations in evaluating the dot product of the element nodal displacements with the element nodal forces for each element.

A disadvantage of the resolution measure is that it provides no useful information for the simplex element models. We must have more behavior states than the constant-strain states to get a measure of the discretization accuracy for an element. We discount this deficiency by arguing that the simplex models are relatively inefficient compared with other models. We have noted the proof that for the torsion element, the hyperbolic rectangle cannot be less efficient than the simplex model. Practitioners have virtually abandoned simplex modeling of membranes, plates, and shells based on experiences with the inefficiencies of simplex elements compared with other models.

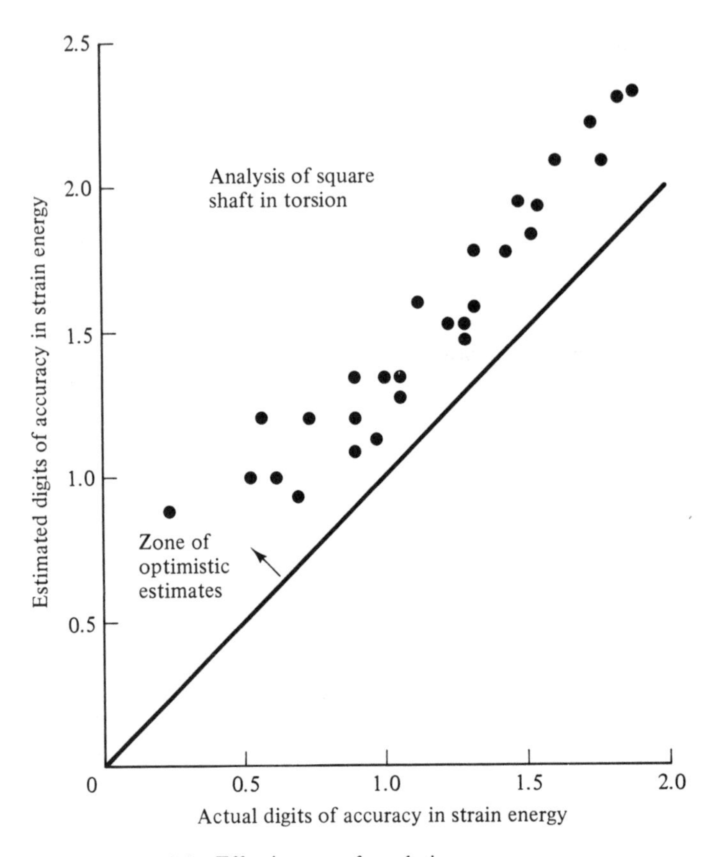

Figure 9.9 Effectiveness of resolution accuracy measures.

Figure 9.9 illustrates the effectiveness of the resolution ratio of Eq. (9.17). It displays estimated and actual discretization errors, measured in terms of the number of digits of accuracy, for a number of analyses of the square shaft in torsion. The display indicates a reduction in scatter as the modeling accuracy increases. The fact that all points lie above the 45-degree line is evidence that the resolution accuracy measure is always an optimistic estimate of the actual discretization accuracy in potential energy analyses. The data of Fig. 9.9 shows that the measure is at most a half digit optimistic.

9.6 ADAPTIVE LOCAL GRID REFINEMENT

Adaptive local grid refinement is a strategy for redesigning FEA grids to develop solutions with known discretization accuracy. In discussing the strategy, we adopt the attitude of the analyst who is depending only on FEA for assessment of the accuracy of the FEA.

Adaptive grid refinement is based on hereditary grids. Each child grid is constructed using measures of discretization accuracy based on the FEA of its parent. The refinement strategy is h refinement because the grid interval h is changed in the child grid but the same element models are used for the same volume of the structure. The refinement strategy is local because not all elements of the parent grid need be subdivided in creating the child grid.

For example, Fig. 9.10 shows the sequence of hereditary grids generated in adaptive local grid refinement of the square shaft in torsion using the hyperbolic shape function rectangular element. Notice that the process has produced grids that favor finer grid intervals near the edge of the square. (This is where our knowledge of the exact solution indicates that the gradient of the warping function is highest.)

The starting grid involves only one finite element for the octant. This grid is the simplest that represents the original geometry adequately. For this problem, it represents the geometry with no error. Since the antisymmetric boundary conditions prescribe all the nodal warping variables to be zero, there are no active nodal variables for the starting grid. The estimate of the reduction in stiffness caused by warping is zero; hence the estimate of torsional stiffness is the polar moment of inertia of the square.

The second grid is created from the first by bisections. The parent grid analysis provides trivial results. Therefore, we have no criteria for regridding, other than the constraint of hereditary grids. We choose bisection in these cases because it is simplest, avoids extreme geometries, and tends to minimize the maximum-minimum grid interval.

Table 9.2 lists data of the second grid analysis. Since the resolution error indicates 0.9000 (90 percent) accuracy, we observe first that the results of analysis of the second grid do not meet our goal of less than 1 percent error in the FEA. We need a third grid. Since the optimality ratios are not equal (see the optimality variance and the data in the second column of Table 9.2), we will develop the third grid by local grid refinement of the second grid.

The column 3 data indicates that element b is acting only in a constant-strain state. This occurs because of the antisymmetric boundary conditions for element b. It will be the case for every element that straddles the diagonal of the square when we use the linear or hyperbolic displacement shape rectangular model. Since element c straddles the diagonal, it must be in a constant-strain state, but since antisymmetric constraints require that the warping be zero at each of its nodes, the constant must be zero.

The third column of Table 9.2 lists the locality residuals for the elements. This data recommends subdivision of element a but, because of the lack of non-constant-strain participation for elements that straddle the diagonal, provides no basis for choosing to subdivide the first or second internodal interval. This unusual case arises because the model involves so few elements. We can subdivide either the first nodal interval along the axis or the second. We choose the first, perversely.

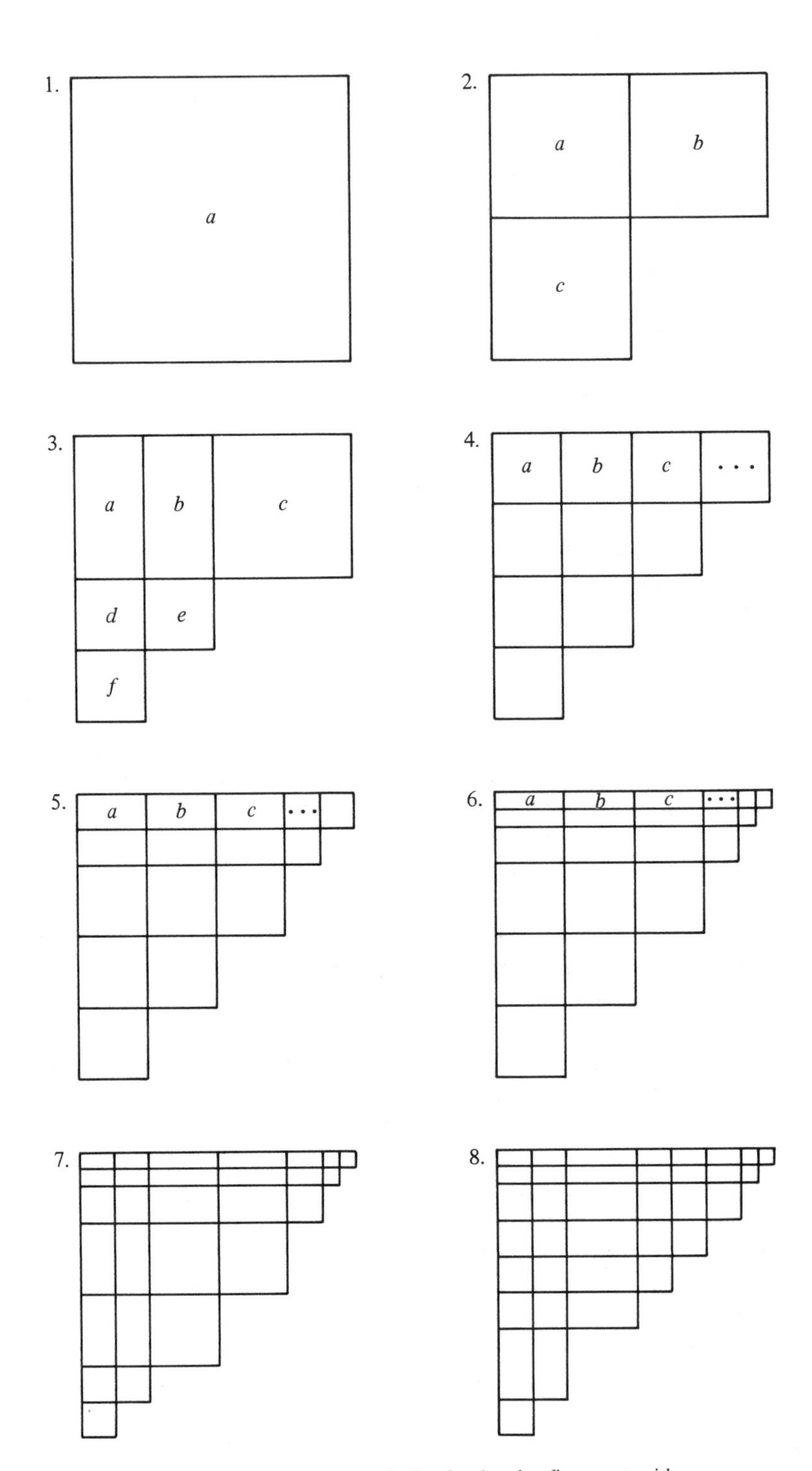

Figure 9.10 Sequence of adaptive local refinement grids.

TABLE 9.2 TORSION ANALYSIS RESULTS: THREE HYPERBOLIC SHAPE ELEMENTS

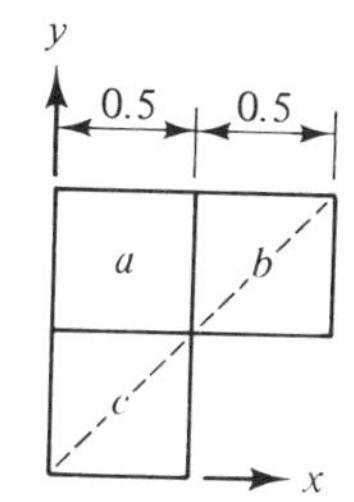

Strain energy = 0.03750
Resolution accuracy = 0.9000
Variance from optimum = 0.0622

Grid Status

Element	Constant SE/Total SE	Residual SE: Total − Constant
a	0.0095	0.7500
b	0	1.0000
c	0	0

Table 9.3 lists data that is more typical in adaptive local grid refinement of the square because there are more finite elements than associate with Table 9.2. We note that the resolution measure recommends a fourth grid and that the data in the third column of the grid status table indicates that the grid is still not optimal. To establish which x-axis nodal interval to bisect, we compare the sum of the locality measures for each of the elements in each candidate subdivision. Among candidate subdivisions, we select that one that has the maximum locality residual sum. As the data in the second column of the remesh assessment portion of Table 9.3 shows, bisection of the first interval along the x axis is recommended. Our perversity is confirmed.

Table 9.4 gives the accuracy measures for the first grid, which indicates an accuracy estimate of more than 99 percent. (The actual accuracy is 98 percent.) The column 2 data demonstrates the predominance of the constant-strain states in this acceptable model. The optimality variance is two orders of magnitude better than that of the Table 9.2 grid.

Figure 9.11 displays the estimated and actual accuracy of the torsional stiffness prediction for the square shaft as a function of the number of active generalized displacements. Active displacements are nodal variables evaluated by solving the stiffness equations. This data exhibits the improving reliability of the resolution measure as solution accuracy increases. Monotonic convergence of solution estimates is also evident.

To compare the convergence rate of the adaptive regridding with the rate for a sequence of optimum grids, we compare the slopes of the dashed curve of Fig. 9.11 with the slope of the lower envelope curve of Fig. 9.7. This comparison reveals that convergence of locally refined grids can be much more rapid than that of optimum grids. This superconvergence rate occurs because the adaptive process is improving the grid optimality while it is increasing the number of nodal variables. Furthermore, the comparison confirms that the sequence of adaptive local grid refinements tends to hover around the optimum grids envelope when many levels of refinement have taken place.

TABLE 9.3 TORSION ANALYSIS RESULTS: SIX HYPERBOLIC SHAPE ELEMENTS

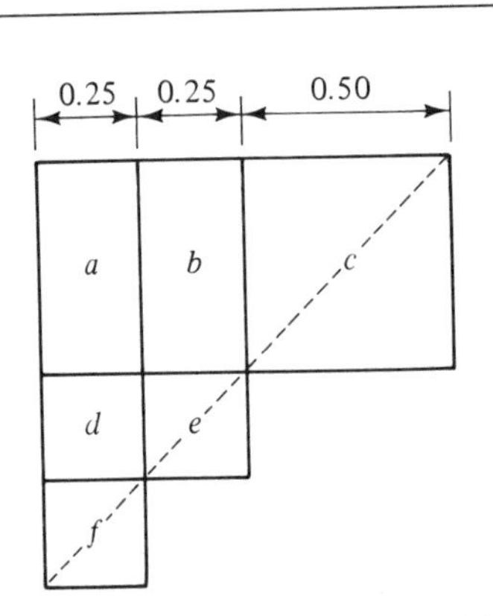

Strain energy = 0.03820
Resolution accuracy = 0.93800
Variance from optimum = 0.04360

Grid Status

Element	Constant *SE*/Total *SE*	Residual *SE*: Total − Constant
a	0.8140	0.001266
b	0.8797	0.001067
c	1.0000	0
d	0.7500	0.000020
e	1.0000	0
f	0	0

Remesh Assessment of Intervals for Grid Refinement

Interval	Elements to Cut	Sum of Residuals	Cut Priority
0–0.25	*a, d, f*	0.00128	2
0.25–0.50	*b, e, d*	0.00109	3
0.50–1.0	*c, b, a*	0.00233	1

Grid line coordinates for next grid = 0.25, 0.50, 0.75, 1.00

TABLE 9.4 DISCRETIZATION MEASURES: 36-ELEMENT GRID

Element Value	Constant *SE*/Total *SE*	Residual *SE*: Total − Constant
a	+9.862531E − 01	+1.142401E − 05
b	+9.877140E − 01	+1.135818E − 05
c	+9.712063E − 01	+7.529557E − 05
d	+9.951911E − 01	+1.050415E − 05
e	+9.970875E − 01	+9.740703E − 06
f	+9.985274E − 01	+7.962808E − 06
g	+9.997491E − 01	+1.013279E − 06
h	+1.000000E + 00	+0.000000E + 00
i	+9.840688E − 01	+9.062584E − 06
j	+9.861191E − 01	+8.888426E − 06
k	+9.702649E − 01	+5.545828E − 05
l	+9.958539E − 01	+6.776187E − 06
m	+9.980145E − 01	+5.092472E − 06
n	+9.994882E − 01	+2.148561E − 06
o	+1.000000E + 00	+0.000000E + 00
p	+9.688811E − 01	+1.957768E − 05
q	+9.740191E − 01	+1.870655E − 05
r	+9.650556E − 01	+7.626531E − 05
s	+9.953932E − 01	+9.513460E − 06
t	+9.987859E − 01	+4.085712E − 06
u	+1.000000E + 00	+0.000000E + 00
v	+9.569482E − 01	+1.044014E − 05
w	+9.671162E − 01	+9.482988E − 06
x	+9.683758E − 01	+3.039365E − 05
y	+9.983650E − 01	+1.753215E − 06
z	+1.000000E + 00	+0.000000E + 00
a'	+9.382120E − 01	+4.769245E − 06
b'	+9.590792E − 01	+3.984984E − 06
c'	+9.787383E − 01	+8.082978E − 06
d'	+1.000000E + 00	+0.000000E + 00
e'	+8.074737E − 01	+4.921223E − 06
f'	+8.978103E − 01	+3.028381E − 06
g'	+1.000000E + 00	+0.000000E + 00
h'	+7.500000E − 01	+4.927090E − 08
i'	+1.000000E + 00	+0.000000E + 00
j'	+0.000000E + 00	+0.000000E + 00

Total external work = 0.0513231
(Total constant *SE*)_(total strain energy) = 0.9918208
Estimated error in strain energy = 0.8179247
Variance from optimum grid = 7.390308E − 04

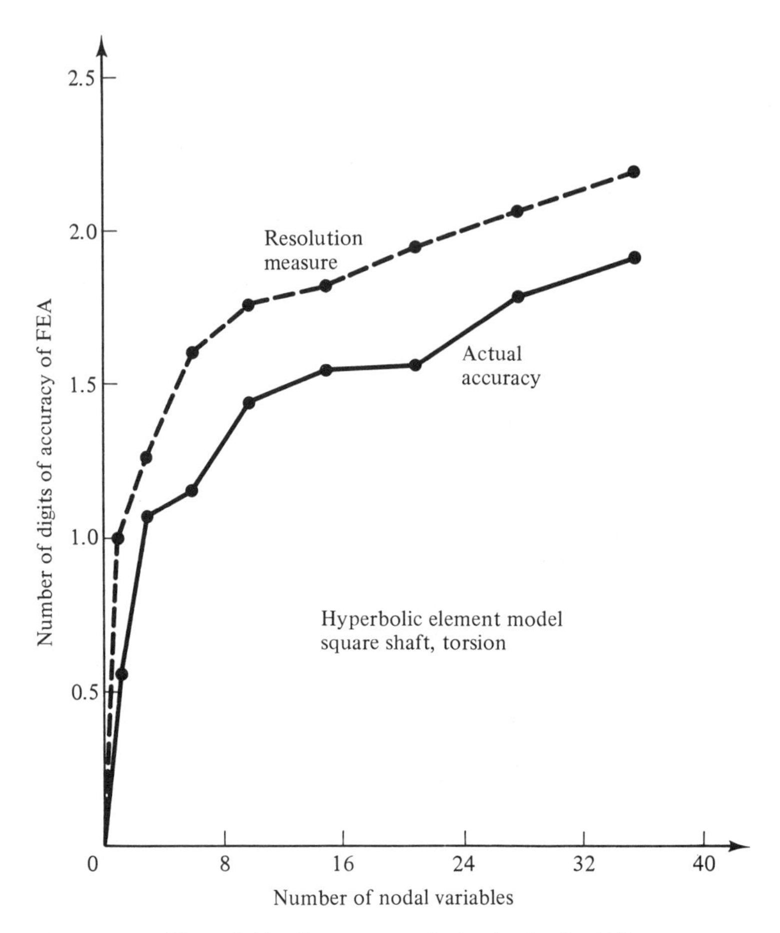

Figure 9.11 Convergency of adaptive local gridding.

9.7 ADAPTIVE POLYNOMIAL GLOBAL REFINEMENT

Polynomial refinement, p refinement, occurs when the order of the interpolating polynomial is increased while there is no change in the nodal lines. The number of nodes may increase, or the number of generalized displacements at a given node may increase. Though local refinement may be used with polynomial changes, we limit this discussion to the case where refinement occurs everywhere in the mesh as the element model is replaced by one with more polynomial terms.

Before proceeding with adaptive polynomial refinement, we wish to limit our choice of element models to those that guarantee monotonic convergence in the energy norm. We can prove monotonicity by the inclusion principle: We need only show, for potential energy models, that the higher-order model has the capability of behaving like the lower-order model for element geometries of interest.

TABLE 9.5 PROOF OF MONOTONIC CONVERGENCE: HYPERBOLIC TO SUPERQUADRATIC

Given:
$$\begin{aligned} 4f = & -(1+\zeta)(1+\eta)(1-\zeta-\eta)f_1 - (1-\zeta)(1+\eta)(1+\zeta-\eta)f_2 \\ & -(1-\zeta)(1-\eta)(1+\zeta+\eta)f_3 - (1+\zeta)(1-\eta)(1-\zeta+\eta)f_4 \\ & +2(1-\zeta^2)(1+\eta)f_5 + 2(1-\zeta)(1-\eta^2)f_6 \\ & +2(1-\zeta^2)(1-\eta)f_7 + 2(1+\zeta)(1-\eta^2)f_8 \end{aligned} \quad \text{(a)}$$

Substitute:
$$f_5 = \frac{f_1+f_2}{2};\quad f_6 = \frac{f_2+f_3}{2};\quad f_7 = \frac{f_3+f_4}{2};\quad f_8 = \frac{f_1+f_4}{2} \quad \text{(b)}$$

Yielding:
$$4f = (1+\zeta)(1+\eta)f_1 + (1-\zeta)(1+\eta)f_2 + (1-\zeta)(1-\eta)f_3 + (1+\zeta)(1-\eta)f_4 \quad \text{(c)}$$

But Eq. (c) is the same as Eq. (a) of Table 7.9. So the superquadratic interpolation function possesses the capability of reproducing the behavior of the bilinear element. Since both are potential energy elements, including the extra generalized displacements in superquadratic case guarantees response predictions cannot be worse than the hyperbolic, that is, convergence must be monotonic.

Table 9.5 illustrates the process for the superquadratic element and the hyperbolic model for rectangular geometries. Here we show that if each midside node implies a linear variation of displacements along its side, the interpolation function reduces to the hyperbolic interpolation function. Therefore, the superquadratic element model must have the same or more strain energy when it replaces the hyperbolic model.

Given the set of element models to be used, the pure strategy of adaptive polynomial global refinement involves the following steps:

1. Choose a starting mesh that will represent the geometry of the structural system as accurately as desired for the element models to be used.

2. Perform an FEA of the system using the starting mesh and the element model with the lowest-order polynomial. In the process of analyzing the structure, evaluate the discretization accuracy measures.

3. If the mesh requires refinement, redesign the mesh, replacing every element by the next-level element model.

4. Perform an FEA using the redesigned mesh.

5. Repeat steps 3 and 4 until the desired solution accuracy is attained.

Is the adaptive polynomial global refinement approach superior to the adaptive local refinement approach? Both approaches can attain any required level of discretization accuracy; both can be guaranteed to produce monotonically converging strain energy estimates. Thus the question is which of the strategies associates with least computer resources.

Babuska proves that when the grid is fine enough, the rate of convergence of polynomial remeshing is twice that of h grid refinement.[6] This useful piece of information is part of the story.

To illustrate the effectiveness of this strategy for finding the torsional stiffness of the square shaft, we choose the hyperbolic shape element and the fully integrated superquadratic element. We perform the following analyses:

1. FEA using the hyperbolic torsion element and the optimum grid spacing for the grid of Fig. 9.10, sketch 2.
2. FEA using the hyperbolic torsion element and the optimum grid spacing for the grid of Fig. 9.10, sketch 4.
3. FEA using the eight-node superquadratic torsion element and the grid of analysis 1.

Figure 9.12 provides a graphical display of the results of these three analyses. The data suggests that an improvement of more than a factor of 2 occurs in the convergence rate for this illustration. The dashed line connects a two-interval

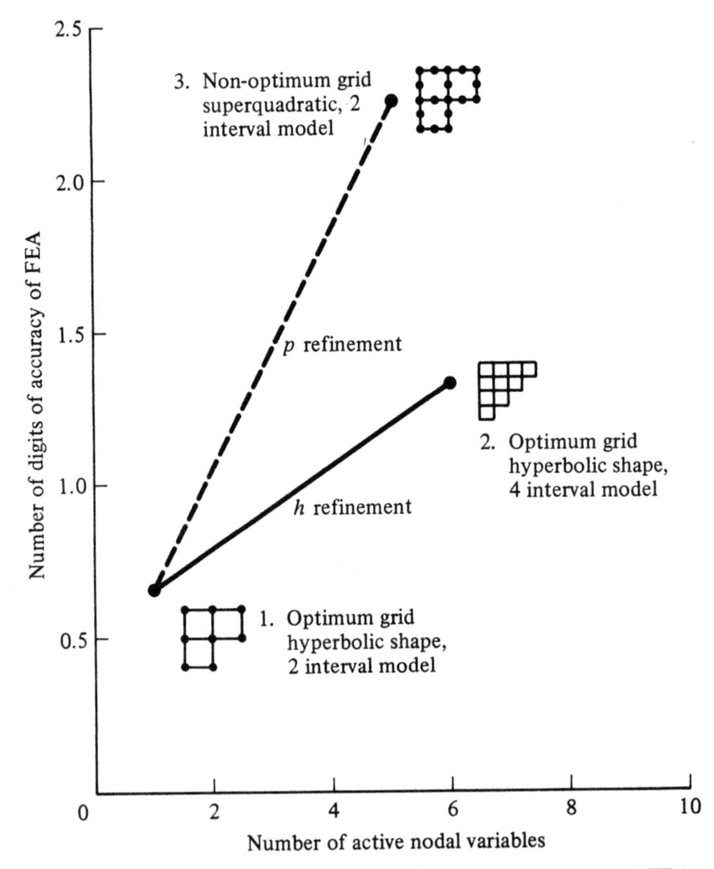

Figure 9.12 Interval versus polynomial refinement for square-shaft FEA.

hyperbolic element model solution to a superquadratic solution that uses the same grid. Since the polynomial convergence rate is retarded by moving from an optimum to a nonoptimum grid, the slope of the line is a pessimistic measure of the convergence rate for polynomial remeshing. The four-interval grid case requires about the same computer solution time as the two-interval superquadratic model, confirming the desirability of using the higher-order model.

These conclusions have been drawn with respect to ideal results. Much of that stated about erratic convergence as a function of grid design has its counterpart in polynomial refinement as a function of element selection. In particular, indiscriminate selection of the element types used in polynomial remeshing can yield a sequence of estimates with monotonically decreasing discretization accuracy even though all element models meet the minimum performance requirements of finite element models. Careful selection of element types and parent grids is the rest of the story if we aspire to exploit the potential improvements of efficiency of p refinement.

9.8 EXTRAPOLATION USING REGULAR GRID REFINEMENT

Extrapolation uses the results of two or more grids as the basis for estimating discretization accuracy, improving the accuracy of the solution or determining the grid needed to obtain results of the desired accuracy.

The general approach to extrapolation is to hypothesize the error function, fit the function to experimental results, and use the fitted function to evaluate unknowns. The key to accurate and efficient extrapolation is selection of the simplest relevant error function.

Let us assume that the discretization error is a continuous function. Without loss of generality, we expand the function in a power series in the grid interval

$$f(0) - f(h) = a_1 h + a_2 h^2 + a_3 h^3 + \ldots \tag{9.21}$$

where

$f(h)$ is the value of the unknown variable when the grid interval is h,
$f(0)$ is the exact value of the function,
h is the grid interval, and
$a_1, a_2, a_3, \ldots$ are the fitting constants.

Note that the constant term of the power series is omitted under the assumption that the error will be zero when the grid interval is zero. As a consequence, we can normalize the interval parameter on the first interval size. More important, we do not need to determine what the grid interval is when nodal spacing has more than one value, as it does for the grids of Fig. 9.13.

Suppose that we have j estimates of structural response. From Eq. (9.21) we can evaluate the a_i by solving the equations

$$\begin{pmatrix} f(h_1) - f(h_2) \\ f(h_2) - f(h_3) \\ f(h_3 - f(h_4) \\ \cdot \\ \cdot \\ \cdot \end{pmatrix} = \begin{bmatrix} h_1 - h_2 & h_1^2 - h_2^2 & h_1^3 - h_2^3 & \cdots \\ h_2 - h_3 & h_2^2 - h_3^2 & h_2^3 - h_3^3 & \cdots \\ h_3 - h_4 & h_3^2 - h_4^2 & h_3^3 - h_4^3 & \cdots \\ \cdot & \cdot & \cdot & \cdots \\ \cdot & \cdot & \cdot & \cdots \\ \cdot & \cdot & \cdot & \cdots \end{bmatrix} \begin{pmatrix} a_1 \\ a_2 \\ a_3 \\ \cdot \\ \cdot \\ \cdot \end{pmatrix} \tag{9.22}$$

where the h subscripts refer to data of experiment j. To eliminate the unknown $f(0)$ from the equations, we have subtracted pairs of equations of the form of Eq. (9.21). The equations require one more piece of experimental data than the number of terms in the power series, so we will have as many equations as unknown a_i.

Solving the linear simultaneous equations (9.22) for the a_i completes the fitting process. Substituting the a_i in Eq. (9.21) yields the estimate of the function when the interval vanishes, $f(0)$. Given $f(0)$, we can solve for $f(h)$ for any value of h of interest reusing Eq. (9.21).

The theory implies that a single-interval parameter characterizes the discretization error. To comply with this assumption, we limit extrapolation to regularly and globally refined grids. By regular grid refinement we mean that there exists a root grid. Each subsequent grid has the same number of equal subdivisions in each element in each dimension referenced to the root grid. Figure 9.13 displays regular subdivision of a root grid for the square shaft. Even though the root grid does not have a uniform grid interval, the subsequent grids meet the requirements of regularly refined grids.

Accurate extrapolation will require more terms in the series when estimates of $f(h)$ oscillate as h decreases. Therefore, we require that hereditary grids be used to support monotonic converging element models. Combined with the need for one parameter refinement, insistence on hereditary grids limits suitable rectangular grids to those illustrated in Fig. 9.13.

An attractive sequence of normalized sizes for extrapolation is 1, 0.5, 0.25, ... This choice minimizes computer resources needed to generate FEA solutions suitable for extrapolation. Even for this choice of intervals, the matrix on the right-hand side of Eqs. (9.22) becomes rapidly and progressively more ill conditioned. The following table shows how rapidly the condition number grows. This growth advises round-off error checks to ensure the accuracy of calculations for the extrapolation process.

Extrapolation Coefficient Matrix Condition Number

Number of fitting equations	2	3	4
Condition number	13.076	69.505	7672.2

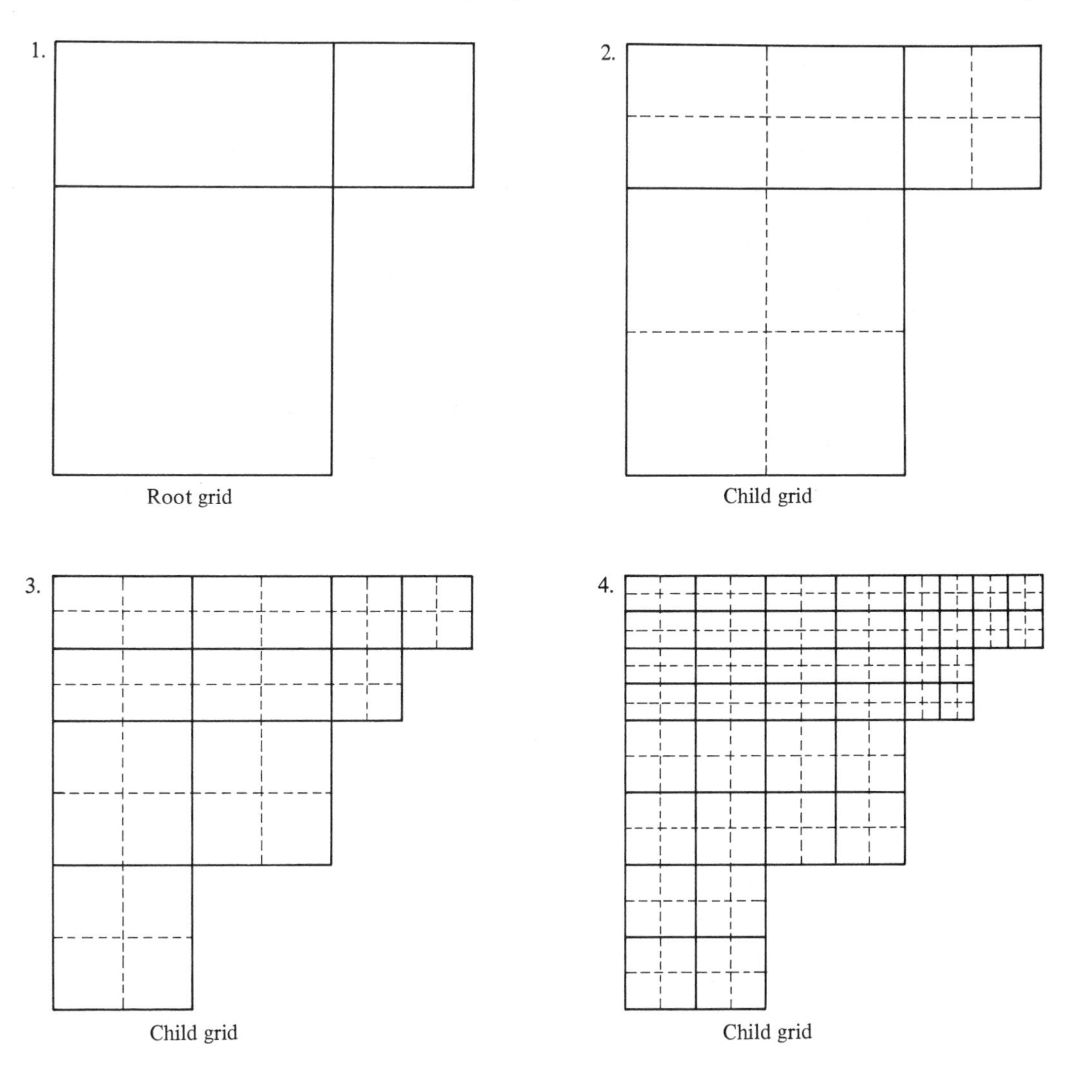

Figure 9.13 Sequence of regular grid subdivisions.

We also observe that the left-hand vector of Eq. (9.20) involves the differencing of results of successive finite element solutions. We conclude that if we want accurate extrapolations, we must be less tolerant of round-off error in the FEA results than otherwise.

Bolkir illustrates the importance of mainframe double-precision FEA and extrapolation calculations for accurate extrapolation.[7] His tests show results of single-precision analysis that are meaningless compared with double. This sensitivity may well explain the erratic success of extrapolation in previous finite element applications.[7]

Ultimately, the success of extrapolation rests on how many and which terms of the series expansion are used in fitting. Richardson observes that the use of the central difference formula often leads to an expression for the error function that commences with the h^2 term.[8] He recommends use of only the h^2 term or the h^2 and h^4 terms.

The popularity of h^2 fitting suggests its use when the actual error function is unknown. However, for finite element analysis, Bolkir recommends adaptive selection of the h terms of the series. The concept is that the error function that associates with any given finite element model can be deduced by pretesting of the model in problems with known solutions. Given the error function, the most efficient extrapolation basis is established.

As an illustration of the effectiveness of this strategy, the literature reports the success of deducing the error functions (h^2, h^4) and finding estimates of strain energy to 12 digits of accuracy for a beam with a triangular loading represented by lumped forces.[9] This type of success was reproduced for five different grids.

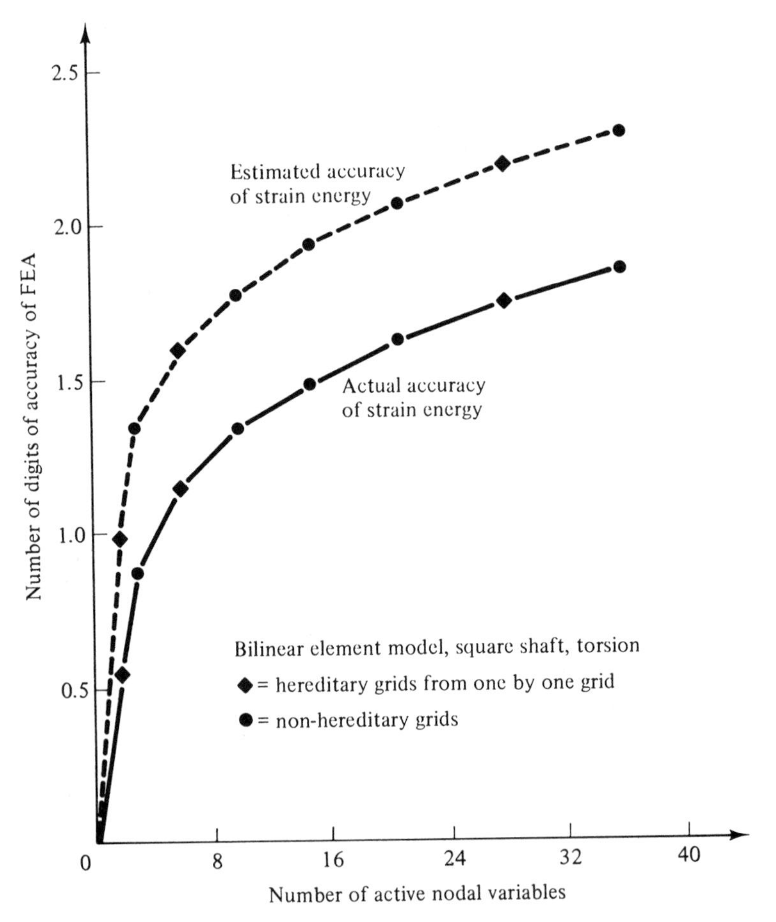

Figure 9.14 Extrapolation of finite element energy estimates.

Were the error function known *ab initio*, the 12-digit accuracy could be obtained by FEA of only three coarse (one-, two-, and four-element) grids.

Figure 9.14 summarizes extrapolation of FEA results for the torsion problem using the hyperbolic element model. This data relates to regular grids based on uniform intervals in the root grid. Note the monotonicity of the accuracy improvement using the regular grids compared with that of the adaptive local gridding of Fig. 9.11. Even results of the nonhereditary grids reflect this trend, though we cannot prove that they must.

The best estimate of the error function for the hyperbolic element considers both even and odd order power terms. This conclusion is based on the best error function for extrapolating to the exact solution of the square shaft torsion problem. We arrive at this conclusion using results of four hereditary-grid FEA solutions. We consider first all terms in the power series through h^4. This analysis suggests that all power series terms make a significant contribution to the extrapolated solution, $f(0)$. Elimination of odd or even terms or reduction of the number of terms degrades the prediction of the exact solution severely.

Using h, h^2, h^3 extrapolation yields an $f(0)$ prediction of the change in torsional stiffness from the nonwarping case with an accuracy of 99.7 percent. The estimated error function indicates the 99.0 percent level is attained with an 11-interval regular grid. Comparing the number of calculations for the 11-interval grid with the sum of the calculations for the $h = 1$, $h = 0.5$, $h = 0.25$, and $h = 0.125$ grids shows that extrapolation is much more efficient than FEA of the 11-interval grid.

9.9 DESIGNING AND REFINING MESHES

The key to self-qualification of FEA with respect to discretization accuracy is the appropriate design and redesign of meshes. The goal is to choose the mesh to adequately represent the original and deformed geometry of the structure. Since the original geometry is usually known, selection of a mesh to define this geometry accurately is primarily a problem of choosing element models and nodes to describe the boundaries of the system to the prescribed accuracy. The central problem reduces to selecting a mesh refinement strategy based on meaningful measures of discretization accuracy that leads to acceptably accurate representation of the deformed geometry.

A broad spectrum of refinement strategies are suitable. In refinement, subdivision of elements may range from one element to all the elements, and one or all the elements may be exchanged for elements of more polynomial terms. We can rely exclusively on discretization sensors from one analysis to establish discretization accuracy, rely exclusively on extrapolation, or involve some combination of sensor and extrapolation data.

In practice, we need to deal with accuracy of predictions of the generalized displacements and stresses as well as with strain energy. In principle, this is not a complication: The strain energy accuracy measures can accommodate these goals.

We need simply to increase our strain energy accuracy requirement to control the accuracy of displacement estimates and increase it further to obtain comparable accuracy in stress estimates.[10]

In practice, obtaining accurate stress predictions may require dealing with singular stresses where changes in geometry are abrupt, places where the use of fillets would be effective in reducing peak stress. Traditionally, singularities are accommodated by using singular-stress finite element models in the neighborhood of the stress risers. Then the treatment of discretization accuracy can be managed just as it is when the infinite stresses do not occur, as long as the system strain energy is finite.

Practical FEA usually engenders more than one loading condition. Direct extension of the ideas of this chapter is appropriate. We find the residuals for each loading condition on a given mesh model and give priority to elements with the highest residuals considering all the loading conditions in mesh redesign.

In practice, few FEA computer programs provide measures of discretization accuracy. Lacking these capabilities, analysts must resort to heroic actions to qualify their analyses. If they choose self-qualification, they can perform multiple analyses, determining the sensitivity of the analysis results to changes in mesh design, and then exercise judgment in interpreting these results with respect to structure.

The mesh redesign strategy is the fabric that holds analysis fidelity together. The ingredients of the strategy include the basis on which finite element model candidates are selected and the strategy of mesh redesign. Selecting elements for which monotonic convergence is assured and using them in a sequence of hereditary meshes guarantees that the strain energy predictions will converge monotonically to an answer. Ultimately, we must use global mesh refinement and the preferentiality capabilities of the elements as the woof and warp of the fabric that guarantees that convergence is to an adequately accurate solution of the relevant structural or elasticity equations.

9.10 HOMEWORK

***9:1.** Torsion FEA boundary conditions.

Given: The square shaft finite element model of Fig. 9.5 with $a = 0.25$, $b = 0.5$, and $c = 0.25$, modeled by simplex elements.

Find: Expressions for the boundary conditions in terms of nodal variables

(a) At node 2

(b) At node 8

9:2. Simplex stiffnes matrix in torsion.

Given: The simplex element matrix of Table 9.1.

Prove: That the stiffness matrix is indifferent to which diagonal divides a square element into two right-angle triangles.

***9:3.** Antisymmetric boundary conditions.

Given: The following stiffness matrix for element 3-4-8-7, which straddles the diagonal of the square shaft of Fig. 9.5.

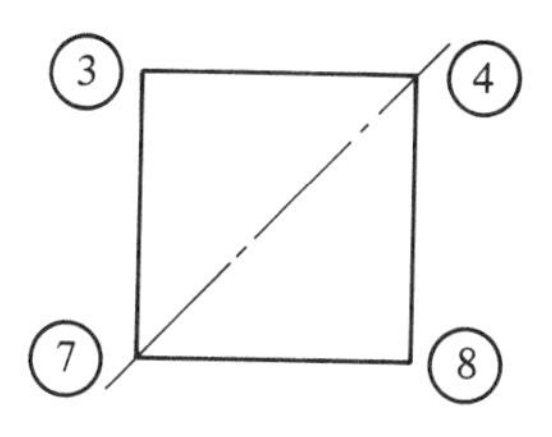

$$\begin{bmatrix} 1 & -0.5 & 0 & -0.5 \\ -0.5 & 1 & -0.5 & 0 \\ 0 & -0.5 & 1 & -0.5 \\ -0.5 & 0 & -0.5 & 1 \end{bmatrix} \begin{pmatrix} \phi_3 \\ \phi_4 \\ \phi_8 \\ \phi_7 \end{pmatrix} = \begin{pmatrix} p_3 \\ p_4 \\ p_8 \\ p_7 \end{pmatrix}$$

Homework Problem 9:3

Find: The element stiffness matrix that implies antisymmetric response, that is, the stiffness matrix which requires that $\phi_8 = -\phi_3$, $\phi_4 = \phi_5 = 0$.

9:4. Monotonicity of polynomial remeshing.

Given: The first and second interpolation functions of Table 7.9.

Determine: If replacing the stiffness matrix associated with the first model with one associated with the second model will result in a monotonic improvement in the estimate of strain energy.

Method: Use the inclusion principle.

9:5. Monotonicity of polynomial remeshing.

Given: The second and third interpolation functions of Table 7.9.

Determine: If replacing the stiffness matrix associated with the second model with one associated with the third model will result in a monotonic improvement in the estimate of strain energy.

9:6. Extrapolation of strain energy.

Given: Simplex model estimates of strain energy for each of the regular grid analyses listed.

Grid Interval	Strain Energy
0.500	0.03125000000000000
0.250	0.04607077205882353
0.125	0.05057623506118045

Find: The value of the strain energy when the interval size is zero using h^2, h^3 extrapolation.

9.11 COMPUTERWORK

9-1. Optimum simplex element grid.

Given: The square shaft of Fig. 9.3 to be modeled by simplex elements.

Do the following:

(a) Plot the strain energy versus x_1, the distance from the y axis to the first grid line for two-interval grids.

(b) Determine the x_1 that associates with the largest estimate of strain energy, to two digits of accuracy.

9-2. Relative accuracy of superquadratic and hyperbolic shape models.

Given: The optimum grid for the hyperbolic element four-interval grid has x_i values of 0, 0.35, 0.64, 0.85, 1.0.

Find: A two-interval superquadratic element grid and a four-interval hyperbolic element grid for which

(a) The hyperbolic element strain energy is more accurate than the superquadratic.

(b) The superquadratic element strain energy is more accurate than the hyperbolic.

9-3. Erratic sequence of strain energy estimates.

Given: The square shaft of Fig. 9.3 modeled by simplex elements in a uniform regular two-interval and a uniform regular three-interval grid.

Find:

(a) An irregular three-interval grid that predicts more strain energy than the regular two-interval grid analysis.

(b) An irregular three-interval grid that gives a more accurate prediction of strain energy than the regular grid.

***9-4.** Discretization accuracy sensors.

Given: The square shaft of Fig. 9.3 modeled by hyperbolic elements in a regular four-interval grid.

Find:

(a) The actual relative accuracy of the finite element prediction of the change in strain energy due to warping.

(b) The estimated relative accuracy as measured by the resolution sensor.

9-5. Adaptive local grid refinement.

Given: The square shaft of Fig. 9.3 modeled by hyperbolic elements and the grid of sketch 4 of Fig. 9.10.

Do the following:

(a) Perform the FEA of the grid.

(b) Determine the best interval to bisect using the element discretization residuals.

(c) Refine the grid by bisecting the best interval.

(d) Refine the grid by bisecting the second best interval.

(e) Perform the finite element analyses of the grids from tasks (c) and (d) and compare results.

9-6. Grid optimality.

Given: The square shaft of Fig. 9.3 modeled by a five-interval grid, $x_i = 0, 0.35, 0.54, 0.70, 0.86, 1.0$.

Do the following:

(a) Determine, by changing the x_i by ± 0.02, whether the grid may be optimum for the bilinear element model.

(b) Determine, by changing the x_i by ± 0.02, whether the grid may be optimum for the simplex element model.

(c) Is the optimum grid dependent on the element model being used, based on the results of these tests?

***9-7.** Interval versus polynomial refinement.

Given: The square shaft problem of Fig. 9.3.

Do the following:

(a) Find the strain energy of a hyperbolic element in a one-interval grid.

(b) Find the strain energy of a superquadratic element in a one-interval grid.

(c) Find the strain energy of a hyperbolic shape element model in the optimum two-interval grid ($x = 0, 0.67, 1.0$).

(d) Find the ratio between the convergence rate of the strain energy of the hyperbolic to superquadratic over the one- to two-interval grids for the hyperbolic element.

9-8. Extrapolation of FEA strain energy.

Given: The square shaft problem of Fig. 9.3.

Do the following:

(a) Obtain the strain energies for simplex element models for uniform grids with intervals of $h = 1, 0.25, 0.20, 0.125$.

(b) Find the best extrapolation function for strain energy using the $h = 0.1, 0.25, 0.125$ analysis results.

(c) Using the best extrapolation function, predict the strain energy for the $h = 0.2$ grid.

(d) Find the actual strain energy of the $h = 0.2$ grid and determine the number of digits of accuracy of the extrapolation.

***9-9.** Extrapolation of displacements.

Given: Results of FEA of torsion for the simplex element in the regular grids listed.

Results of Simplex FEA of Torsion Problem

GRID INTERVAL	STRAIN ENERGY	MIDPOINT WARPING
0.5	0.03125000000000000	−0.1250000000
0.25	0.04607077205882353	−0.1360294118
0.125	0.05057623506118045	−0.1383219066

Do the following:

(a) Find the better of h and h^2 extrapolation basis for the strain energy by comparing the h and h^2 extrapolated results with the exact value, 0.0521793, for an octant of the 2×2 shaft.

(b) Find the exact value of warping at the midpoint of the free edge of the square using your computer code.

(c) Find the better extrapolation basis for the midpoint warping of h and h^2 and compare with the exact warping value of task (b).

9-10. Extrapolation with all-regular grids.

Given: The square shaft strain energy results of Prob. 9-9.

Do the following:

(a) Generate the strain energies for grid intervals of 0.333333333333, 0.2, 0.166666666667, and 0.142857142857 using the simplex element.

(b) Find the best extrapolation function for strain energy using the given data and the results of task (a).

(c) Determine the grid size that will produce estimates of the strain energy with at least 99 percent accuracy.

9.12 REFERENCES

1. Melosh, R.J., "Basis for Derivation of Matrices for the Direct Stiffness Method," *AIAA Journal*, vol. 1, no. 7, July 1963, pp. 1631–1637.
2. Sokolnikoff, I.S., *Mathematical Theory of Elasticity*, New York: McGraw-Hill, 1956, pp. 128–133.
3. Carroll, W.E., "Recent Developments Dealing with Optimum Finite Element Analysis Techniques," in *Finite Element Grid Optimization*, Shephard, M.; Gallagher, R.; eds., New York: ASME, 1979.
4. Melosh, R.J.; Utku, S., "Principles for Design of Finite Element Meshes," in *State-of-the-Art Surveys on Finite Element Technology*, Noor, A.K.; Pilkey, W.D.; eds., New York: ASME, 1983, pp. 57–86.
5. Melosh, R.J.; Marcal, P.V., "An Energy Basis for Mesh Refinement of Structural Continua," *Int. J. for Num. Meth. in Engineering*, vol. 11, no. 7, 1977, pp. 1083–1092.
6. Babuska, I.; Szabo, B.A.; Katz, I.N., *The P-Version of the Finite Element Method*, Report WU/CON-79/1, Center for Computational Mechanics, Washington University, May 1979.
7. Bolkir, A.B., *Extrapolation in Finite Element Analysis*, Ph.D. thesis, Duke University, Apr. 1984.
8. Richardson, L.F.; Gaunt, J.A., "The Deferred Approach to the Limit," *Trans. Royal Society*, vol. 226, Series A, Apr. 1927, pp. 299–361.
9. Melosh, R.J.; Bolkir, A.B., "Self-qualification of Finite Element Analysis Results by Polynomial Extrapolation," *Finite Elements in Analysis and Design*, vol. 1, no. 1, Apr. 1985.
10. Melosh, R.J., "Computer Resources for FEA of Vehicle Crash," paper presented at the annual meeting of the Society of Automotive Engineers, Detroit, Apr. 1984.

Chapter 10

Analyzing Scelernomic Systems

A scelernomic structural system is one that can be analyzed exactly by a finite number of linear analysis steps. Usually, the number of analysis steps can be predetermined, thereby establishing the computer resources needed to support the analysis. In any case, the maximum number of steps can be determined.

Scelernomic structures represent an important class of systems. They include structures whose members fracture, yield, buckle, or slip in their connections. They encompass members that are prestressed or prestrained. In general, they involve structures with members modeled by load deflection curves that are folded lines.

Study of scelernomic structural analysis is technically appealing. This study examines modern modeling and analysis approximations and methods. It introduces some of the analysis concepts embedded in the nonlinear analysis without the need for extensive theoretical development and exhaustive calculations.

Scelernomic analyses can play two roles in structural engineering: a role in providing support for advanced integrity analysis of a structure and a role in assessing whether or not nonlinear simulation is needed. In the first role, low computer resource needs for scelernomic analysis, compared with nonlinear analysis, is the payoff for careful analyst attention to the restrictive approximations of a scelernomic analysis. In the second role, understanding of the approximations of the scelernomic analysis is the basis for establishing the types of nonlinearities that need representation in assuring structural integrity.

10.1 STRUCTURES WITH COUNTERS

A counter is a structural element that has no stiffness for at least one non-null vector of nodal loading.

Counters are often an idealization of a structurally deficient material. We assume that without prestress, concrete and soil have no resistance to tension. The sensitivity of glass to scratches suggests a model that resists only compression and shear. Water is usually assumed to have negligible shear resistance and is therefore a deficient structural material.

Counter elements are also a useful idealization when the geometry of an element results in negligible resistance to one or more loadings. When members are wires, chains, or ropes, their resistance to compressive loads is disregarded. Nonsymmetrical beams may be capable of carrying moment of one sign and virtually incapable of resisting moment of the opposite sign. An open tube may offer essentially no resistance to torsion compared with its closed-section neighbors. Thin panels may be prone to buckle easily in shear, requiring analysis of tension field resistance only. Sheet metal may be so thin that its bending resistance is ignorable.

10.1.1 Counter Models

Figure 10.1 shows the load deflection curve for the simplest counter, a tension-only spring. The elongation of this element increases linearly with the tensile force acting on it. It has no resistance to a compressive force.

The rod counter model is a natural extension of the concept of a spring counter. Figure 10.2 is the schematic of a structure whose cross braces are designed under the assumption that they can only resist tension; the members offer so little resistance to Euler buckling that they become inactive when compressed. The cross braces are rod tension counters.

A beam model has independent resistance to axial extension, torsion, and

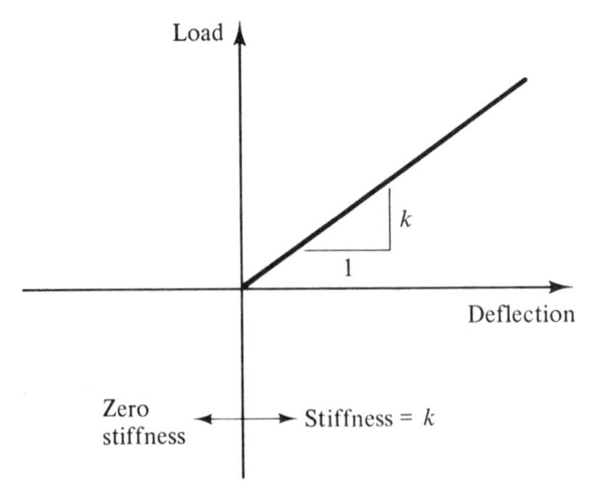

Figure 10.1 Tension-only spring load deflection.

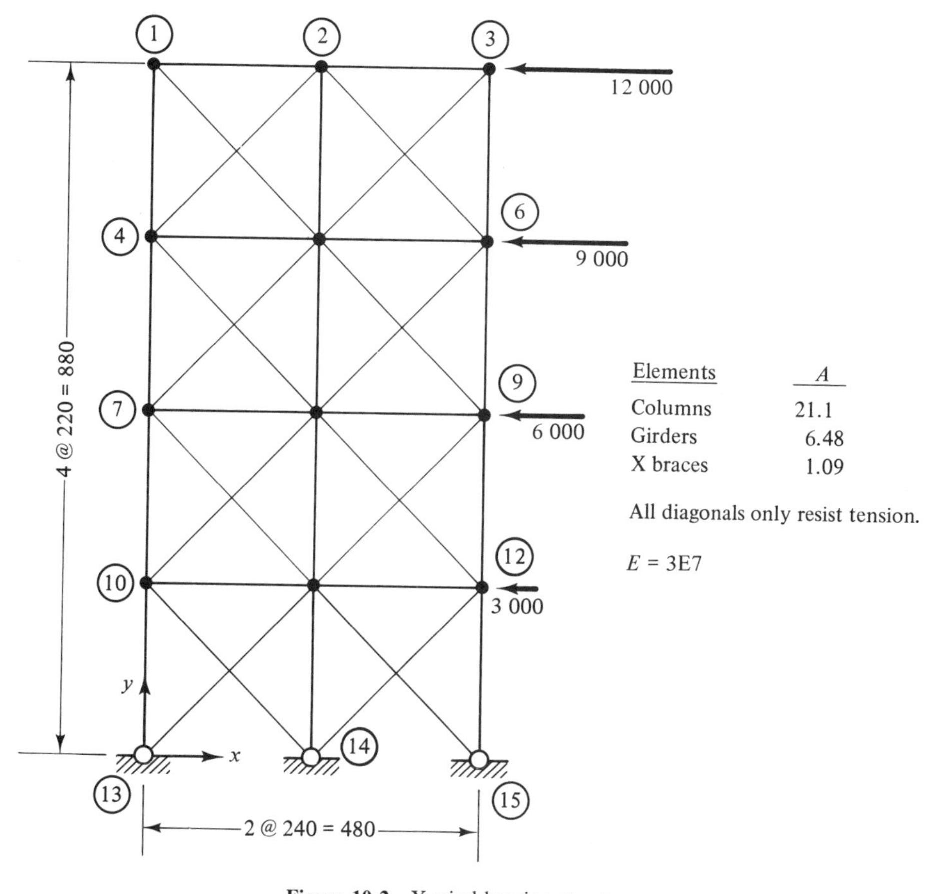

Figure 10.2 X wind-bracing structure.

bending about each of its principal axes. Thus its approximation as a counter must indicate whether axial tension and/or compression is resisted, whether or not twist resistance exists, and whether bending resistance is offered to positive and/or negative moments about each principal axis.

The extension of the concept of counter elements to continuum finite elements requires that we synthesize the model from its component resistances. Since this involves no new concepts, we leave the details to the computer program developer.

For the counter analysis, we assume that the structure behaves as a linear system. This assumption is consistent with the premise that the material is Hookean, strains squared are small compared with strains and negligible compared with 1, and member rotations are so small that the sine of the angle of rotation is sensibly equal to the angle and the cosine equals 1. These are the assumptions of linear structural analysis and hence common to most scelernomic analyses.

We assume that structural changes in geometry take place in pseudotime. This means that we are concerned with the sequence of structural configurations, not with their timing. We neglect inertia and damping effects of the equations of motion and hence only need to consider that the structure is initially unstressed and undeflected.

10.1.2 Analysis Method for Counters

We determine the behavior of the structure that has counters by the following steps.

1. Perform a linear structural analysis assuming all members capable of resisting all loadings.
2. Identify the next member to fail. This is the member that violates its counter load deflection relation by the largest margin. (For the X-bracing, this is the rod that has the largest compressive elongation.)
3. Reanalyze the structure, omitting the element stiffness components of all members previously designated as the next member to fail.
4. Repeat steps 2 and 3 until there are no new member components that fail or the system stiffness matrix becomes singular.

The analysis steps determine if the structure is kinematically stable for the given loading and, if so, what the member forces will be, considering the load deflection relations of the counters. The maximum number of analyses is the total number of counter deficiencies plus 1.

M. al-Mandil has proved seven theorems that bear on this solution method.[1] These ensure that the solution will be unique, that if there is a tie among the next members to fail, the solution will be independent of which is assumed to fail first and that the strain energy of the response will increase monotonically with the analysis number. The theory is predicated on the assumption that all analyses involve the total loading and hence any rod removed is never a candidate for reinstatement.

Table 10.1 lists the response predictions obtained by applying the analysis steps to the wind bracing of Fig. 10.2. Table 10.1 data includes member end force and elongation for the first, second, and last analysis steps. The last step furnishes the counter analysis solution: The forces are in equilibrium at each node, the load deflection relations are satisfied for each member, and members joined at nodes have common nodal displacements. These requirements provide the basis for assessing round-off, input, and computer programming errors.

The response data and theory provides a check on the analysis. The members removed are indicated to have zero force in steps 2 and 9 of Table 10.1. The fact that each of these members, once removed, experiences no negative elongation in the final analysis is evident from the data in the second column of step 2 of Table 10.1. Similar output from each analysis confirms that once removed, an inactive counter would only experience compression in subsequent analysis steps.

TABLE 10.1 COUNTER ANALYSIS RESULTS

Step 1			Step 2		
Analysis Results: Element Axial Responses			Analysis Results: Element Axial Responses		
Bar	Bar Force	Elongation	Bar	Bar Force	Elongation
1	−2.552635+03	−8.87170E−04	1	−2.63490E+03	−9.15762E−04
2	−1.00438E+04	−3.49075E−03	2	−1.03126E+04	−3.58414E−03
3	−2.09529E+04	−7.28221E−03	3	−2.13556E+04	−7.42216E−03
4	−3.38790E+04	−1.17747E−02	4	−3.68094E+04	−1.27932E−02
5	+5.08099E+02	+1.76591E−04	5	+6.82282E+02	+2.37128E−04
6	+2.04873E+02	+7.12039E−05	6	+7.33388E+02	+2.54890E−04
7	−6.28326E+01	−2.18375E−05	7	+8.64445E+02	+3.00439E−04
8	−3.50422E+02	−1.21790E−04	8	−8.72240E+03	−3.03148E−03
9	+3.20626E+03	+1.11434E−03	9	+3.12386E+03	+1.08570E−03
10	+1.06410E+04	+3.69830E−03	10	+1.03769E+04	+3.60652E−03
11	+2.14984E+04	+7.47180E−03	11	+2.09481E+04	+7.28055E−03
12	+3.43949E+04	+1.19540E−02	12	+3.61718E+04	+1.25715E−02
13	−2.78470E+03	−3.43791E−03	13	−2.87443E+03	−3.54868E−03
14	−8.50224E+03	−1.04966E−02	14	−8.59216E+03	−1.06076E−02
15	−1.90046E+03	−2.34625E−03	15	−1.91364E+03	−2.36252E−03
16	−6.05264E+03	−7.47240E−03	16	−6.06099E+03	−7.48271E−03
17	−1.08388E+03	−1.33812E−03	17	−1.02821E+03	−1.26940E−03
18	−4.20797E+03	−5.19502E−03	18	−4.30829E+03	−5.31888E−03
19	−2.11236E+02	−2.60785E−04	19	−2.58423E+03	−3.19041E−03
20	−2.41982E+03	−2.98743E−03	20	+1.81303E+02	+2.23830E−04
21	+3.77762E+03	+3.76117E−02	21	+3.89937E+03	+3.88239E−02
22	−4.25409E+03	−4.23557E−02	22	−4.38308E+03	−4.36400E−02
23	+3.50216E+03	+3.48691E−02	23	+3.37338E+03	+3.35869E−02
24	−4.74493E+03	−4.72427E−02	24	−4.62297E+03	−4.60284E−02
25	+6.83210E+03	+6.80236E−02	25	+6.97902E+03	+6.94864E−02
26	−7.33696E+03	−7.30502E−02	26	−7.47387E+03	−7.44133E−02
27	+6.81840E+03	+6.78871E−02	27	+6.67463E+03	+6.64558E−02
28	−7.50044E+03	−7.46778E−02	28	−7.36037E+03	−7.32833E−02
29	+8.80728E+03	+8.76894E−02	29	+8.86864E+03	+8.83003E−02
30	−9.42137E+03	−9.38035E−02	30	−9.68217E+03	−9.64001E−02
31	+9.14921E+03	+9.10938E−02	31	+9.10687E+03	+9.06722E−02
32	−9.24941E+03	−9.20914E−02	32	−8.96959E+03	−8.93054E−02

(Continued)

TABLE 10.1 (Continued)

Step 1			Step 2		
Analysis Results: Element Axial Responses			Analysis Results: Element Axial Responses		
Bar	Bar Force	Elongation	Bar	Bar Force	Elongation
33	+9.70793E+03	+9.66567E-02	33	+1.31878E+04	+1.31304E-01
34	-1.05346E+04	-1.04888E-01	34	+0.00000E+00	-1.44349E-01
35	+1.05181E+04	+1.04723E-01	35	+1.40866E+04	+1.40252E-01
36	-9.93625E+03	-9.89300E-02	36	-1.34225E+04	-1.33641E-01
Total strain energy = 5772.053			Total strain energy = 6532.381		

Step 9

Analysis Results: Nodal Displacements

No. of analysis steps = 9

Node	Direction 1	Direction 2
1	-9.43431E-01	-3.17501E-02
2	-9.51436E-01	-1.29890E-02
3	-9.66251E-01	+1.60662E-02
4	-7.92499E-01	-2.96843E-02
5	-8.05816E-01	-1.12317E-02
6	-8.23737E-01	+1.60662E-02
7	-5.58526E-01	-2.41821E-02
8	-5.75249E-01	-8.28631E-03
9	-5.95265E-01	+1.43089E-02
10	-2.74671E-01	-1.43645E-02
11	-2.92291E-01	-4.49081E-03
12	-3.12605E-01	+9.29770E-03
13	+0.00000E+00	+0.00000E+00
14	+0.00000E+00	+0.00000E+00
15	+0.00000E+00	+0.00000E+00

Analysis Results: Element Axial Responses

Bar	Bar Force*	Elongation
1	-5.94378E+03	-2.06577E-03
2	-1.58313E+04	-5.50220E-03
3	-2.82478E+04	-9.81757E-03
4	-4.13306E+04	-1.43645E-02
5	-5.05621E+03	-1.75729E-03
6	-8.47479E+03	-2.94543E-03
7	-1.09207E+04	-3.79550E-03
8	-1.29213E+04	-4.49081E-03
9	+0.00000E+00	+0.00000E+00
10	+5.05621E+03	+1.75729E-03
11	+1.44186E+04	+5.01120E-03
12	+2.67520E+04	+9.29770E-03
13	-6.48412E+03	-8.00508E-03
14	-1.20000E+04	-1.48148E-02
15	-1.07865E+04	-1.33166E-02
16	-1.45159E+04	-1.79209E-02
17	-1.35453E+04	-1.67226E-02
18	-1.62134E+04	-2.00166E-02
19	-1.42721E+04	-1.76199E-02
20	-1.64546E+04	-2.03144E-02
21	+8.79617E+03	+8.75788E-02
22	+0.00000E+00	-1.05879E-01
23	+7.48265E+03	+7.45007E-02
24	+0.00000E+00	-9.98194E-02
25	+1.46325E+04	+1.45688E-01
26	+0.00000E+00	-1.73540E-01
27	+1.38553E+04	+1.37950E-01
28	+0.00000E+00	-1.66718E-01
29	+1.83751E+04	+1.82951E-01
30	+0.00000E+00	-2.17465E-01
31	+1.82522E+04	+1.81729E-01
32	+0.00000E+00	-2.10636E-01
33	+1.93611E+04	-1.92768E-01
34	+0.00000E+00	-2.18493E-01
35	+2.13353E+04	+2.12429E-01
36	+0.00000E+00	-2.24155E-01

Total strain energy = 11758.98

*A positive axial bar force is tension.

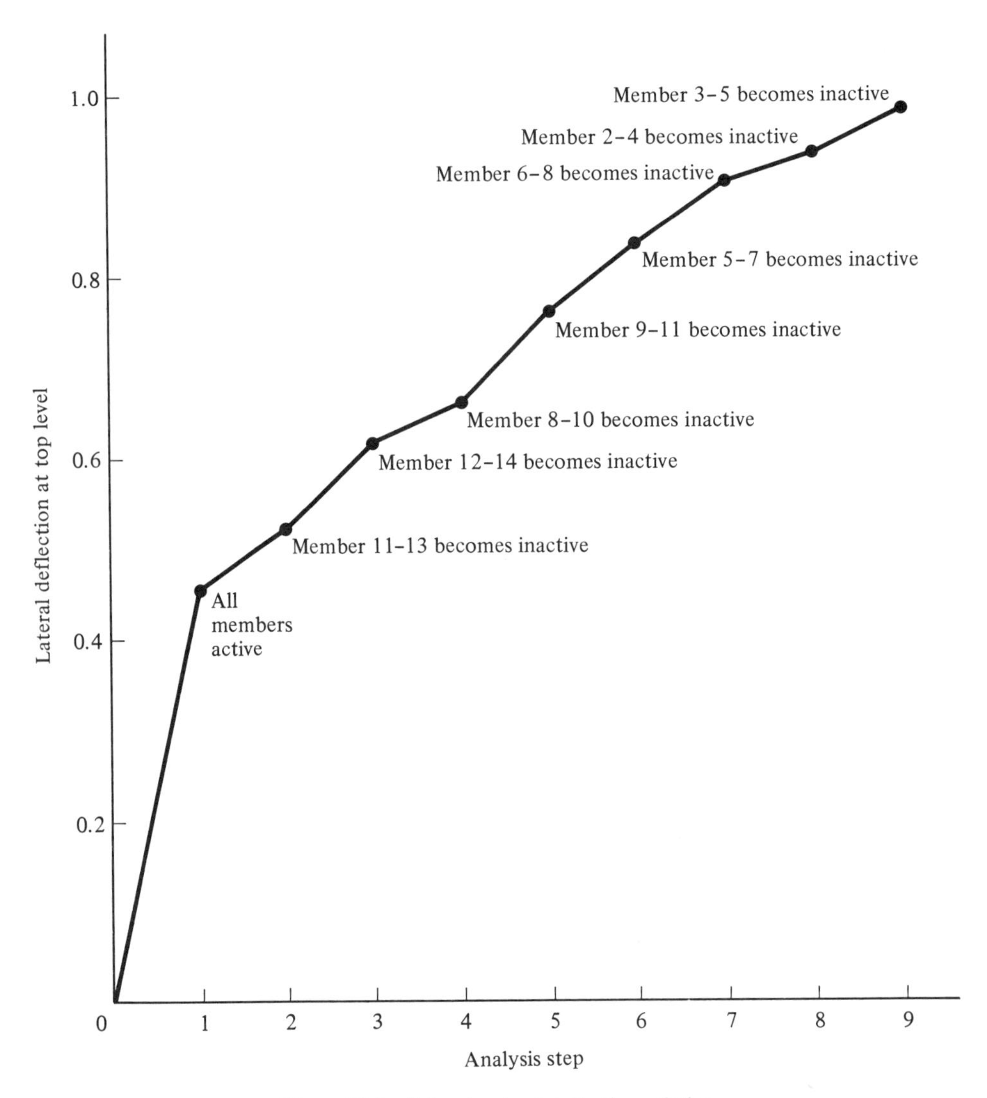

Figure 10.3 X-bracing load deflection relation.

Figure 10.3 displays the analysis step deflection-relation for the X-bracing. The step results are connected by straight lines consistent with the assumptions of scelernomic analysis. The ninth step terminates the analysis because, as the data in Table 10.1 indicates, all the load deflection relations for the counters are satisfied; that is, all the counters that remain active are in tension. The fact that the strain energy increases monotonically with increases in the analysis number is required by the theory. It provides the basis for another check on analysis results.

10.2 LIMIT ANALYSIS

Limit analysis provides an efficient means for estimating the nonlinear behavior of ductile structures.[2] It is usually used to determine the load at which a given structure will collapse. It is also used to determine the sequence in which members go plastic and the deflections that associate with irreversible slip strain in the material.

10.2.1 Limit Analysis Models

The dotted curve of Fig. 10.4 is typical of the engineering stress-strain data available for uniaxial tension tests of steel components. The bilinear curve (shown as a continuous folded line on the plot) is the model used in limit analysis. The left leg of this curve, whose slope quantifies Young's modulus, defines the elastic portion. The horizontal leg defines the plastic. To honor the zero slope, this leg is called perfectly plastic. The dashed curve represents the results of physical experiments.

The elastic perfectly plastic model can be a good representation of curves for construction steels for strains less than about 0.3. It represents the material well up to the proportional limit, approximates the curve continuation to the yield stress by linear extrapolation, and idealizes the postyield behavior with an error increasing as the stress approaches fracture stress.

The model curve is characterized by two parameters, the yield stress and the

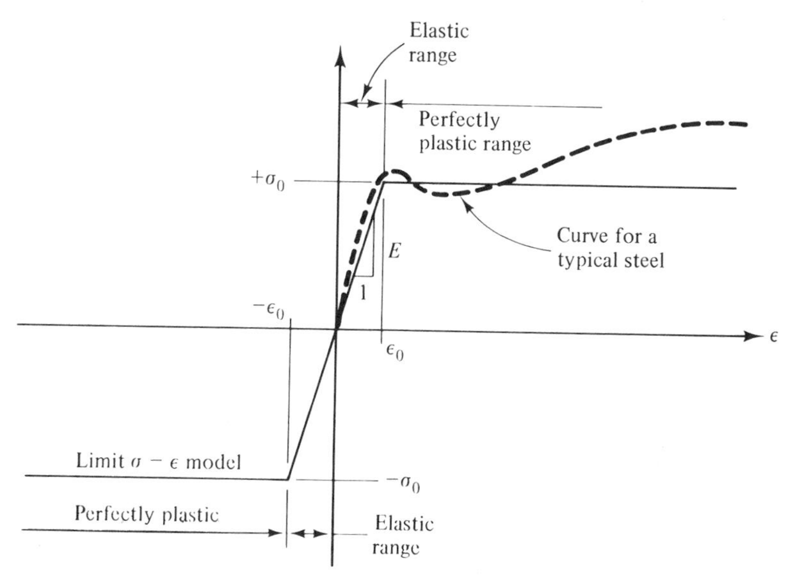

Figure 10.4 Uniaxial stress-strain curves.

yield strain. The elastic leg always passes through the origin and extends to the yield point, where it joins the plastic leg. The plastic leg is always horizontal and extends to infinity. The compression branch of the material model is assumed to be a mirror image of the tension branch.

The use of the bilinear model can be interpreted to mean that the material behaves elastically over the entire strain range as long as there is no unloading. Thus the material can be represented by strain energy and the stress-strain relations can be dealt with as piecewise linear. In summary, despite the fact that yielding implies macroscopic flow of the material, the material model does not depend on time explicitly: It is modeled in pseudotime. Since the material model does not represent unloading, we must reject the results of limit analysis that involve unloading at any point in the structure during any analysis loading step.

One common loading that involves no local unloading is proportional loading. In this case, the load increment for a step of the loading sequence is expressible by

$$\mathbf{P}_{s+1} = \mathbf{p}_s + \lambda_s \, \mathbf{p}_s; \qquad s = 1, 2, \ldots, n_s \geq 0 \tag{10.1}$$

where

$\mathbf{p}_{s+1}$ is the total loading at the end of loading step $s + 1$

$\mathbf{p}_s$ is the loading at the end of step s

$\mathbf{p}$ is the applied loading,

λ_s is the positive scalar load multiplier for step s, and

n_s is the number of loading steps.

Besides the loading defined by Eqs. (10.1), there are many other loadings that will not involve unloading at a point. Since the proportional loading is the most common of such loading sequences, the term proportional loading is applied to the set of all such possible loadings.

Figure 10.5 illustrates the load deflection relation for the limit analysis of a rod. The rod is straight, of uniform cross-sectional area along its length, and loaded with equal and opposite collinear forces at its ends. Since every point of the rod undergoes the same stress and strain for any given value of the end load, the load deflection curve of the member echoes the limit analysis material stress-strain model.

Figure 10.6 illustrates the load deflection relation for limit analysis of a beam. The beam has a straight neutral axis and an unchanging cross-sectional geometry along its length. Since its end loads must satisfy two equilibrium equations, the collection of shear and moment forces involves only two independent variables.

We define a limit beam model as a member that can yield only at a nodal point. This assumption does not limit the number of yield points in a frame structure; it only determines how many beam elements we must use to model a given beam segment as a lattice subsystem. Rather than modify the definition of

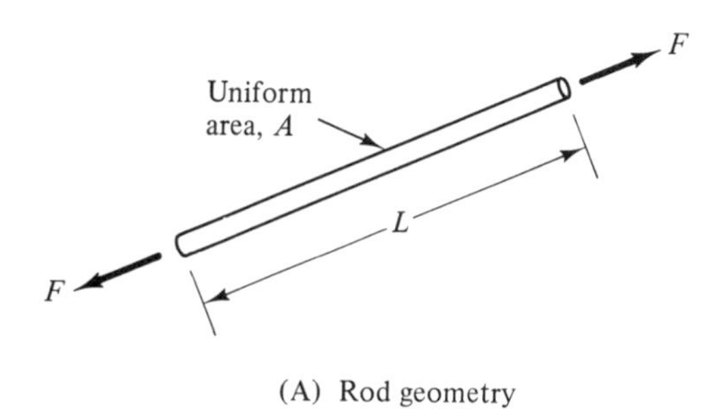

(A) Rod geometry

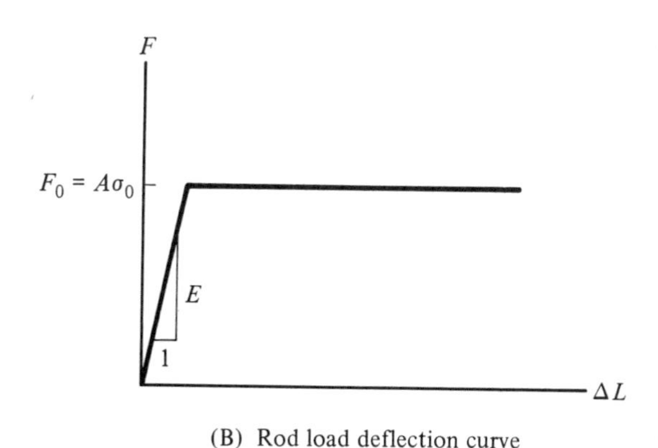

(B) Rod load deflection curve

Figure 10.5 Rod member limit model.

equivalent load, we chose to require that we model our frame with a node at every point where the bending moment may be a local maximum.

The yielding at an end is ignored until the bending moment reaches the plastic moment capability M_0 of the cross section. This is the bending moment the cross section can bear under under the assumption that every fiber over the cross section is either at the tensile or compressive yield stress of the material. The model thereby idealizes the plastic behavior as if a hinge forms in the beam. The model substitutes the hinge model for the gradual development of yield stress across the section and over the span, depicted in Fig. 10.6(B). This gradual development would result in the dotted load deflection curve of Fig. 10.6(B) rather than the folded line of the hinge model.

Table 10.2 shows the usual method of calculating the plastic moment capability as applied to a beam of triangular cross section. This calculation is based on the following assumptions:

1. Only bending stresses are significant; that is, the mean normal stress on the cross section is zero, and shear stresses do not effect yielding.
2. The stress at every point in the cross section has a magnitude of σ_0, the yield stress of the material.
3. The bending neutral stress line is the line in the cross section, parallel to the

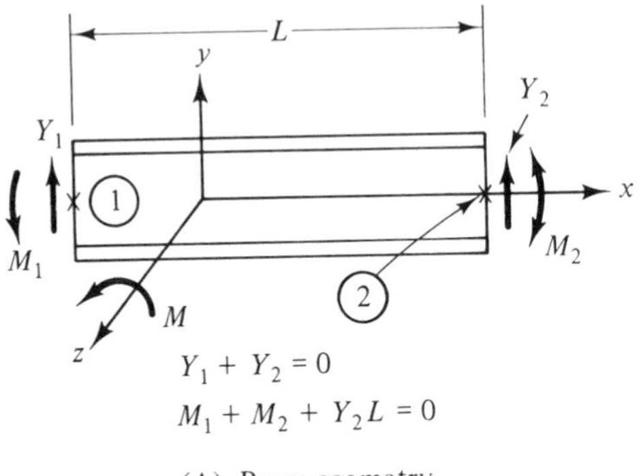

(A) Beam geometry

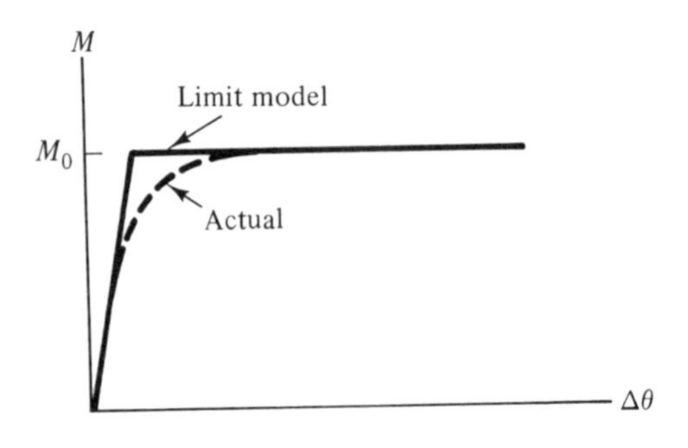

(B) Beam load deflection curve

Figure 10.6 Beam member limit model.

vector of the section bending moment, that divides the section so that all tensile yield stress on the area on one side of the line is in equilibrium with all compression yield stress on the other side.

4. The plastic moment capability is the moment of the yield stresses about the bending neutral stress line.

In summary, the calculation of plastic moment capability requires locating the bending stress line, the line that bisects the cross-sectional area, and calculating the moment of the yield stresses about that line.

If shear stresses were considered in calculating the yield moment, the moment capability would be reduced. Neglect of shear stresses is acceptable as long as the depth of the beam is a small fraction, say, < 0.1, of the beam span.

If the nonzero axial force is considered, the plastic moment capability may be greater or less than its value for the zero-force case. Regardless of the change, the plastic moment capability becomes a function of the magnitude of the axial force. The moment is readily calculated once the force is known.

Continuum models for limit analysis[3] are complicated by the use of approx-

TABLE 10.2 ANALYSIS OF PLASTIC MOMENT CAPABILITY

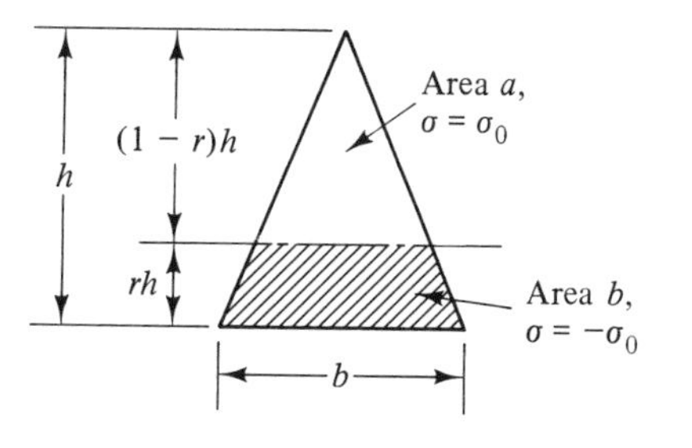

$$\text{area } a = \frac{1}{2}(\text{area } a + \text{area } b)$$

$$\text{so } \frac{1}{2}(1 - r)^2 bh = \frac{1}{2}\left(\frac{1}{2}bh\right)$$

$$\therefore \quad r = 1 - \frac{1}{\sqrt{2}}$$

$$M_0 = \text{moment (area } 1\sigma) + \text{moment (area } 2\sigma)$$

$$M_0 = bh^2\sigma_0 \left\{\frac{1}{6}(1 - r)^3 + \left[\frac{1}{2}(1 - r)r^2 + \frac{2}{3}r^3\right]\right\}$$

imate behavior states and the need to consider functions of up to three parameters defining the yield onset.

To examine the continuum case in more detail, consider the simplex element model of the membrane under plane stress. Here displacements are taken to be

$$\begin{aligned} u &= a_0 + a_1 x + a_2 y \\ v &= b_0 + b_1 x + b_2 y \end{aligned} \tag{10.2}$$

where a_i and $b_i i = 0, 1, 2$ are constants. Setting the displacements and coordinates to nodal values and inverting the relationship, the displacements may be expressed as a linear function of the nodal displacements. It follows from the definition of strains, Eqs. (7.1), that the membrane strains are constants over the element volume. We can evaluate the stresses using these strains and the stress-strain equations of the material.

Using the Von Mises yield criterion, the onset of yielding at a point is defined as occurring when

$$(\sigma_{xx} - \sigma_{yy})^2 + 6\sigma_{xy}^2 = 2\sigma_\sigma^2 \tag{10.3}$$

where σ_0 is the value of the yield stress of a coupon in uniaxial tension. The insistence on proportional loading is that the left-hand side of this equation never be greater than the right and at no stage of the loading sequence can it be less, once the equality is satisfied.

When yield occurs in the simplex membrane, it occurs simultaneously throughout the element. If we use higher than first-order displacement functions in the spatial coordinates, yield need not occur at all points in the element at the same loading. (This fact is dealt with approximately in the numerically integrated element models by considering the stress states at each Gauss point independently.)

Other continuum element models will evoke different yield criteria. A plate model uses a yield function expressed in terms of moments and twists; a shell model must admit the interaction of normal forces, bending moments, and shears. A model of a three-dimensional solid will use the three-dimensional form of a yield criterion.

Just as in the analysis of counters, limit analysis assumes that linear structural analysis assumptions prevail. This is necessary for scelernomic analysis. An important consequence is that up to the loading at which the structure is doomed to collapse, superposition of stresses and deflections is valid as an analysis tool.

10.2.2 Incremental Load Analysis

We observe that all limit analysis element models involve piecewise linear behavior. Either an element is behaving in the linear elastic range, or it behaves in the plastic range where its stiffness (or Gauss point stiffness) is zero. This fact invites incremental scelernomic analysis.

The concept of incremental load analysis is to deal with the finite difference representation of the stiffness equations

$$\mathbf{K}_t \, \Delta\mathbf{g} = \Delta\mathbf{p} \tag{10.4}$$

where

$\mathbf{K}_t$ is the tangent stiffness matrix,
$\Delta\mathbf{g}$ is the vector of displacement changes, and
$\Delta\mathbf{p}$ is the change of external loads.

Accordingly, the loading and responses are applied in a series of steps, calculating the tangent stiffness and applying the finite element process at the beginning of each step and adding step responses:

$$\mathbf{p} = \Sigma\Delta\mathbf{p}_s; \qquad \mathbf{g} = \Sigma\Delta\mathbf{g}_s; \qquad s = 1, 2, \ldots, n_s \tag{10.5}$$

For lattice system linear analysis, the incremental load analysis process arrives at the collapse load by the following steps:

1. Determine the incremental deflections of the structure by a linear analysis. (In this analysis, each element is given the stiffness corresponding to the current estimate of stresses in the element. If the stresses are below the yield strength of the element, the usual stiffness is selected. If the stresses are at the yield strength, the component is assumed yielded, and the plastic stiffness is selected. As a convenience, the incremental load may be taken to be the total load.)

2. Determine the minimum positive scalar multiplier of the loading λ_s such that the load increment will induce yielding at one new location. (This can be found by determining the multipliers each of which induces yielding at a different location and selecting the smallest positive multiplier of this set of multipliers.)

3. Add the incremental load scalar, displacements, and stresses to their previously accumulated values.

4. Repeat tasks 1–3 until the stiffness matrix becomes singular. Then the load scalar is the load multiplier at which collapse occurs and accumulated stresses and displacements define the state of the structure at the onset of kinematic collapse.

In applying this process, the most frequent mistake is neglecting the sum of the previous increments in finding the incremental load multiplier.

For the rod (or torque tube), the plastic stiffness is zero and the element's stiffness matrix is null. For the bending component of the beam, the plastic stiffness is represented by implying a hinge at the end of the element that is at the limiting yield strength.

Table 10.3 illustrates the analysis procedure in application to a four-bar truss. Note that once the yield strength of an element is attained, subsequent steps cannot change its stress.

Figure 10.7 exhibits the load deflection relation for the truss. The loading scalar necessarily increases monotonically with each analysis step until the structure becomes kinematically unstable. The small sketches indicate the structures analyzed for each load increment.

If we study lattice structures, we conclude that collapse occurs when the yielding is active at $R + 1$ locations, where R is the number of force redundants in the system. Accordingly, there will be at most $R + 1$ linear analyses using the analysis process just described. Fewer than $R + 1$ are needed when the yield strength is attained at two or more locations for the same loading or when yielding divides the structure into distinct substructures.

The interpretion of analysis results is enriched by the lower-bound theorem.[2] This theorem states that the load, in equilibrium with internal forces, which nowhere implies exceeding the yield strength, must be less than or equal to the collapse load of the structure.

In accordance with this theorem, the result of each cycle of tasks 1–3 of the analysis process yields a lower-bound (hence conservative) estimate of the collapse load. Furthermore, since the modeling involves lattice models, if we persist in cycling until collapse is indicated, we will arrive at the greatest lower-bound estimate—the exact solution.

The theorem also applies when limit continuum models are used, but only when these are equilibrium models. For most continuum elements, limit analysis provides an estimate of the collapse load that becomes increasingly more accurate, within the restrictions of the limit idealization of the material model, as the number of nodal displacement variables increases.

In general, limit analysis furnishes estimates of the structural collapse load in a finite number of analysis cycles. It provides an efficient means of determining the importance of inelastic material effects. Traditionally, it provides a rational basis for facile assessment of structural integrity based on system plastic collapse.

TABLE 10.3 SEQUENTIAL COLLAPSE LOAD ANALYSIS

Member	A	L	$\frac{AE}{L^*}$	λ	μ	$\frac{AE}{L}\lambda^2$	$\frac{AE}{L}\mu$	$\frac{AE}{L}\lambda\mu$
1	1	30	0.0333	−1	0	0.0333	0	0
2	2	50	0.0400	−0.6	0.8	0.0144	0.0256	−0.0192
3	3	40	0.0750	0	1	0	0.0750	0
4	4	50	0.0800	0.6	0.8	0.0288	0.0512	0.0384
			Totals			0.0765	0.1518	0.0192

Original Geometry of the Truss

1. Analysis for first yield zone

   Using the finite element method, $\mathbf{K}\,\mathbf{u} = \mathbf{P}$ with

$$\begin{bmatrix} \sum_{i=1}^{4}\left(\frac{AE}{L}\right)_i \lambda_i^2 & \sum_{i=1}^{4}\left(\frac{AE}{L}\right)_i \lambda_i\mu_i \\ \text{Sym.} & \sum_{i=1}^{4}\left(\frac{AE}{L}\right)_i \mu_i^2 \end{bmatrix} \begin{pmatrix} u_1 \\ v_1 \end{pmatrix} = \begin{pmatrix} 0 \\ \Delta P_5 \end{pmatrix} \tag{a}$$

Taking a unit value of $\Delta P_1 = -1$ and evaluating Eq. (a),

$$\begin{bmatrix} 0.0765 & 0.0192 \\ 0.0192 & 0.1518 \end{bmatrix} \begin{pmatrix} u_1 \\ v_1 \end{pmatrix} = \begin{pmatrix} 0 \\ -1 \end{pmatrix}; \qquad \text{thus} \begin{pmatrix} u_1 \\ v_1 \end{pmatrix} = \begin{pmatrix} 1.708 \\ -6.804 \end{pmatrix} \tag{b}$$

Member stresses resulting from Eqs. (b), scaling ΔP_1:

Member	$\Delta L\dagger = \lambda \Delta u + \mu \Delta v'$	$\Delta\sigma = E\Delta L/L$	$\Delta P_1 \times \Delta\sigma$	
1	1.708	0.0569	1.0035	
2	6.468	0.1294	2.2822	
3	6.804	0.1701	3.0000	← First yield zone,
4	4.418	0.8836	1.6584	$\Delta P_1 = -17.637$

2. Analysis for second yield zone
   Repeating Eq. (a) omitting member 3 in the summation,

$$\begin{bmatrix} 0.0765 & 0.0192 \\ 0.0192 & 0.0768 \end{bmatrix} \begin{pmatrix} u_1 \\ v_1 \end{pmatrix} = \begin{pmatrix} 0 \\ -1 \end{pmatrix}; \quad \text{thus,} \begin{pmatrix} u_1 \\ v_1 \end{pmatrix} = \begin{pmatrix} 3.487 \\ -13.892 \end{pmatrix} \tag{c}$$

   Member stresses resulting from Eqs. (c) and (b), scaling AP_2:

Member	ΔL	$\Delta\sigma$	σ_1	$\sigma_2 = \sigma_1 + \alpha_2 \Delta\sigma$	
1	3.487	0.1162	1.0035	1.3193	
2	13.21	0.2641	2.2822	3.0000	← Second yield zone,
4	9.021	0.1804	1.5584	2.0487	$\Delta P_2 = -2.7179$

3. Analysis for third yield zone
   Repeating Eq. (a) omitting members 2 and 3,

$$\begin{bmatrix} 0.0621 & 0.0384 \\ 0.0384 & 0.0512 \end{bmatrix} \begin{pmatrix} u_1 \\ v_1 \end{pmatrix} = \begin{pmatrix} 0 \\ -1 \end{pmatrix}; \quad \text{thus,} \begin{pmatrix} u_1 \\ v_1 \end{pmatrix} = \begin{pmatrix} 22.52 \\ -36.42 \end{pmatrix} \tag{d}$$

   Member stresses resulting from Eqs. (b), (c), and scaled 60:

Member	ΔL	$\Delta\sigma_1$	σ_2	$\sigma_3 = \sigma_2 + \alpha_3 \Delta\sigma_2$	
1	22.52	0.7507	1.3193	3.0000	← Third yield zone,
4	15.62	0.3124	2.0487	2.7481	$\Delta P_3 = -2.2388$

4. Collapse load $= \sum_{\delta=1}^{3} \Delta P_s = 22.59 = 7.531\ \sigma_0$

   Only member 4 survives without plasticity.

*Assume $E = 1$, $\sigma_0 = 3$.

†For example, $\Delta L_1 = \lambda_1 (u_2 - u_1) + \mu_1 (v_2 - v_1)$.

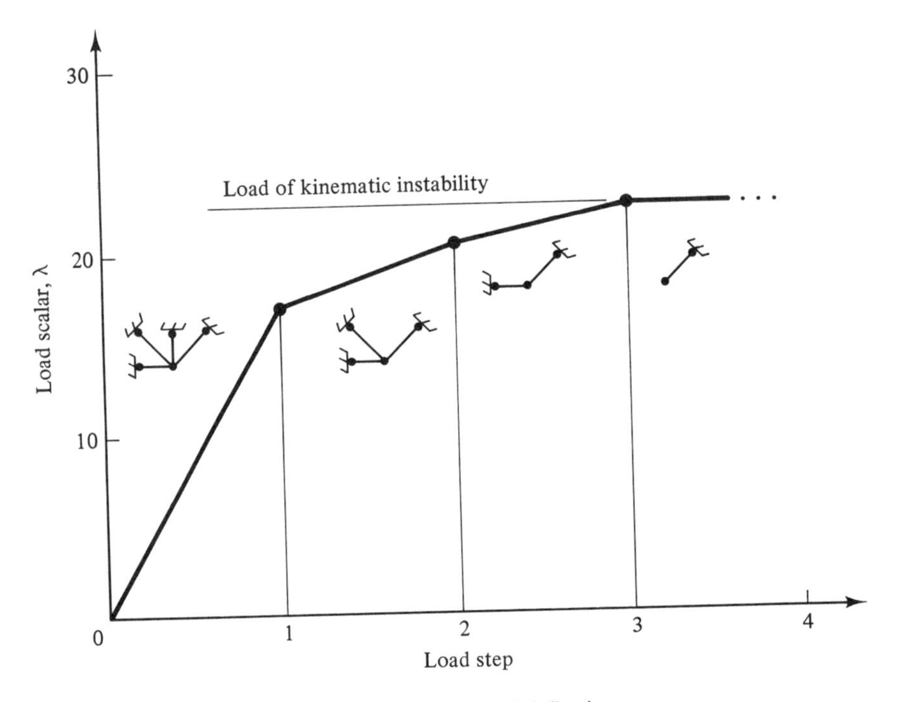

Figure 10.7 Four-bar truss load deflection curve.

10.3 STRUCTURES OF HYPERELASTIC MATERIALS

A hyperelastic material is an elastic material whose stress-strain equations are derivable from a positive-definite strain energy function. Though hyperelastic materials include linearly elastic models as a subset, we will be concerned here with materials that are modeled, in uniaxial response, by a folded line, no segment of which has a negative slope. Figure 10.8 shows the kind of stress-strain relation of interest.

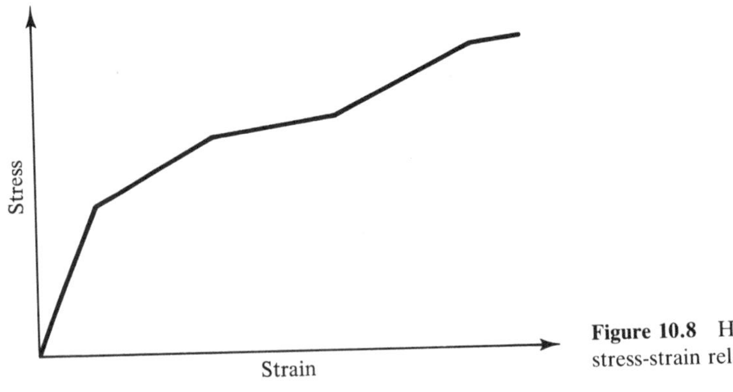

Figure 10.8 Hyperelastic material stress-strain relation.

Like the limit analysis model, this material model guarantees to result in a unique solution of the structural equations as long as unloading is precluded. As for limit analysis, we can find that solution by incremental loading in pseudotime.

We extend the limit analysis process to the piecewise continuous hyperelastic material by following the progress of stress changes at each point we wish to monitor in the structure. As a consequence, the number of analysis cycles may be as many as

$$n_c = n_\sigma(n_B - 1) + 1 \tag{10.6}$$

where

- n_c is the number of analysis cycles,
- n_σ is the number of points where stress is monitored, and
- n_B is the maximum number of branches in a quadrant of the stress-strain curve.

Figure 10.9 and Table 10.4 illustrate a typical analysis problem. We wish to determine the deflection of the tip of this tower for the structural configuration of the figure.

Table 10.5 lists the final results of the incremental analysis. The analysis cycles are terminated when the residual forces

$$\mathbf{r} = \mathbf{K}_t\,\Delta\mathbf{g} - \Delta\mathbf{p} \tag{10.7}$$

have a relative norm of zero compared with the norm of the total applied load.

The second column of Table 10.5 lists the internal force in each rod caused by the loading of Table 10.4. These are the internal forces when the unloaded truss has no internal forces and the folded stress-strain curve of Fig. 10.9 is active. The third column cites the change in each rod internal force induced by the Table 10.4 loading assuming the unloading Young's modulus is the same as the loading modulus. Subtracting the entries of the third column from those of the second produces the entries of the fourth column of Table 10.5. This superposition simulates unloading of the tower. Therefore, the fourth column defines the residual internal forces in the rods induced by the plastic straining under the applied loading. Similarly, the displacements can be superimposed to evaluate the permanent set in the tower due to plastic straining.

The table that follows cites the number of analysis cycles required to find the solution of this problem as a function of a loading multiplier. It provides both the number of cycles for incremental load analysis and for general iterative nonlinear analysis. The scelernomic analysis solution is exact to five digits of the maximum displacement. The nonlinear solution is accurate to two digits in the largest displacement. Nevertheless, this data suggests that computer resources are conserved by using the incremental loads approach for the piecewise linear structural model, particularly when the number of response branches to explore is low.

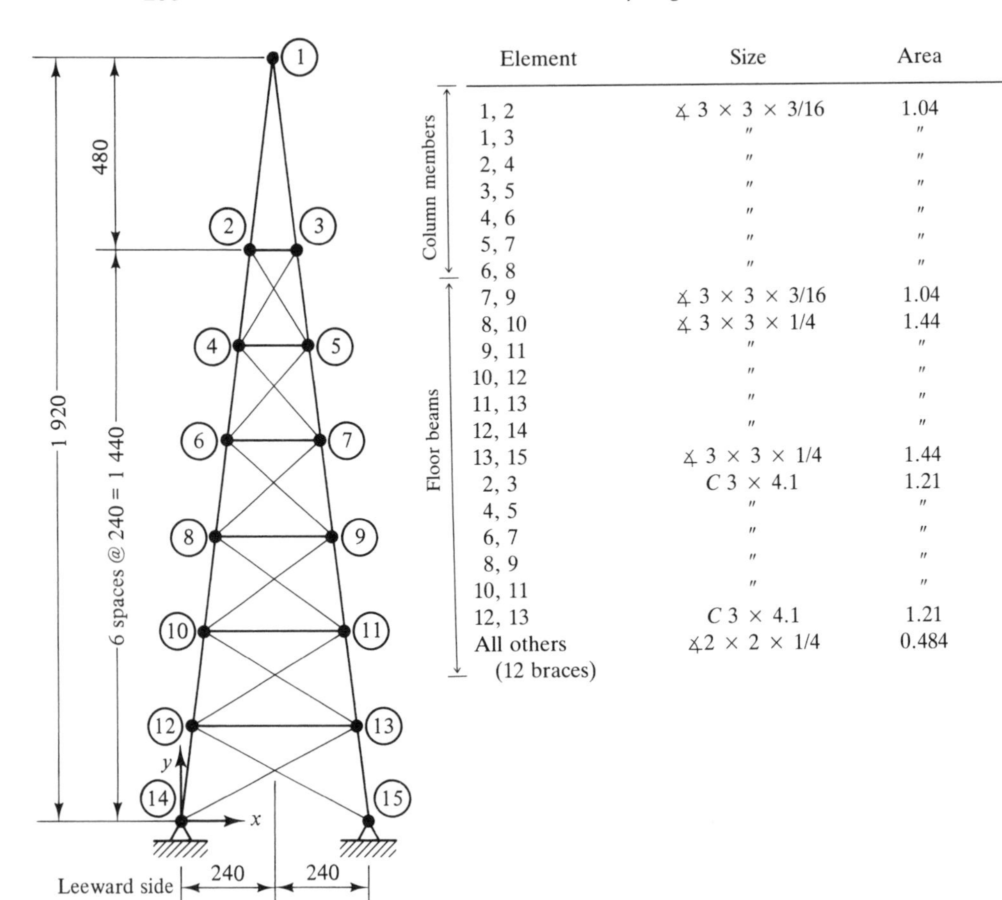

Element	Size	Area	$\sqrt{I/A}$
1, 2	∡ 3 × 3 × 3/16	1.04	0.929
1, 3	"	"	"
2, 4	"	"	"
3, 5	"	"	"
4, 6	"	"	"
5, 7	"	"	"
6, 8	"	"	"
7, 9	∡ 3 × 3 × 3/16	1.04	0.929
8, 10	∡ 3 × 3 × 1/4	1.44	0.930
9, 11	"	"	"
10, 12	"	"	"
11, 13	"	"	"
12, 14	"	"	"
13, 15	∡ 3 × 3 × 1/4	1.44	0.930
2, 3	C 3 × 4.1	1.21	0.404
4, 5	"	"	"
6, 7	"	"	"
8, 9	"	"	"
10, 11	"	"	"
12, 13	C 3 × 4.1	1.21	0.404
All others (12 braces)	∡2 × 2 × 1/4	0.484	0.626

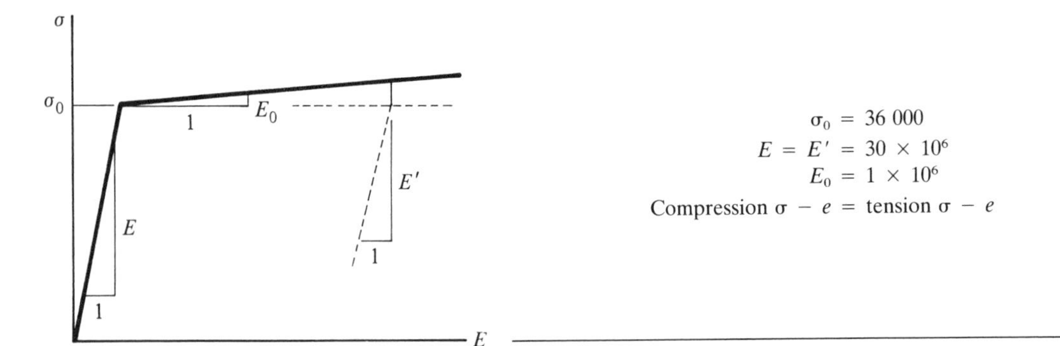

$$\sigma_0 = 36\ 000$$
$$E = E' = 30 \times 10^6$$
$$E_0 = 1 \times 10^6$$
Compression $\sigma - e$ = tension $\sigma - e$

Figure 10.9 Navigation tower configuration.

TABLE 10.4 NAVIGATION TOWER EXTERNAL LOADING

Node No.	Vertical Load*	Lateral Load†	Node No.	Vertical Load*	Lateral Load†
1	−308	−9650	8	−304	−87.5
2	−273	−50	9	−304	−438
3	−273	−250	10	−374	−100
4	−292	−62.5	11	−374	−500
5	−292	−312	12	−378	−112
6	−294	−75	13	−378	−563
7	−294	−275			

*Includes dead loads and floor loads.

†Includes wind load and airplane impact load (based on peak accelerations).

TABLE 10.5 NAVIGATION TOWER INTERNAL FORCES

Rod	Applied Load	Unloading	Residual
1, 2	−39 100	−39 100	0
1, 3	38 700	38 700	0
2, 4	−39 100	−39 400	300
3, 5	39 000	−38 800	200
4, 6	−39 600	−40 200	600
5, 7	39 400	39 000	400
6, 8	−40 400	−41 100	700
7, 9	40 000	39 300	700
8, 10	−42 300	−42 200	−100
9, 11	39 700	39 800	−100
10, 12	−43 500	−43 500	0
11, 13	40 300	40 300	0
12, 14	−44 900	−44 900	0
13, 15	40 900	40 900	0
2, 3	150	0	150
4, 5	628	0	628
6, 7	1 250	0	1 250
8, 9	641	0	641
10, 11	0	0	0
12, 13	0	0	0
2, 5	−211	120	−331
3, 4	−683	−351	−332
4, 7	−415	224	−639
5, 6	−1 180	−539	−641
6, 9	−656	351	−1 007
7, 8	−1 717	−709	−1 008
8, 11	623	554	69
9, 10	−772	−840	68
10, 13	727	733	−6
11, 12	−1 045	−1 040	−5
12, 15	945	945	0
13, 14	−1 251	−1 250	−1

Analysis Steps and Cycles for the Navigation Tower

LOAD SCALAR VALUE	SCELERNOMIC ANALYSIS	GENERAL NONLINEAR ANALYSIS
0.5	2 steps	2 cycles
1.0	9 steps	55 cycles
2.0	15 steps	20 cycles
16.0	20 steps	15 cycles

10.4 LINEAR BUCKLING ANALYSIS

Prediction of buckling loads of structures is the most common use of scelernomic analysis. These estimates of the loading scalar at which the tangent stiffness becomes zero involve consideration of nonlinear strain displacement equations. The solution process involves two equation-solving steps.

We will develop the concepts by focusing the discussion on structures whose displacements are confined to the xy plane. We attend structures that may buckle without stresses exceeding the proportional limit.

10.4.1 Buckling Models

We will need the nonlinear two-dimensional strain displacement equations

$$\begin{aligned} e_{xx} &= \delta u/\delta x + 0.5[(\delta u/\delta x)^2 + (\delta v/\delta x)^2] \\ e_{yy} &= \delta v/\delta y + 0.5[(\delta u/\delta y)^2 + (\delta v/\delta y)^2] \\ e_{xy} &= 0.5(\delta u/\delta y + \delta v/\delta x) + 0.5[(\delta u/\delta y)(\delta u/\delta x) + (\delta v/\delta y)(\delta v/\delta x)] \end{aligned} \qquad (10.8)$$

and the corresponding strain energy expression

$$SE = 0.5 \iiint \boldsymbol{\sigma}^T \mathbf{e}\, dx\, dy\, dz \qquad (10.9)$$

where

$\boldsymbol{\sigma}$ are the stresses evaluated by replacing $\mathbf{e}$ by $\mathbf{E}^{-1}\boldsymbol{\sigma}$ in the stress-strain equations (7.9),

$\mathbf{e}$ is the vector whose components are given by Eq. (10.8), and

integration extends over the volume of the structure.

Consider the rod element of Fig. 10.10. The rod is prismatic and formed of a homogeneous and isotropic Hookean material. Its right end is free to assume any position in the xy plane. Its left end is pinned at the origin of the coordinate system. The total strain in the rod is

$$e_{xx} = \frac{L_f}{L_o} - 1 \qquad (10.10)$$

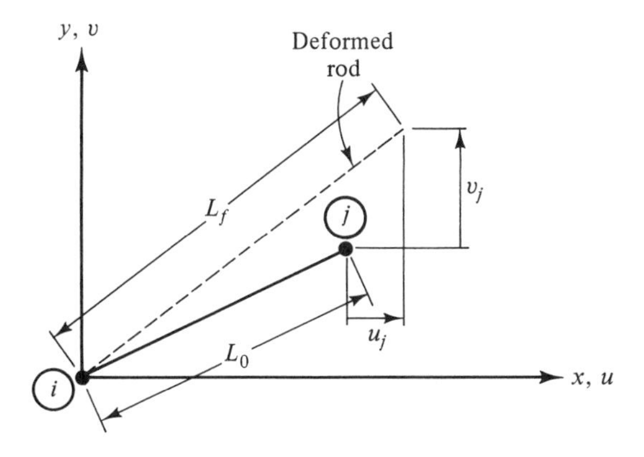

Figure 10.10 Rod with large deflections.

where

L_f is the final length of the rod and
L_o is its original (unstrained) length.

The strain is assumed to be constant over the rod. Predicating uniaxial stress and recognizing the rod's uniform cross section, the strain energy becomes

$$SE = 0.5\, AEL_o e_{xx}^2 \tag{10.11}$$

where

A is the cross-sectional area and
E is Young's modulus.

Substituting Eq. (10.10) in Eq. (10.11) and expressing L_o and L_f in terms of the displacements of the free node furnishes

$$SE = 0.5AEL_o \frac{[(x_o + u_1)^2 + (y_o + v_1)^2]^{0.5}}{(L_o - 1)^2} \tag{10.12}$$

Evaluating the stiffness coefficients as the second derivatives of the strain energy with respect to the nodal displacements supplies the tangent stiffness matrix:

$$\mathbf{K}_t\,\mathbf{g} = (\mathbf{K}_o + \lambda\mathbf{K}_i)\,\mathbf{g} \tag{10.13}$$

where

$$\mathbf{K}_o = \frac{AE}{L_f^3}\begin{bmatrix} (x_o+u_1)^2 & & & \text{Sym.} \\ (x_o+u_1)(y_o+v_1) & (y_o+v_1)^2 & & \\ -(x_o+u_1)^2 & -(x_o+u_1)(y_o+v_1) & (x_o+u_1)^2 & \\ -(x_o+u_1)(y_o+v_1) & -(y_o+v_1)^2 & (x_o+u_1)(y_o+v_1) & (y_o+v_1)^2 \end{bmatrix} \tag{10.14}$$

$$\mathbf{K}_i = \frac{P_a}{L_f}\begin{bmatrix} 1 & & & \text{Sym.} \\ 0 & 1 & & \\ -1 & 0 & 1 & \\ 0 & -1 & 0 & 1 \end{bmatrix} \quad \text{and} \quad \mathbf{g} = \begin{pmatrix} u_1 \\ v_1 \\ u_2 \\ v_2 \end{pmatrix} \tag{10.15}$$

where

λ is the buckling scale factor,
$\mathbf{K}_o$ is the conventional stiffness matrix,
$\mathbf{K}_i$ is the induced stiffness matrix, and
P_a is the axial load on the rod.

These stiffnesses differ from the nonbuckling linear analysis model in two important ways:

1. The conventional stiffness matrix is a function of the final length and the orientation of the rod.
2. The induced stiffness matrix has been added to the conventional stiffness matrix. The stiffness changes induced by the internal forces in the rod depend on the final axial force and length of the rod. (This induced stiffness matrix has the unusual property that it does not change with rotations of the rod or the coordinate axes.)

In deriving the stiffnesses for the rod, we have implied that the rod can undergo large rigid rotations (and rigid translations) but only infinitesimal changes in length. As an alternative procedure for deriving the stiffnesses, we can develop the equations in the local axes aligned with the final orientation of the rod and use a contragradient transformation of the stiffness matrices to refer them to the original axes. This is the corotational coordinates approach.

For the beam model, we begin by constraining the two-dimensional strain field by the Kirchhoff hypothesis:

$$e_{xx} = \delta u/\delta x - y\, \delta^2 v/\delta x^2 + 0.5(\delta v/\delta x)^2 \tag{10.16}$$

Evaluating stresses for the uniaxial stress state and substituting stress and strain in Eq. (10.9) gives

$$SE = 0.5\iiint E[\delta u/\delta x - y\,\delta^2 v/\delta x^2 + 0.5(\delta v/\delta x)^2]^2\, dx\, dy\, dz \tag{10.17}$$

$$SE = 0.5\iiint E(\delta u/\delta x)^2\, dx\, dy\, dz + 0.5\iiint E(y\delta^2 v/\delta x^2)^2\, dx\, dy\, dz$$
$$+\, 0.5\iiint E\, \delta u/\delta x\, (\delta v/\delta x)^2\, dx\, dy\, dz \tag{10.18}$$

where Eq. (10.17) is simplified to Eq. (10.18) by assuming that the cross section is symmetric and the integral of 0.25 $(\delta v/\delta x)^4$ is negligible. Under the supposition that u displacements vary linearly with the spanwise coordinate and v varies cubically, upon integration and differentiation with respect to nodal displacements, the first integral of Eq. (10.18) yields the conventional rod stiffness matrix and the second, the conventional Bernouilli-Euler beam matrix. The third integral gives the induced stiffness matrix of the beam:

$$\mathbf{K}_i\,\mathbf{g} = \frac{P_a}{30L_f}\begin{bmatrix} 36 & & & \text{Sym.} \\ 3L_f & 4L_f^2 & & \\ -36 & -3L_f & 36 & \\ 3L_f & -L_f^2 & -3L_f & 4L_f^2 \end{bmatrix}\begin{pmatrix} v_1 \\ t_1 \\ v_2 \\ t_2 \end{pmatrix} \tag{10.19}$$

The cogradient transformation of the conventional beam stiffness matrix and the induced stiffness matrices of Eqs. (10.15) and (10.19) imply rotation of the coordinate axes from one aligned with the axes of the rigidly rotated beam to one aligned with global axes. As Eq. (10.14) suggests, this transformation, when it allows large rigid rotation angles, results in stiffness coefficients that are nonlinear in the generalized displacements.

Induced stiffness matrices for membranes, plates, shells, and three-dimensional solids can be developed in a similar fashion. Thus by using the shape functions of linear FEA models, the nonlinear strain displacement equations of elasticity, and the strain energy form and by neglecting higher-order deformations than linear, we arrive at acceptable models. These models reduce to the conventional linear models when rigid rotation angles of the elements are small enough so that the sines of the angles are equal to the angles and cosines equal to 1.

The induced stiffness matrices are like conventional element stiffness matrices. They are symmetric. They imply force equilibrium. They reflect rotation of axes by cogradient transformation. They can be dealt with by the FEA process without modifying the process. They are positive semidefinite when all stresses are tensile or zero.

These matrices also differ from the conventional in important ways. They are independent of material properties. They are a function of the gross dimensions of the element (like span) but independent of local dimensions (like thickness, area, and moments of inertia). They will be indefinite or negative semidefinite for some compressive stress states.

10.4.2 Evaluation of Buckling Loads

The buckling problem is to evaluate the buckling load scalar—the multiple of the stress state that causes the tangent stiffness matrix to be singular.

To estimate the buckling scalar by scelernomic analysis, we linearize the tangent stiffness matrix. We assume that the orientation of the member is given by its original orientation. We assume that the final geometry is adequately approximated by the original. For the truss and beam, for example, we replace L_f by L_o. We assume that the internal force distribution, motivated by P_a in Eqs. (10.15) and (10.19), does not change as the load changes to the buckling load.

It is convenient to recast the requirement that the tangent stiffness matrix be singular as an eigenproblem. Then the mathematical problem is to solve the equations

$$\mathbf{K}_o \, \mathbf{g} = -\lambda \, \mathbf{K}_i \, \mathbf{g} \tag{10.20}$$

where λ is the buckling load scalar. Usually, our interest is in finding the smallest positive λ and its associated eigenvector **g**. (Many algorithms are available to evaluate the eigensolution.)

Therefore, the tasks of scelernomic buckling analysis are as follows:

1. Perform a conventional linear FEA, determining nodal displacements by solving the simultaneous equations and finding the values of the nodal forces on each finite element.
2. Generate and exercise the finite element process for the induced stiffness matrix, using the element nodal forces of task 1 in evaluating element-induced stiffness matrices.
3. Find the lowest positive buckling load scalar and its vector by solving the eigenproblem of Eqs. (10.20) for the smallest positive λ.

The tasks of this process involve two equation-solving steps: solving the linear simultaneous equations and finding a solution of the eigenproblem.

The characteristics of the stiffness matrices determine when a positive buckling scalar will occur. For truss and frame analysis, buckling cannot occur unless at least one of the elements is undergoing compression. In the all tension system, all the element matrices will be positive semidefinite or positive definite and therefore only kinematic instability is possible. Membranes, plates, and shells can buckle in shear without the presence of compressive stresses.

In the general case, the tangent stiffness matrix will be indefinite. Accordingly, the lowest-magnitude buckling load scalar may be positive, negative, or zero. The zero-mode eigenvector may be a kinematic mode, but positive and negative scalars imply elastic buckling. Negative scalars can be interpreted as buckling when the stress state implies a loading vector of opposite sign to that assumed.

Analysis of truss buckling involves no discretization error. The truss model idealizes members, assuming that they remain straight. Therefore, this analysis requires separate calculations to ensure that no rod buckles in Euler buckling at a lower force than would excite the truss element buckling of Eq. (10.15), the *P*-delta buckling.

In the case of a frame, the induced stiffness includes element rod behavior as a special case, so this check is not needed. The beam model, based on the shape functions of Eq. (10.2), is not a lattice model, however. A measure of the need for additional grid refinement can be based on comparison of the buckling mode in an element with the number of nodes needed to represent the element mode for the accuracy desired.

For example, experiments indicate that twice as many equal-span elements are needed for a beam segment as the number of changes of curvature of the mode

TABLE 10.6 DISCRETIZATION ERROR IN COLUMN BUCKLING LOAD

Grid of Model	Error λ_1 (%)	Error λ_2 (%)
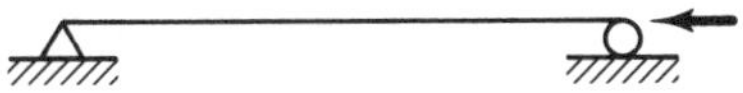 1 element	21.6	52.0
 2 elements	0.753	21.6
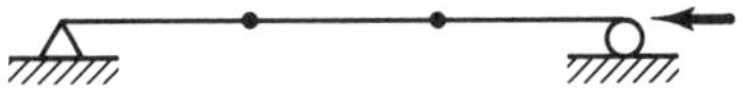 3 elements	0.158	2.17
 4 elements	0.0492	0.787
 5 elements	0.0191	0.384
 6 elements	0.00740	0.214

plus 1 to attain engineering accuracy in predicting the buckling scalar.[4] This criterion can also be used for estimating whether discretization error will be excessive for linear buckling analyses of plates and shells.

Table 10.6 summarizes results of a study of the relation between the number of finite elements and the accuracy of Euler buckling predictions for the pinned-pinned beam. This data reflects the discretization rule and demonstrates monotonic convergence to the Euler value of the buckling load.

For beams, it is possible to develop the induced stiffness matrix using the exact solution of the Euler differential equation. Then the frame buckling load can be calculated without discretization error.[5] This approach evokes the need to solve a nonlinear eigenproblem. The approach is guaranteed to minimize computer resource needs when highly accurate linear buckling analysis is justified.

Mathematical modeling error is a more subtle source of error. Loss of stiffness due to overstraining the material may cause buckling at load scalars much below the scelernomic estimates. The early triggering of buckling due to imperfections in the geometry is known to be important in curved plates and shells.[6]

One of the most significant discoveries of the 1940s was the discovery that cylindrical shell buckling requires a nonlinear buckling theory. The scelernomic predictions of buckling loads of axially loaded cylindrical shells were found to be as much as three times those experienced in the laboratory.[6] This history should caution the engineer against accepting the results of scelernomic buckling analysis without circumspection. The linear buckling theory is elegant. It is widely used in the design of frames, where it produces satisfactory estimates of critical load when strains are small. If it indicates buckling of any articulated or continuum model of a structure, it may simply mandate a nonlinear analysis that simulates the effects of large geometry changes more realistically.

10.5 STRUCTURES WITH PRESTRESSES

Prestressing is designed into prestressed concrete structures, guyed towers, stayed bridges, and masonry arches. Prestressing also occurs in casting, stamping, forging, rolling, ball-peaning, and welding operations of fabrication. It arises in shoring, jacking, buttressing, or any force fitting during assembly or erection. It may be caused by heating or cooling during service.

Regardless of the source, prestressing means that the initial stress in the structure is not zero everywhere, though there may be no apparent source of loading. We define it to mean that a self-equilibrating stress state exists within the structure.

We need no new finite element models to deal with prestress analysis. We do need to modify the analysis process, using Duhamel's superposition principle to establish the effects on structural response. These effects include changing the stiffness of the structure as well as its state.

To particularize explanation of the process, consider analysis of a truss that is heated above its stress-free temperature. The analysis tasks are as follows:

1. Assume that the elements of the structure are unconnected and determine the change in geometry of each element due to heating. This task requires integrating the thermal strain over the element geometry.

2. Find the nodal forces that would cause the deflections calculated by task 1. These are given by

$$\mathbf{p}_e = \mathbf{K}_e \, \mathbf{g}_e \tag{10.21}$$

where $\mathbf{g}_e$ is the vector of nodal displacements of task 1. Note that $-\mathbf{p}_e$ are the element forces that will restore the element original geometry so the structure can be pieced back together with no gaps.

3. Assemble the element load vectors $\mathbf{p}_e$ and determine the responses of the connected structure to this loading.

4. Superimpose responses from tasks 1–3 to find the change in state caused by the prestresses introduced in task 2.

The tasks involve two analyses: one to determine the compensating forces on each element to make the structure fit back together in the original geometry and one to evaluate the system changes caused by the force fitting.

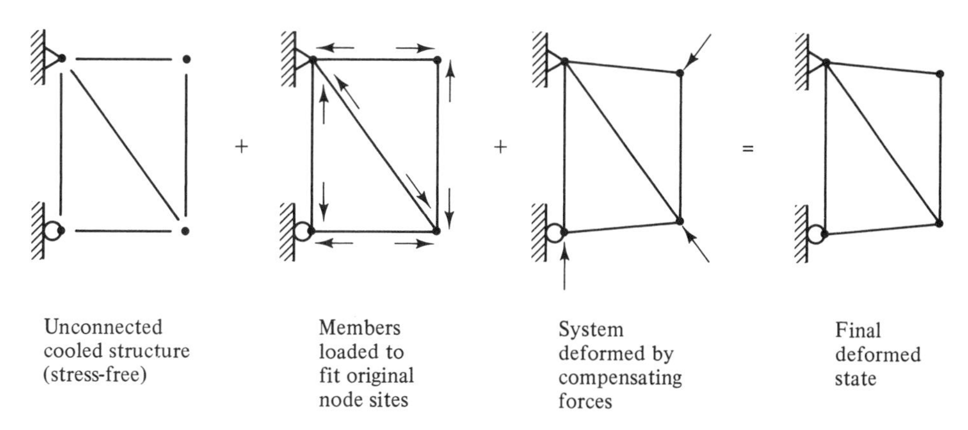

Figure 10.11 Superposition states for thermoelastic analysis.

Figure 10.11 graphically illustrates the process for a simple truss. The final state of the structure is determined by superimposing three states: the state of each element of the unconnected structure under heating, the state of each element when it is loaded to fit the original geometry, and the changes in response due to the compensating nodal forces. The final deflections are equal to the deflections of the third state because the deflections of the unconnected structure cancel those of the force fitting. The final stresses are the sum of the stresses when the elements are forced to fit together in the original geometry and the stresses caused by the compensating forces. For the heated truss, all the stresses of the second state are compressive, whereas most of the stresses of the third state are tensile.

If we wish to include the stiffness changes induced by prestress, we simply include the stiffness induced by the prestress in calculating the tangent stiffness. Then singularity of the stiffness matrix reflects either kinematic or elastic instability. Nonpositive definiteness, without singularity of the matrix, indicates that the prestress magnitude is greater than that to cause thermoelastic buckling.

Duhamel's principle states that we can always do linear elastic analysis using compensating nodal forces, like those of the thermoelastic case, to predict responses. This principle is valid even in the face of nonlinear stress-strain relations. Therefore, the same analysis process is appropriate regardless of the source of prestress and whether the model is a lattice or a continuum. If we know only the prestress state, instead of prestrains, the first analysis task must include calculating equivalent nodal forces.

10.6 SLOTTED-JOINT STRUCTURES

Metal structures are often bolted together. When these structures are an integral part of a scientific instrument, such as a large steerable antenna or a telescope, accuracy of performance may be unduly degraded by joint slip.

We will need a special model to represent joint stiffness changes and will use

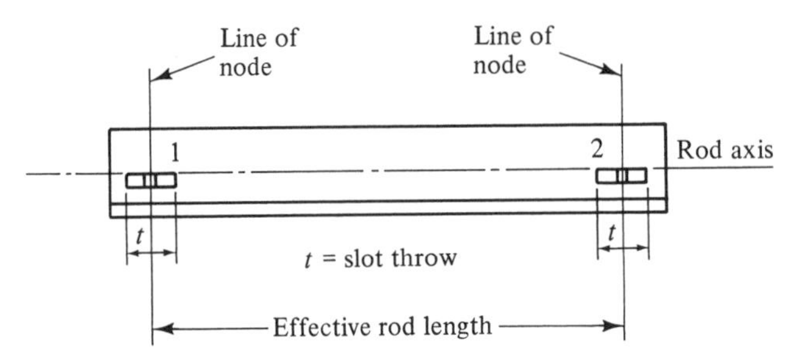

Figure 10.12 Slotted-rod geometry details.

pseudotime step analysis to determine the worst-case deflections of articulated structures with slotted joints.[7]

10.6.1 Rod-Joint Model

Consider a truss. The model of each rod involves the following assumptions:

1. The coordinates of the nodes of the structure define the original location of the center of the single equivalent pin that connects all rods at a joint of the structure.
2. During any displacement change, the stiffness of each element is constant. When the rod is not slipping, its stiffness is the tangent stiffness. When the rod is slipping, its stiffness is zero.
3. The slot, as Fig. 10.12 suggests, is assumed to be collinear with the rod axis. No forces motivate movement of the rod normal to its undeflected axis.

The first and third assumption imply that the only special data needed to describe the slotted joints of an element is the throw of the slot. The throw is the maximum distance a rod can slip. The sign of the throw is disregarded under the assumption that the slot freedom for element movement matches the direction of motivated movement. Thus if the distance between nodes of a rod's ends is reduced, the rod may offer no resistance until the slip equals the throw. Alternately, if the nodes separate, there may be no resistance until the separation exceeds the throw.

Assumptions 2 and 3 imply that the load deflection relation for a rod is trilinear, as Fig. 10.13 shows. In the initial and full throw states, the stiffness of the rod is its tangent stiffness when motivation presages nonslip. At intermediate length changes, the stiffness is zero when the rod is free to move in slip and the tangent stiffness otherwise. The joint trilinear load deflection relation is the only nonlinearity of the mathematical model.

10.6.2 Pseudotime Analysis

We use a pseudodynamic approach for finding the solution to the joint slip problem. We assume that the structural changes follow a prescribed sequence with negligible inertial and damping contributions to the equations of motion. This approach

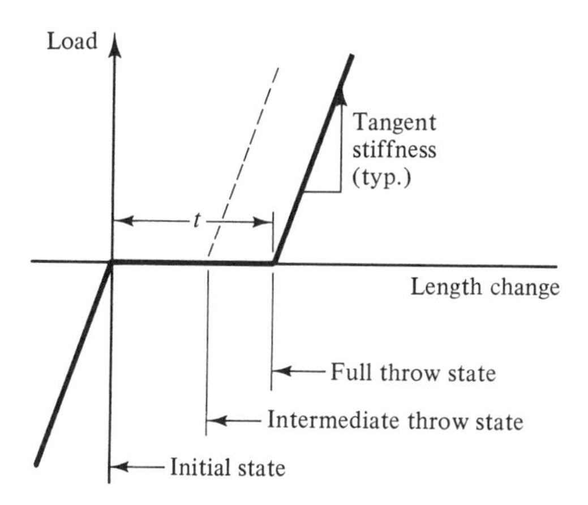

Figure 10.13 Rod load deflection relation.

ensures that the final structural configuration is realizable by slowly changing from the initial state.

The simulation involves the following steps:

1. Perform a conventional linear static structural analysis to establish a first-order estimate of rod internal forces. Thus we assume that the original geometry is perturbed by the loading into an offset state. This change in geometry takes place with no slip.

2. Identify the rod whose slip will contribute the most to the deflection change of the current configuration, the slipping rod.

3. Reanalyze the structure, predicting the change in displacements induced by neglecting the stiffness of the slipping rod and considering the induced stiffnesses associated with the offset state.

4. Establish a new equilibrium configuration considering movement of the slipping rod. If the slip predicted in step 3 is less than the slot throw for the slipping rod, the desired equilibrium configuration is the result of step 3. If not, the displacement increment of step 3 is scaled to exploit the throw of the joint, and another analysis is performed with the slipping rod locked to establish the equilibrium geometry.

5. Repeat steps 2–4 until the responses do not change.

This analysis process simulates a relatively high starting resistance against element slip but zero resistance while the element is slipping. It represents static equilibrium transitions from one set of slot movements to the next. Given the original geometry, material properties, and boundary conditions, the process can lead to only one solution.

The number of steps needed to complete the solution search will be finite in the sense that convergence is monotonic. That is, as long as the criterion that identifies the next element to slip never results in changing the direction of slip of an element, convergence is monotonic. For example, if the criterion depends only on rod internal force, the process will be monotonically improving.[1]

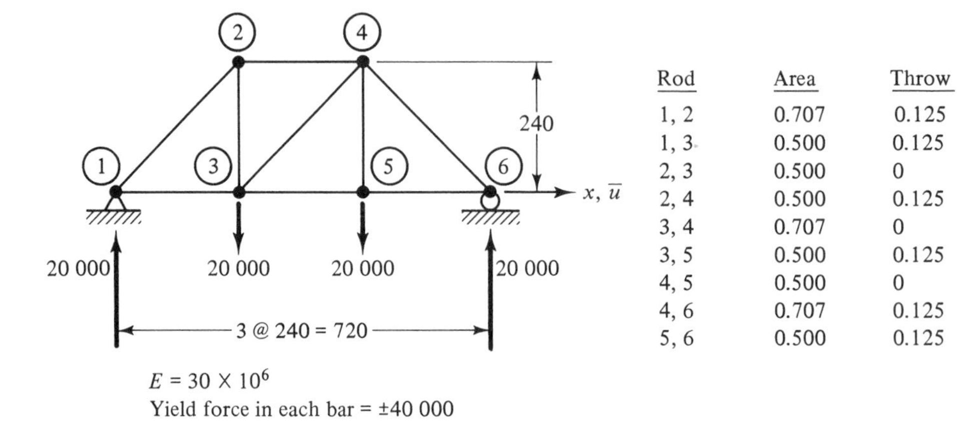

Rod	Area	Throw
1, 2	0.707	0.125
1, 3	0.500	0.125
2, 3	0.500	0
2, 4	0.500	0.125
3, 4	0.707	0
3, 5	0.500	0.125
4, 5	0.500	0
4, 6	0.707	0.125
5, 6	0.500	0.125

Figure 10.14 Determinate truss problem.

The maximum number of cycles through steps 2 and 3 of the analysis process depends on the number of slotted rods and structural redundants. For determinate trusses it must be n_s, the number of slotted rods. For indeterminate structures with as many or more redundants than slotted rods, it must be $\leq n_s!$ Fortunately, for most trusses the number of redundants n_r will be much less than the number of slotted elements, and the maximum number of cycles is then $n_s!/(n_s - n_r)!$

Figure 10.14 describes a two-dimensional determinate truss. It is assumed

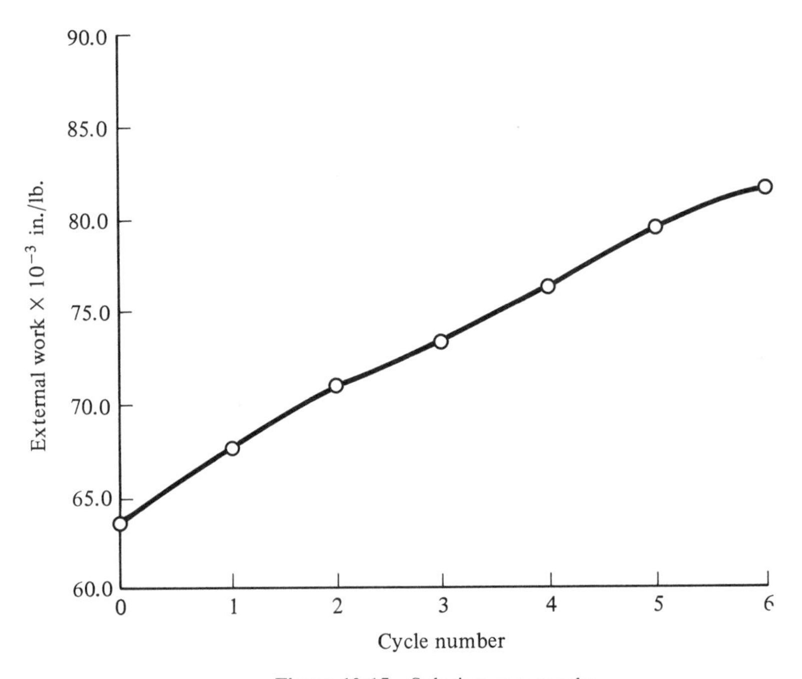

Figure 10.15 Solution step results.

TABLE 10.7 EQUILIBRIUM DISPLACEMENTS (inches)

State	u_2	v_2	u_3	v_3	u_4	v_4	u_5	v_5
No slip	0.5333	−1.1734	0.3200	−1.4934	0.2133	−1.3868	0.6400	−1.7068
Slipped	0.7834	−1.6002	0.4451	−1.9202	0.3385	−1.8138	0.8901	−2.1335

TABLE 10.8 EQUILIBRIUM ELONGATIONS (inches)

Rod	Cycle 0	1	2	3	4	5	6
1, 2	−0.4526	−0.5770	−0.5775	−0.5777	−0.5776	−0.5780	−0.5776
1, 3	0.3200	0.3195	0.3202	0.4453	0.4452	0.4454	0.4451
2, 3	0.3200	0.3192	0.3199	0.3196	0.3197	0.3203	0.3200
2, 4	−0.3200	−0.3198	−0.3199	−0.3205	−0.4449	−0.4451	−0.4449
3, 4	0	−0.0001	0.0001	0.0003	−0.0001	−0.0003	−0.0001
3, 5	0.3200	0.3199	0.3198	0.3202	0.3200	0.3205	0.4450
4, 5	0.3200	0.3200	0.3192	0.3200	0.3200	0.3199	0.3197
4, 6	−0.4526	−0.4525	−0.5769	−0.5779	−0.5777	−0.5777	−0.5777
5, 6	0.3200	0.3202	0.3195	0.3204	0.3202	0.4453	0.4453

that bolt holes have been punched, resulting in a worst-case slip tolerance of one-eighth. All chord elements may slip as much as one-eighth; other members cannot slip. Figure 10.15 is a plot of the external work implied in each step of the solution search. The curve illustrates the monotonic increase in work in each step as the elements are simulated as slipping. The final solution is found in six steps, the number of members that were permitted to slip.

Table 10.7 facilitates comparison of nonslip and slip displacements of truss nodes. The accumulation of slip is evident, especially in the vertical displacements.

Table 10.8 lists rod elongations for each analysis cycle. In every case, these increase monotonically with cycle number, within the limitations of computer accuracy of the results.

The slotted-joint structural model and analysis procedure obtains a solution that satisfies the element load deflection equations, considering the slip limits, node equilibrium, and displacement boundary conditions. It develops worst-case estimates of geometry changes by assuming slip of the member that has the maximum slip-motivating length change in each analysis cycle.

10.7 FEATURES OF SCELERNOMIC ANALYSIS

Scelernomic analysis is stepwise linear structural analysis. The models are simplified nonlinear and dynamic models created by approximations that include linearization. Solution search is effected by analysis in pseudotime. Usually this means using incremental loading or incremental deflection steps.

Scelernomic analysis is much more economical than nonlinear or dynamic analysis and is often equally accurate. The analyst must exercise special care in interpreting these analysis results because of the special approximations. Rather than being an end in itself, the scelernomic analysis may only serve to justify general nonlinear or dynamic analysis.

Assurance of accuracy of a scelernomic analysis requires study of the results of the analysis steps and the final solution. Evaluation of the step results confirms whether the loading is proportional, as implied by the theory. Examination of the results of the last step establishes whether the structural or elasticity equations are satisfied.

Proportional loading bears directly on the relevance of scelernomic analysis of counters, limit analysis, and hyperelastic material effects. Once a counter becomes inactive, subsequent steps cannot require its activity. Once a member becomes plastic, subsequent steps assume that the member does not unload. Unloading is prohibited in scelernomic hyperelastic material effects analysis to avoid the possibility of multiple equilibrium solutions.

The proportional-loading concept bears indirectly upon the buckling and joint-slip analyses. The accuracy of buckling load predictions depends on the relevance of the stress state associated with the first analysis step. The joint-slip analysis steps may continue endlessly if joint-slip direction changes sign from one analysis step to the subsequent step.

The last-step results must satisfy the fundamental equations of structures or elasticity. For structures, the internal forces in elements must be consistent with the element load deflection relations. The member internal forces at the nodes must be in equilibrium with the applied external forces. All member endpoints sharing a node must undergo the same endpoint movements.

The concept of scelernomic analysis as a special analysis category is useful. It identifies a class of problems that evoke approximations beyond those of conventional linear structural and elasticity analysis and short of those of nonlinear static and dynamic analysis.

Scelernomic analyses play an important role in structural engineering, a role unusually dependent on the involvement of a knowledgeable engineer.

10.8 HOMEWORK

10:1. Counter theorem.

Given: The structure defined by the following finite element input.

Element	AE	Counter	Node	$\underline{x}$	$\underline{y}$
1-2	3E6	+1	1	0	0
1-4	5E6	+1	2	120	0
2-3	2E6	+1	3	200	0
2-4	1E6	+1	4	120	100
3-4	1E6	+1			

Fixities: 1, 1; 1, 2; 3, 1; −3, 2
Loads: −2, 1, 1200

Show:

(a) Removal of rod 2-3 leads to response predictions that satisfy all the structural equations.

(b) Removing any other rod leads to response predictions that violate the load deflection relations of at least one element.

***10:2.** Beam counters.

Given: The continuous beam of the sketch, with $P_1 = P_2$, and the fact that the section can resist only bending that causes compression in the top fibers.

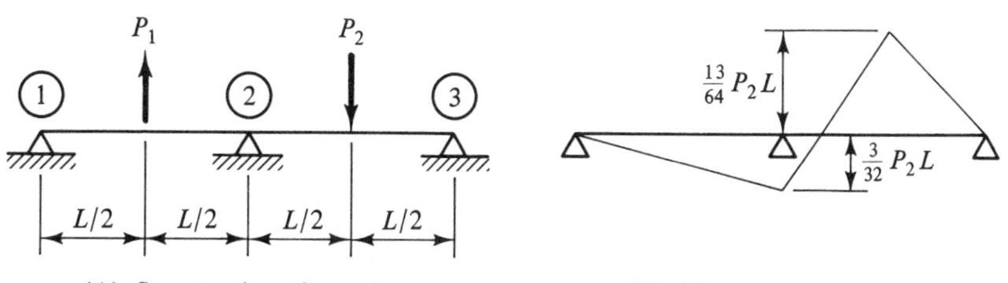

(A) Structural configuration (B) Moment diagram for P_1

Homework Problem 10:2

Determine:

(a) Whether either of the beam elements becomes inactive under the loading shown.

(b) Why the structural design is unacceptable.

***10:3.** Plastic moment capability, nonsymmetric cross section.

Given: The beam cross section shown, formed of an elastic perfectly plastic material with yield stresses of $\pm 20\ 000$.

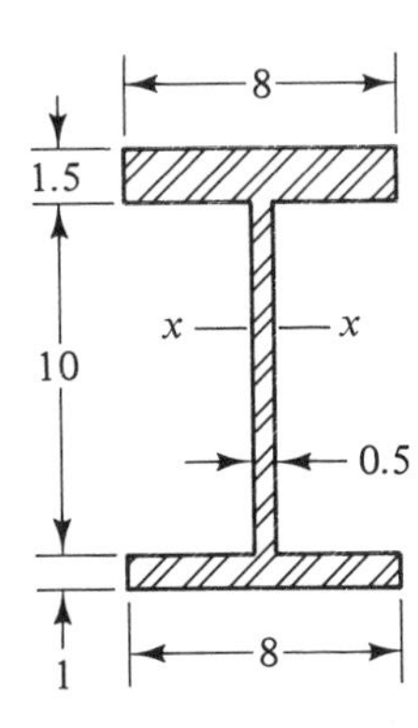

Homework Problem 10:3

Find: The plastic moment capability for bending about the *xx* axis.

10:4. Plastic moment capability, nonasymmetric stress-strain relation.

Given: The tensile yield stress for the material of the rectangular cross section shown is 40 000 and the compression yield stress is $-20\ 000$. Both the tension and the compression branches of the stress-strain relation are elastic perfectly plastic.

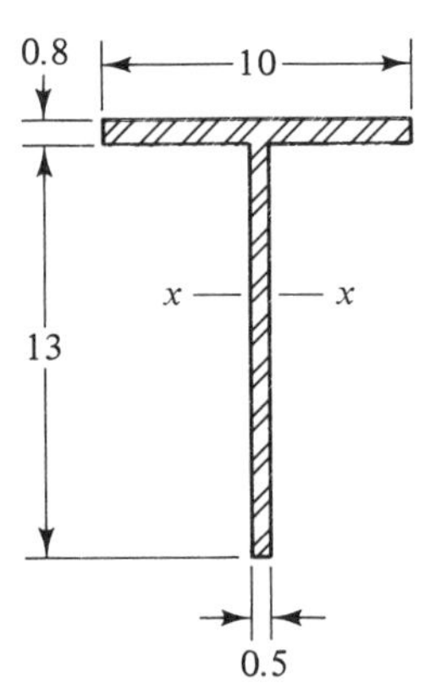

Homework Problem 10:4

Find: The positive and negative plastic moment capabilities of the cross section defined.

***10:5.** Three-bar truss of strain-hardened material.

Given: A three-bar truss and the stress-strain curve of its members, as shown.

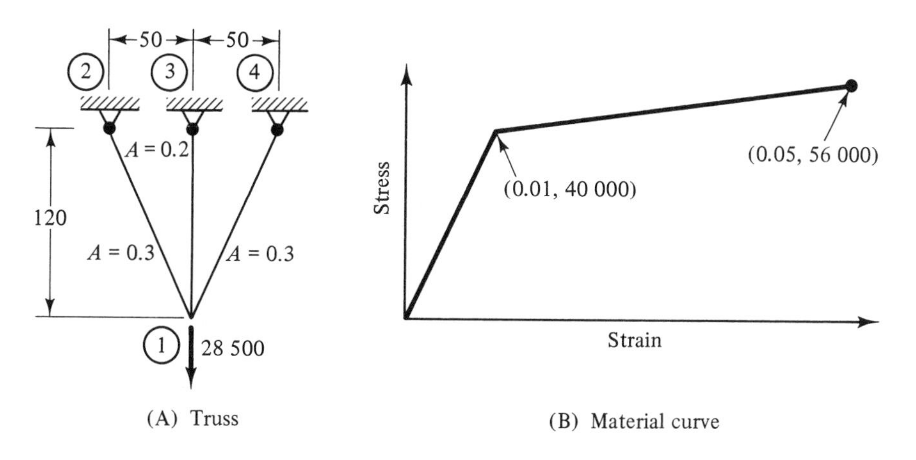

(A) Truss (B) Material curve

Homework Problem 10:5

Find: The stress in each rod.

10:6. Hyperelastic two-bar truss.

Given: The truss shown is loaded with a vertical force whose magnitude will vary. The stress-strain curve for the rod material is shown.

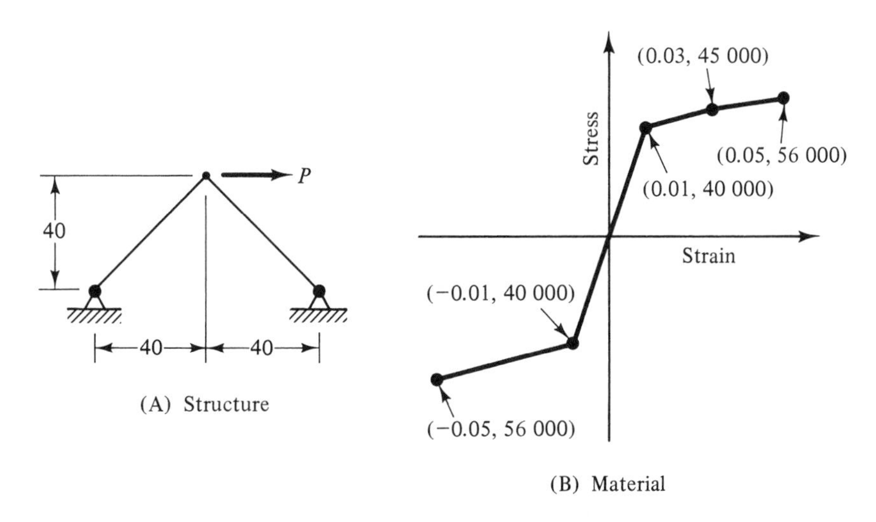

(A) Structure (B) Material

Homework Problem 10:6

Find: The load deflection curve for the system.

10:7. Rod induced stiffness.

Given: The induced stiffness matrix for a rod in two-dimensional space.

Show: That coefficients of the matrix are unchanged by rotation of the reference axes through any given angle.

10:8. Vanishing induced stiffness.

Given: The truss of the figure.

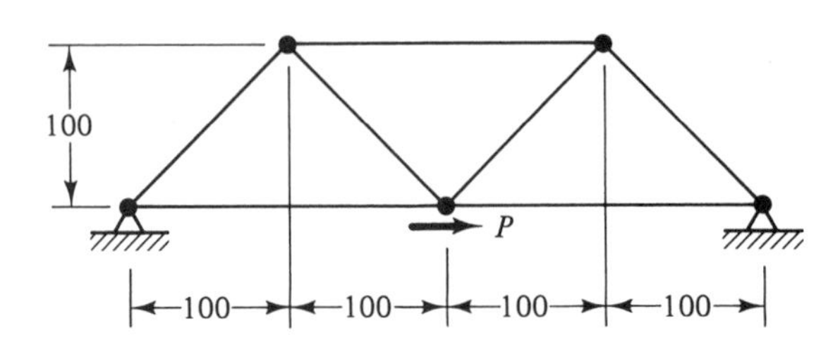

Homework Problem 10:8

Show: That the induced stiffness matrix is null for this problem.

10:9. Heating of a determinate structure.

Given: A statically determinate truss or frame resting on determinate supports and heated in any way.

Show: That thermoelastic stresses will be zero.

10:10. Heating of an indeterminate structure.

Given: An internally statically indeterminate structure resting on determinate supports. The structure is formed of a single material and everywhere has the same rise in temperature from the zero-stress stage.

Show: That thermoelastic stresses will be zero.

10.9 COMPUTERWORK

10-1. Building wind bracing.

Given: The structural geometry of Fig. 10.1, designed with X-bracing tension-only counters resisting the wind loads.

Determine:

(a) If the counter analysis method is relevant to this structure, that is, if all elements of the truss experience proportional loading.

(b) If the structure is kinematically stable or not.

(c) If the bracing is strong enough for the loading.

***10-2.** Tensionless foundation.

Given: The two-dimensional truss model of a nuclear reactor building resting on a sandy soil and loaded by dead loads and the D'Alembert forces of a major earthquake shock.

Determine: If the building will overturn.

Method: Find out if compression-only counters, placed between the soil and the structure, result in a kinematically stable response.

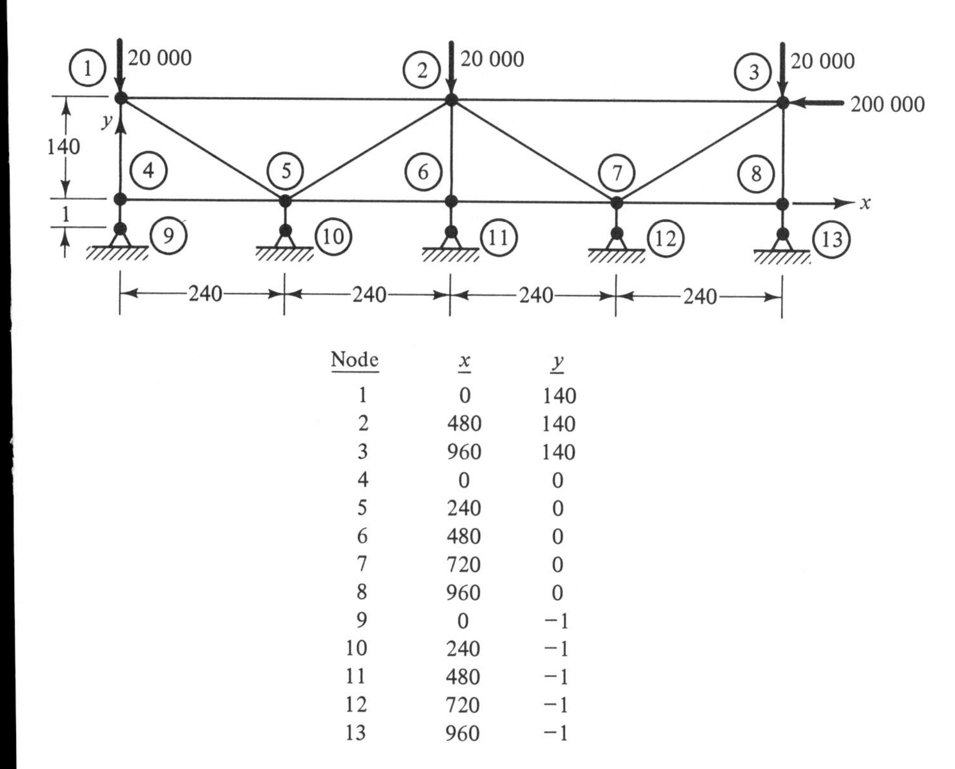

Node	x	y
1	0	140
2	480	140
3	960	140
4	0	0
5	240	0
6	480	0
7	720	0
8	960	0
9	0	−1
10	240	−1
11	480	−1
12	720	−1
13	960	−1

Elements
1, 2
2, 3
4, 5
5, 6
6, 7
7, 8
1, 4
2, 6
3, 8
1, 5
5, 2
2, 7
7, 3
4, 9
5, 10
6, 11
7, 12
−8, 13

Fixities: ?

Loads: 1, 2, −20 000; 2, 2, −20 000; 3, 2, −20 000; −3, 1, −200 000

Computerwork Problem 10-2

10-3. Inverted arch.

Given: The designer claims that the structure shown will act like an inverted arch, that is, that all members will be in tension for the loading shown.

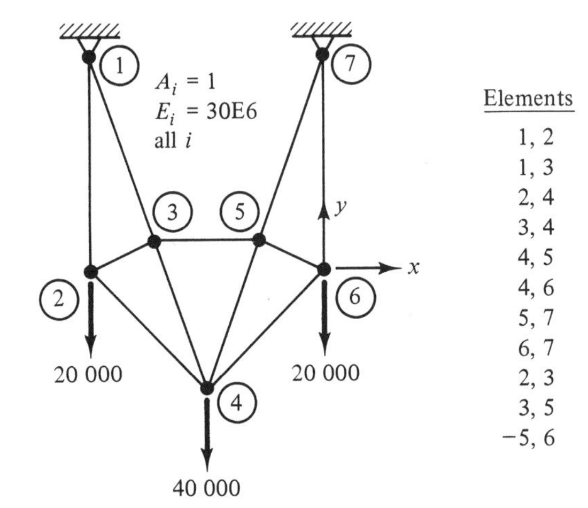

Node	x	y
1	−120	120
2	−120	0
3	−86	20
4	−60	−60
5	−34	20
6	0	0
7	0	120

Fixities: 1, 1; 1, 2; 7, 1; −7, 2

Loads: 2, 2, −20 000; 4, 2, −40 000; −6, 2, −20 000

Computerwork Problem 10-3

Determine:

(a) If his claim is valid.

(b) Whether some members can be removed to make his claim valid and retain kinematic stability.

10-4. Truss collapse.

Given: The three-dimensional truss shown.

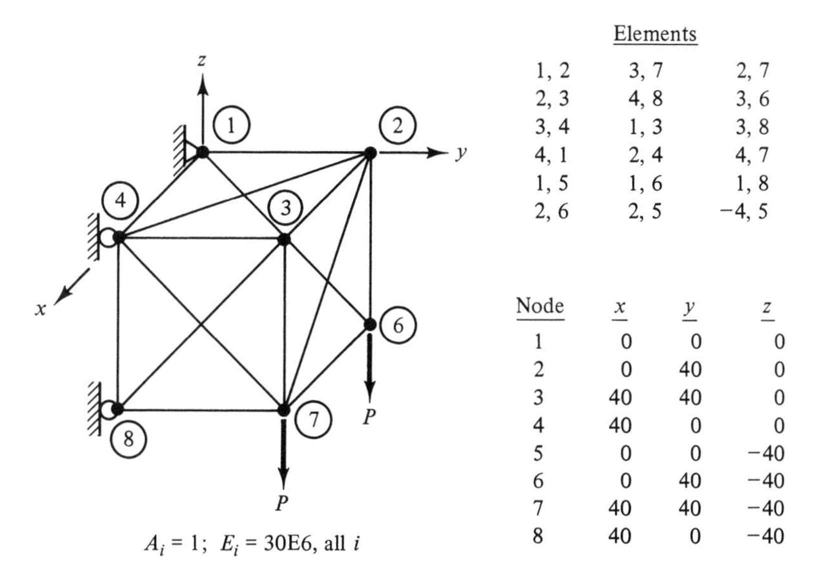

Elements

1, 2	3, 7	2, 7
2, 3	4, 8	3, 6
3, 4	1, 3	3, 8
4, 1	2, 4	4, 7
1, 5	1, 6	1, 8
2, 6	2, 5	−4, 5

Node	x	y	z
1	0	0	0
2	0	40	0
3	40	40	0
4	40	0	0
5	0	0	−40
6	0	40	−40
7	40	40	−40
8	40	0	−40

$A_i = 1$; $E_i = 30\text{E}6$, all i

Fixities: 1, 1; 1, 2; 1, 3; 4, 1; 4, 2; −8, 2

Computerwork Problem 10-4

Find: The collapse load when the yield stress is $\pm 40\ 000$.

10-5. Collapse of a hoist beam.

Given: The two-bay continuous beam loaded by a trolley hoist, as shown.

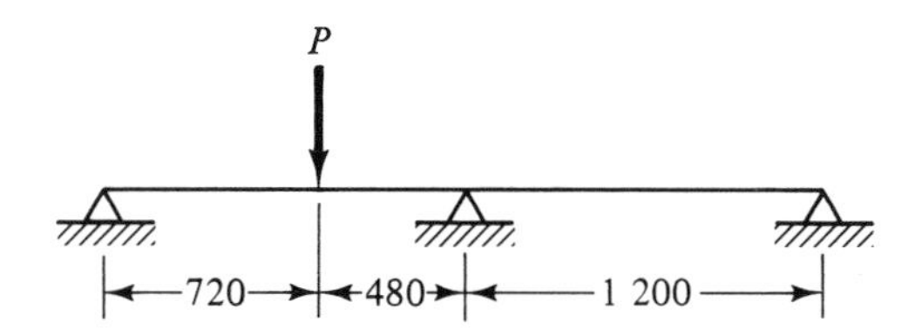

Computerwork Problem 10-5

Find: The load factor at which collapse will occur, as a function of the yield moment capability of the cross section.

***10-6.** Kayak floatation tank structure.

Given: The wooden truss structure shown. It supports the canvas wrapping for a bow flotation tank of a kayak. The stress-strain relation is approximated by the curve shown. The loads are caused by the hydrostatic pressure at a depth of h, where h is large compared with the dimensions of the tank.

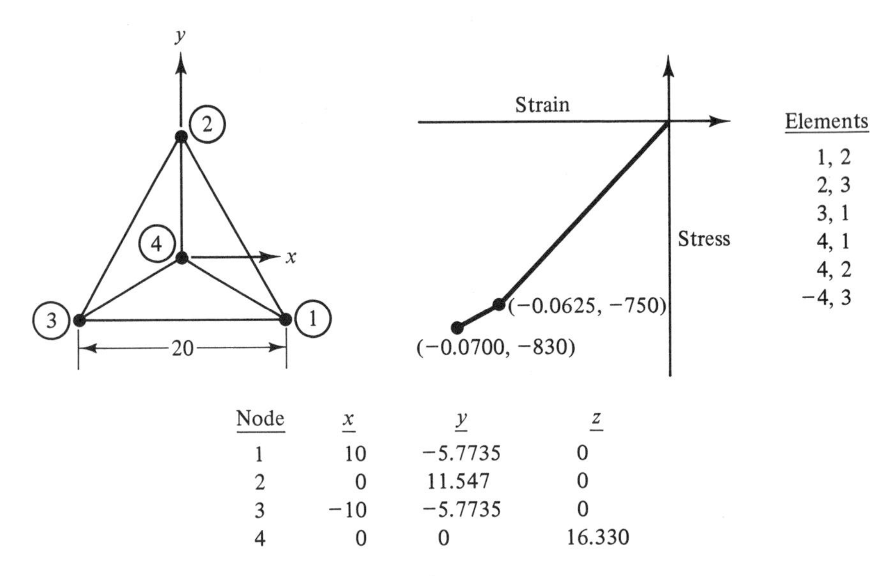

Node	x	y	z
1	10	−5.7735	0
2	0	11.547	0
3	−10	−5.7735	0
4	0	0	16.330

Fixities: 1, 1; 1, 2; 1, 3; 2, 1; 2, 3; −3, 2

Loading: 2, 2, $-3.77h$; 3, 1, $2.95h$; 3, 3, $2.41h$; −4, 3, $-4.17h$

Computerwork Problem 10-6

Determine: The value of h at which the structure will fracture (strain less than −0.07).

10-7. Car-barrier low-speed quasi-static impact analysis.

Given: The simple truss model of car structure shown. The model is two-dimensional. Loads are due to inertia forces and the forces from the vertical barrier in a low-speed head-on crash.

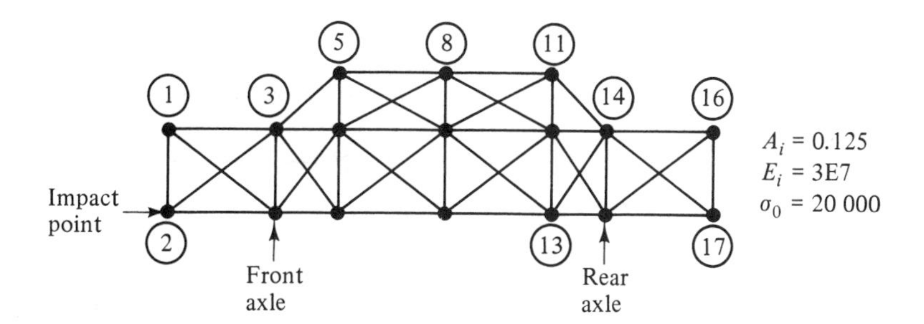

Elements		
1, 2	6, 7	10, 13
1, 3	6, 8	11, 12
1, 4	6, 9	11, 14
2, 3	6, 10	12, 13
2, 4	7, 9	12, 14
3, 4	7, 10	12, 15
3, 5	8, 9	13, 14
3, 6	8, 11	13, 15
3, 7	8, 12	14, 15
4, 6	9, 10	14, 16
4, 7	9, 11	14, 17
5, 6	9, 12	15, 16
5, 8	9, 13	15, 17
5, 9	10, 12	−16, 17

Node	x	y
1	0	36
2	0	9
3	36	36
4	36	9
5	54	54
6	54	36
7	54	9
8	90	54
9	90	36
10	90	9
11	126	54
12	126	36
13	126	9
14	144	36
15	144	9
16	180	36
17	180	9

Fixities: ?

Loads: 1, 1, −20; 2, 1, 600;
3, 1, −100; 4, 1, −100
5, 1, −20; 6, 1, −30;
7, 1, −40; 8, 1, −30;
9, 1, −40; 10, 1, −30;
11, 1, −20; 12, 1, −20;
13, 1, −30; 14, 1, −20
15, 1, −30; 16, 1, −30;
−17, 1, −40

Computerwork Problem 10-7

Determine: If the impact forces result in permanent deformation of the structure.

10-8. Buckling of a beam.

Given: The pinned-roller beam of the figure. The Euler buckling load of this beam is $n^2\pi^2\ EI/L^2$, where n is the number of the buckling mode.

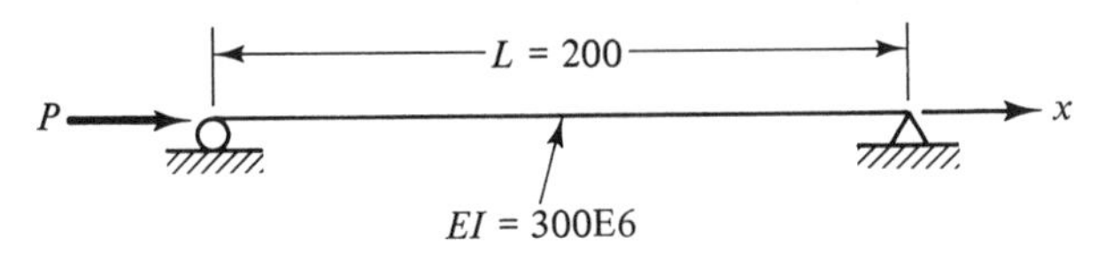

Computerwork Problem 10-8

Determine: The number of equal-span finite elements to predict the buckling load for the first and second mode, with less than 5 percent error compared with the Euler load.

***10-9.** Buckling of a Verendeel truss.

Given: The Verendeel deck truss of the figure. (In analysis terms, this would be called a frame.)

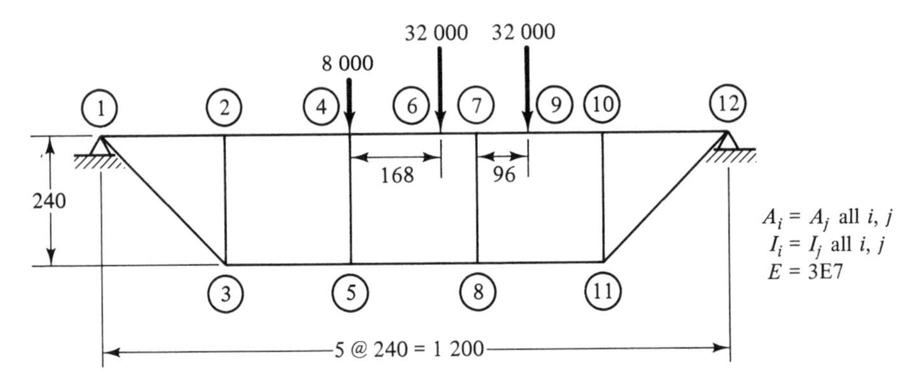

Computerwork Problem 10-9

Determine:

(a) The loading multiplier that associates with first-mode buckling.

(b) Whether discretization is fine enough to expect results of engineering accuracy (95% accuracy).

10-10. Force fitting a truss.

Given: Rod 3, 4 of the truss of the figure is stretched 0.6 units in force-fitting it into the undeformed geometry.

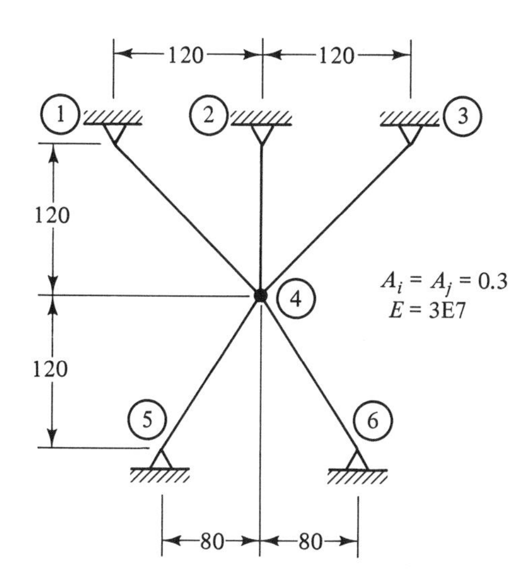

Computerwork Problem 10-10

Determine: The rod stresses and nodal deflections caused by the force fitting.

10-11. Turbine blade stiffness degradation.

Given: The truss model of a turbine blade designed to operate in a high-temperature environment. The tables define the problem data.

Element Data

Element	Area	Class
1, 2	0.2	1
1, 3	0.2	1
1, 4	0.2	2
1, 5	0.2	3
1, 6	0.2	1
1, 7	0.2	1
2, 3	0.2	4
2, 4	0.2	1
2, 5	0.2	1
2, 6	0.2	4
2, 7	0.2	4
2, 8	0.2	1
3, 4	0.2	1
3, 5	0.2	1
3, 6	0.2	4
3, 7	0.2	4
3, 8	0.2	1
4, 6	0.2	1
4, 7	0.2	1
−4, 8	0.2	3

Nodal Data

Node	x	y	z
1	10	0.15	1
2	10	0.30	0
3	10	0	0
4	10	0.15	−1
5	0	0.15	1
6	0	0.30	0
7	0	0	0
8	0	0.15	−1

Fixities: 5, 1; 5, 2; 5, 3;
6, 1; 6, 2; 6, 3;
7, 1; 7, 2; 7, 3;
8, 1; 8, 2; −8, 3

Loads: 1, 2, 1; −4, 2, −1

Material Properties

Mean Temperature (°)	Relative E*	Expansion Coefficient (× 10E6)
70	1.00	4.65
200	0.96	4.70
300	0.92	4.75
400	0.90	4.80
500	0.88	4.90
600	0.86	5.00
700	0.84	5.15
800	0.80	5.30
900	0.76	5.50
1000	0.70	5.70

Element Temperature States (°)

Class 1	Class 2	Class 3	Class 4
70	70	70	70
250	250	400	100
312	320	500	125
375	380	600	150
438	450	700	175
500	520	800	200
562	580	900	225
625	650	1000	250

*E @ 70° = 16E6 (titanium).

Find: The torsional stiffness of the model as a function of temperature, with and without considering induced stiffness.

10-12. Joint-slip truss geometry.

Given: The indeterminate truss described.

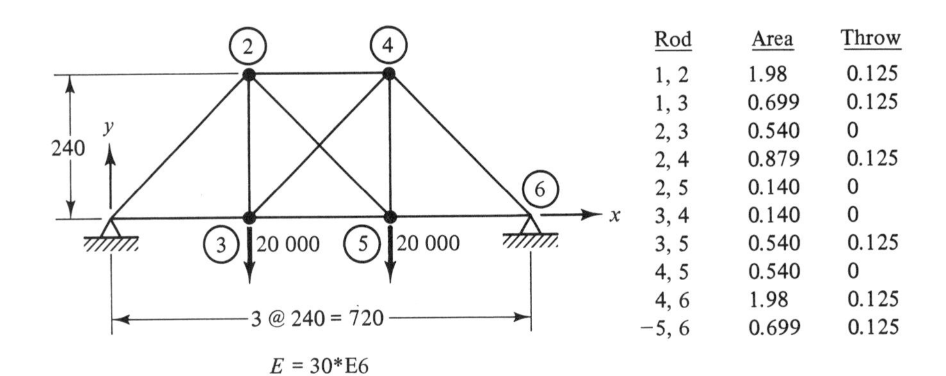

Rod	Area	Throw
1, 2	1.98	0.125
1, 3	0.699	0.125
2, 3	0.540	0
2, 4	0.879	0.125
2, 5	0.140	0
3, 4	0.140	0
3, 5	0.540	0.125
4, 5	0.540	0
4, 6	1.98	0.125
−5, 6	0.699	0.125

Computerwork Problem 10-12

Find: The displacements of the joints:

(a) When slip does not occur.

(b) When worst slipping occurs.

10.10 REFERENCES

1. Al-Mandil, M.Y., "Analysis and Behavior of Structures on Discrete Tensionless Foundations," Ph.D. dissertation, Duke University, 1981.
2. Hodge, P.G., *Plastic Analysis of Structures*, New York: McGraw-Hill, 1959.
3. Mallett, R.W., ed., *Limit Analysis Using Finite Elements.* New York: ASME, 1976.
4. Melosh, R.J., "Finite Element Approximations in Transient Analysis," *J. Astron. Sci.*, vol. 31, no. 3, 1983, pp. 343–358.
5. Williams, F.W., Wittrick, W.H., "Exact Buckling and Frequency Calculations Surveyed," *J. Struct. Eng.*, vol. 110, no. 1, Jan. 1984.
6. Donnell, L.H., "New Theory of Buckling of Thin Shells under Axial Compression and Bending," *Trans. ASME*, vol. 56, 1934, pp. 795–806.
7. Melosh, R.J., et al., "Scelernomic Analysis of Structures Considering Connection Slip," *Finite Elements in Analysis and Design*, New York: North-Holland, 1986.

Appendix A

Cross-Chapter Equation Notation

Principal Variables

b = semibandwidth of a square matrix
$\mathbf{C}$ = a constraint matrix of a contragradient transformation
e = a strain component
$\mathbf{E}$ = matrix of material property constants
E = Young's modulus
EW = external work
$\mathbf{g}$ = vector of generalized displacements
$\mathbf{K}$ = the stiffness matrix
n = the number of items of a class of items
$\mathbf{p}$ = vector of forces conjugate to $\mathbf{g}$, such that $\mathbf{p}^T\mathbf{g}$ is the external work
P = an applied load component
$\mathbf{q}$ = vector of equivalent forces at nodes
r, s, t = rotations about the x, y, and z axes, respectively
$\mathbf{s}$ = vector of stresses at a set of points in the structure
SE = strain energy
u, v, w = displacements along the x, y, and z axes, respectively

Superscripts

-1 = the matrix inverse
T = the transposed matrix

Subscripts

c = constrained
e = related to a finite element
i, j = matrix coefficients

Operators

$|\cdot\cdot\cdot|$ = the absolute value or determinant operator on . . .
δ = the partial differentiation or variational operator
Δ = the finite change operator
$\int$ = the integration operator
$\|\cdot\cdot\cdot\|$ = the norm operator on . . .
Σ = the summation operator

Appendix B

Answers and Comments on Selected Homework and Computerwork Problems

CHAPTER 1 HOMEWORK

1:1. a, b, d

1:3. b, c, d

1:5. b, c

1:7. a, d

1:9. a, b, d, e

1:11. c, d

CHAPTER 2 HOMEWORK

2:4.

$$\begin{pmatrix} t_3 \\ v_4 \\ t_4 \\ v_5 \\ t_5 \end{pmatrix} = \begin{bmatrix} 0 & 0 & 0 & 0 \\ 0 & 0 & 1 & 0 \\ 0 & 0 & 0 & -1 \\ 1 & 0 & 0 & 0 \\ 0 & -1 & 0 & 0 \end{bmatrix} \begin{pmatrix} v_1 \\ t_1 \\ v_2 \\ t_2 \end{pmatrix} \quad \text{or} \quad \begin{pmatrix} v_1 \\ t_1 \\ v_2 \\ t_2 \\ t_3 \end{pmatrix} = \begin{bmatrix} 0 & 0 & 1 & 0 \\ 0 & 0 & 0 & -1 \\ 1 & 0 & 0 & 0 \\ 0 & -1 & 0 & 0 \\ 0 & 0 & 0 & 0 \end{bmatrix} \begin{pmatrix} v_4 \\ t_4 \\ v_5 \\ t_5 \end{pmatrix}$$

or selected equations from each set of constraints, depending on which generalized displacements are desired for the constrained equations.

2:7. $v_1 = 0$

2:8.

$$\begin{pmatrix} u_3 \\ v_3 \\ t_3 \end{pmatrix} = \begin{bmatrix} 1 & 0 & -\mathrm{b} \\ 0 & 1 & \mathrm{a} \\ 0 & 0 & 1 \end{bmatrix} \begin{pmatrix} u_2 \\ v_2 \\ t_2 \end{pmatrix}$$

2:11. **b.** Nodal sparsity = 190/256
c. Bandwidth = 4

CHAPTER 3 COMPUTERWORK

3-1.

Element Data		Nodal Data		
Element	AE	Node	x	y
1, 2	43.5E5	1	0	80
1, 3	39.0E6	2	60	80
2, 3	28.5E6	3	60	0
2, 4	57.0E6	4	0	0
−3, 4	75.0E6			

Boundary Conditions	
Fixities	Loads
1, 1	2, 1, 10000
1, 2	−3, 2, 23000
4, 1	
−4, 2	

The user may enter the element data lines and boundary condition lines in any order, but the last line must start with a minus sign. The nodal data lines must be ordered by the nodal number.

3-5. **a.** 1. Negative AE value (last line of element data) is not acceptable.
2. No element connected to node 7.
3. Minus sign appears on the third from last fixity rather than the last.
4. Nodal variable 5, 1 has both been fixed and given a prescribed settlement. (A zero settlement may be prescribed.)

CHAPTER 4 HOMEWORK

4:1.

$$\begin{bmatrix} 8\frac{7}{8} & -\frac{3}{4} \\ -\frac{3}{4} & 3\frac{1}{2} \end{bmatrix} \begin{pmatrix} u_1 \\ u_2 \end{pmatrix} = \begin{pmatrix} U_1 \\ U_2 \end{pmatrix}$$

4:4. **a.** $V_1 = q(L^3a - La^3 + a^4/2)/L^3$

$T_1 = q(12a^2/2 - 2La^3/3 + a^4/4)/L^2$

b. $V_1 = q(L^3a - La^3 + a^4/2 + Lc^3 - c^4/2)/L^3$

4:5.

Structure Number	$\mathbf{K}_c$ Order	$\mathbf{K}_c$ Max. Rank	$\mathbf{K}_c$ Actual Rank	Degeneracy Source
1	4	2	2	v
2	2	2	1	v
3	9	9	9	—
4	10	10	9	u

4:7.

$$\begin{pmatrix} v_1 \\ t_1 \end{pmatrix} = \begin{bmatrix} 0 & 1 & 0 \\ 0 & 0 & 1 \end{bmatrix} \begin{pmatrix} u_1 \\ v_2 \\ t_2 \end{pmatrix}$$

4:11.

$$\frac{GJ}{L}\begin{bmatrix} 1 & 0 & -0.5\sqrt{3} & -0.5 \\ & 0 & 0 & 0 \\ & & 0.75 & 0.25\sqrt{3} \\ \text{Sym.} & & & 0.25 \end{bmatrix} \begin{pmatrix} t_{x1} \\ t_{y1} \\ t_{x2'} \\ t_{y2'} \end{pmatrix} = \mathbf{p}$$

CHAPTER 4 COMPUTERWORK

4-4. The analyses for the test cases lead to the following results and conclusions.

Test	Computer Result	Possible Rank
$u_1 = 0$	Matrix singular	$<$ or $= 5$
$u_1 = t_1 = 0$	Matrix singular	$<$ or $= 4$
$u_1 = t_1 = t_2 = 0$	Matrix singular	$<$ or $= 3$
$u_1 = u_2 = v_1 = 0$	Matrix singular	$<$ or $= 3$
$u_1 = v_1 = v_2 = 0$	Matrix nonsingular	$>$ or $= 3$

CHAPTER 5 HOMEWORK

5:4.

$$\mathbf{L} = \begin{bmatrix} 3 & 6 & 6 \\ & -1 & -3 \\ & & 8 \end{bmatrix}$$

By inspection, in accordance with the rules of definiteness, the matrix is indefinite.

CHAPTER 5 COMPUTERWORK

5-2. Since the reactions at node 4 of the left figure structure equal the loads on node 4 of the right figure structure and the fixities define determinate supports, the forces in all rods of the left truss must equal those of the right truss. Computer analyses furnish the data in the following table, confirming the consistency of the computer configuration for this problem.

Left Truss Rod No.	Right Truss Rod No.	Internal Force
1, 2	2, 1	−9 201.4
1, 3	2, 4	5 098.2
1, 4	2, 3	11 502.
2, 3	1, 4	−8 498.2
2, 4	1, 3	5 098.9
3, 4	3, 4	6 798.6

CHAPTER 6 HOMEWORK

6:1. The analyses lead to the results listed in the table.

Singularity Ratios for a Beam

NODAL NUMBERING	NODE n, ROTATION DEGREE OF FREEDOM	NODE n, TRANSLATION DEGREE OF FREEDOM
Tip to root	0.5	0.5
Root to tip	n^{-3}	$0.25n^{-2}$

6:3. $\|\Delta\mathbf{C}\| \leq (\|\Delta\mathbf{A}\|\ \|\mathbf{B}\| + \|\Delta\mathbf{B}\|\ \|\mathbf{A}\| + \|\Delta\mathbf{A}\|\ \|\Delta\mathbf{B}\|)\|\mathbf{B}^{-1}\|/\|\mathbf{B}\|$

6:7. **a.** The answers follow. Because the matrix is symmetric, the zero and infinity norms are equal.

Condition Number of Wilson's Matrix

BASIS OF THE NORM	FORMULA TO CALCULATE	$\|K\|$	$\|K^{-1}\|$	CONDITION NUMBER
One norm	max. $j\Sigma_i\|K_{ij}\|$	33.00	136.0	4488
Infinity norm	max. $i\Sigma_j\|K_{ij}\|$	33.00	136.0	4488
Euclidean norm	$(\Sigma_i\Sigma_j\|K_{ij}\|^2)^{1/2}$	30.54	98.53	3009

b. Assuming that the error in a coefficient must be less than 2^{-t}, the maximum error in arithmetic, using the lesser condition number, is 3009×2^{-t}.

CHAPTER 6 COMPUTERWORK

6-2. The singularity ratio associates with rotation at node 2, which can be determined by calculating all the ratios of D_{ii}/K_{ii}. For example, the ratio for rotation at node 2 is

$$\frac{2EI_{2,3}/L_{2,3}}{(4EI/L)_{1,2} + (4EI/L)_{2,3}} = 0.2499993235$$

The IEEE single-precision computer analysis finds the singularity ratio to be 0.2499994.

6-5. Computer results for the analyses are as follows.

Internal Bending Moments at the Nodes

NODE	BASELINE ANALYSIS	REVISED PROBLEM	DOUBLE PRECISION	EXACT SOLUTION
1	6.00000E0	−14.500E0	0	0
2	−0.94800E3	−1.05300E3	−1.00000E3	−1.00000E3
3	−8.41850E3	−8.81688E3	−9.00000D3	−9.00000D3
4	−67.9800E3	−71.1167E3	−73.0000D3	−73.0000D3

Tabulating the relative error measures gives the following.

Relative Error in Beam-Bending Moments

NODE	BASELINE VS. REVISED (%)	BASELINE VS. EXACT (%)	REVISED VS. EXACT (%)
2	11.1	5.2	5.3
3	4.7	6.5	2.0
4	4.6	6.9	2.6

This data reflects that when the round-off self-consistency tests exhibit errors, they estimate the round-off errors.

CHAPTER 7 HOMEWORK

7:3. $(1 + \nu) > 0; (1 - 2\nu) > 0; (1 - \nu) > 0$

7:5. Taking the strain energy of the resisting foundation as

$$SE = 0.5 \int_0^L k\left[\frac{u_1 + (u_2 - u_1)x}{L}\right]^2 dx$$

and the external work as

$$EW = U_1 u_1 + U_2 u_2$$

the boundary element matrix can be developed by the potential energy minimization.

7:7. Since $\begin{pmatrix} u_1 \\ u_2 \end{pmatrix} = \begin{bmatrix} 1 & 0 \\ 1 & L \end{bmatrix}\begin{pmatrix} a_0 \\ a_1 \end{pmatrix}$, the Vandermode matrix is $\begin{bmatrix} 1 & 0 \\ 1 & L \end{bmatrix}$, and the interpolation function is $u = u_1 + (u_2 - u_1)x/L$.

7:11. **a.** $x = 0.5$
b. $x = 0.75$
c. estimated error $= 0.75 - 0.50 = 0.25$
d. remainder at $\theta = \pi/4$ is -0.17814
e. actual error $= -0.04289$

7:14. Since the highest-order integrand is x^3, we need at least a two-point rule. Using two-point Gauss quadrature, we find the integral to give

$$\begin{bmatrix} 2 & 1.66667 \\ 1.66667 & 0 \end{bmatrix}$$

7:16. Since the integrand is at worst a 10-parameter polynomial and since each Gauss point is defined by two coordinates and one weight value, the least number of Gauss points for exact integration is 4.

7:19. $J = 0.5\begin{bmatrix} x_1 - x_2 & y_1 - y_2 \\ x_1 - x_4 & y_1 - y_4 \end{bmatrix}$

CHAPTER 7 COMPUTERWORK

7-2. The program should detect that the stiffness matrix is singular and the equation for node 7, component 2 is dependent and consistent.

7-5. **b.** The problem is solved by imposing settlements as follows.

$$
\begin{aligned}
u_1 &= b/(a^2 + b^2)^{0.5} & v_1 &= -a/(a^2 + b^2)^{0.5} \\
u_2 &= -b/(a^2 + b^2)^{0.5} & v_2 &= -a/(a^2 + b^2)^{0.5} \\
u_3 &= b/(a^2 + b^2)^{0.5} & v_3 &= a/(a^2 + b^2)^{0.5} \\
u_4 &= -b/(a^2 + b^2)^{0.5} & v_4 &= a/(a^2 + b^2)^{0.5}
\end{aligned}
$$

CHAPTER 8 HOMEWORK

8:4. **a.** Any set of u and v that satisfy the equations

$$(y - b)u_1 - (y + b)u_2 + (y + b)u_3 - (y - b)u_4 = \text{constant}$$

$$(x - a)v_1 - (x - a)v_2 + (x + a)u_3 - (x + a)v_4 = 0$$

$$(x - 1)u_1 - (x - a)u_2 + (x + a)u_3 - (x + a)u_4$$
$$+ (y - b)v_1 - (y + b)v_2 + (y + b)v_3 - (y - b)v_4 = 0$$

For example, $u_1 = u_2 = 0$; $u_3 = u_4 = 1$; $v_1 = v_2 = v_3 = v_4 = 0$

b. Possible solutions:

$$u_1 = u_2 = u_3 = u_4 = 0; \quad v_1 = v_2 = 0; \quad v_3 = v_4 = 1$$

$$u_1 = u_2 = u_3 = u_4 = 1; \quad v_1 = v_4 = 1; \quad v_2 = v_3 = 0$$

$$u_1 = u_2 = u_3 = u_4 = 1; \quad v_1 = v_4 = 2; \quad v_2 = v_3 = 1; \quad u_3 = u_4 = 1$$

8:7. **a.** The least squares solution is $(u_1, u_2) = (-1, 2)$.

b. The residuals are $(0, 0, 0)^T$.

8:11. **a.** The model and corresponding stiffness equations are

$$\mathbf{K\,g} = \begin{bmatrix} 2 & -1 & -1 & 0 \\ -1 & 2 & 0 & -1 \\ -1 & 0 & 2 & -1 \\ 0 & -1 & -1 & 2 \end{bmatrix} \begin{pmatrix} g_1 \\ g_2 \\ g_3 \\ g_4 \end{pmatrix}$$

b. The model and corresponding stiffness equations are

$$\mathbf{K\,g} = \begin{bmatrix} 2 & -1 & -1 & 0 \\ -1 & 2 & 0 & -1 \\ -1 & 0 & 2 & -1 \\ 0 & -1 & -1 & 2 \end{bmatrix} \begin{pmatrix} g_1 \\ g_2 \\ g_3 \\ g_4 \end{pmatrix}$$

c. Since **K** of part (a) equals **K** of part (b), the torsional rigidity is independent of the direction of the diagonal.

CHAPTER 8 COMPUTERWORK

8-4. The constant-strain state is evident by the fact that all Gauss points indicate zero direct stresses and shear stresses are all 8.01282E3.

8-7. **a.** The amount of constant-strain strain energy is (1 − %local error/100) × total work. Therefore, the constant-strain strain energies are 0.4313589 × 92.09695 = 39.72684 for (B) and 0.7046497 × 148.8565 = 104.8917 for (C).

b. Total work is calculated by imposing the displacements of model in (B) on the one-point Gauss element. This gives total work = 39.72681, essentially the same value as found in the analysis of (B) in part (a).

CHAPTER 9 HOMEWORK

9:1. For the simplex model, the warping varies linearly along node lines. Since the node lines of the sides of the rectangular elements are parallel to the coordinate axes, the boundary conditions are expressed by

$$\delta\phi/\delta n = y \cos(x, n) - x \cos(y, n)$$

a. At node 2, $\delta\phi/\delta n = (\phi 2 - \phi 6)/0.25 = -0.25$

b. At node 8, $\delta\phi/\delta n = (\phi 8 - \phi 7)/0.25 = 0.75$

9:3. Since $\phi_4 = \phi_7 = 0$, the stiffness equation for p_8 gives

$$p_8 = -0.5\phi_4 + \phi_8 - 0.5\phi_7 = \phi_8$$

Therefore, the stiffness equations can be represented by

$$\begin{bmatrix} 1 & -0.5 & 0 & -0.5 \\ & 1 & -0.5 & 0 \\ & & 0 & -0.5 \\ & \text{Sym.} & & 1 \end{bmatrix} \begin{pmatrix} \phi_3 \\ \phi_4 \\ \phi_8 \\ \phi_7 \end{pmatrix} = \begin{pmatrix} p_3 \\ p_4 \\ 0 \\ p_7 \end{pmatrix}$$

with the boundary conditions $\phi_4 = \phi_7 = 0$; and $\phi_8 = -\phi_7$ by a contragradient transformation, as usual.

CHAPTER 9 COMPUTERWORK

9-4. The exact torsional rigidity $J = I_{xx} + I_{yy} - SE_{\text{warping}}$. Thus $J = 2.666667 - SE_{\text{warping}} = 2.249232$.

a. The four-grid interval bilinear element estimates $J = 2.666667 - 0.0483734 \times 8 = 2.279679$. Therefore, the actual relative error of SE is 1.36 percent and the relative accuracy, 98.6 percent.

b. The constant-strain energy estimates a discretization error of 2.51 percent, giving a relative accuracy of 97.49 percent.

9-7. **a.** $SE = 0.0$

b. $SE = 0.0480769$

c. $SE = 0.0420813$

d. Convergence rate of hyperbolic models (h refinement) = 0.0420813

Convergence rate of hyperbolic to superquadratic (p refinement) = 0.0480769

In this case, p refinement involves a 14 percent higher rate of convergence than h.

9-9. The exact $SE = 0.0521793$ for an octant of the shaft.

a. h extrapolation yields $SE = 0.055082$
h^2 extrapolation yields $SE = 0.052078$
h, h^2 extrapolation $SE = 0.053145$

h^2 extrapolation is closest to the exact SE.

b. ϕ at midpoint on free edge $= -0.138914$

c. ϕ from h extrapolation $= -0.140614$
ϕ from h^2 extrapolation $= -0.139086$

CHAPTER 10 HOMEWORK

10:2. **a.** The bending moments of elastic analysis are as follows.

Site	Moment	Top Fiber Stress
at P_1 load	$PL/4$	tension
at node 2	0	—
at P_2 load	$-PL/4$	compression

b. The structural design is unacceptable because beam segment 1, 2 will fail in overstress no matter how small the value of P.

10:5. The stresses are as follows:

Rod	Value of Stress
1, 2	36 900
1, 3	40 300
1, 4	36 900

Index

Accelerated Gauss Seidel iteration, 89–94
analysis basis, 267
illustrative problem, 267–69
material model, 266
number of analysis steps, 267
Analysis interpretation, 12–13
Arithmetic relative error, 106

Bandwidth minimization algorithms, 89
Boundary conditions:
displacement, 9, 18, 19, 21–23
force, 16
Boundary finite elements, 141, 151
Buckling analysis. (*see* Linear buckling analysis)

Choleski's method, 89
Cogradient transformation, 60–65, 113
Comparison test of stiffness matrices, 205–6
Compatibility equations of elasticity, 141
Computer arithmetic error, 105–8
Computer configuration, 8, 11–12, 32–33, 71
consistency checking, 37
validation, 36–37
Computer errors, 12, 102–24
Computer number representation, 103–5
Computer tests:
advantages and disadvantages, 184–85
of arithmetic mode, 105
comparison test of stiffnesses, 205–6
of computer arithmetic, 112–14
Condition number of a square matrix, 110
Conjugate loads, 19, 296
Consistent equations, 88, 92
Constant strain:
discretization sensors, 227–30
evaluation, 199
Continuum element models, 130–83
derivation of, 152–56, 175–76
equivalent loads, 172–73
field and boundary, 141, 151
membrane, 142–43
plate, 144–47
selection of models, 176–77
solid, 5, 136, 148–50
types, 131–36
Continuum element model tests:
beam element test data, 190–94
constant strain states, 171
eigendata test, 195–96
eigenvalue test, 187–89
equivalent load test, 172–73
extended lattice test, 189–90
membrane element test data, 199–205
potential energy model test, 163
subdivisibility test, 196–99
Contragradient transformation, 23, 59–60
Coordinate transformation. (*see* Cogradient transformation)
Co-rotational coordinates, 272
Counters in structures, 250–56
analysis checks, 252, 256
analysis method, 252–56
load deflection relation, 256
rod element model, 250–51
theorems on analysis results convergence, 252
Crout's method, 89

Definiteness of a matrix, 49, 85
evaluation by triangularization, 85
evaluation from eigenvalues, 187–88
relation to strain energy, 188
Definiteness of material properties matrix, 138, 141
Degeneracy of a matrix, 57, 86
Differencing error, 106–8, 113, 123
Digits lost by round-off, 106, 111
Direct error measure, 90
Discretization 9–11, 49–50
Double-precision analysis, 103, 190–91, 241
Duhamel's principle, 276, 277

EASE:
input data, 34–36
input data checks, 35–36
installation, 33
minimum hardware, 33
model types, 33
program scope, 32–33
Eigendata test, 187–89, 195–96
Eigenvalue test, 187
Elasticity equations, 9, 137–40
Element energy, 19–23
Element potential energy test, 163
Engineer's responsibilities, 12–13, 82
Engineering system, 26–27
Equation solving, 8, 11
data processing, 82, 89
direct method, 82–89
iterative method, 82, 89–94
N-step iterative method, 82, 94–95
selection of, 95–96
validation of, 96
Equivalent loading:
continuum elements, 204, 205
lattice elements, 52–54, 186
superposition of, 54
Extended lattice test, 189–90, 192–93
External work, 19–21
Extrapolation of FEA results, 239–43
Richardson h^2 and h^2, h^4, 241–42

Sensitivity to round-off error, 241

FEA. (*see* Finite element analysis)
Field finite element models, 131
Finite element analysis:
analysis scope, 1
history of, 2, 96
on-going research, 1
process, 11, 16–28
steps, 9–11
technology, 2–8
widespread use of, 1
Finite element process:
assembly, 16–23
imposing boundary conditions, 18–19, 21–23

Gauss elimination, 83
Gauss integration rules, 165–67
Gauss point material yielding, 261
Gauss quadrature, 163–68
Gauss Seidel iteration, 89–94
convergence of solution estimates, 90–93
number of calculations, 93
optimization of, 91, 93
sensing equation consistency with, 92

Hereditary grids, 225–27
Hermite polynomials, 157–61
Hestenes-Stiefel equation solving method, 94
Higher precision arithmetic, 114–15, 122, 124
Hyperelastic materials structures, 266–70

Idealized model, 3
Implicit variables, 27
Inconsistent equations, 88, 92
Incremental load analysis, 262–63, 266–67
Incremental load analysis method, 262–66, 267, 278–79
Indirect error measure, 90
Induced stiffness matrix, 272–74, 277, 278
Interpolation function, 157–61
order of error, 160
remainder, 157

Jacobian, 169

Kinematic instability, 88

LaPlace's equation, 216–17
Lattice element models, 42–71
derivation from test data, 65–69
equivalent load, 52–54
microscopic models, 54–57
model formulation, 69
stiffness characteristics, 46–49
types, 42–46
unbraced and braced, 46, 57–59, 65–69
uniqueness symmetry, 68
Lattice element models, 43–46
Lattice element tests:
equivalent load test, 51–52
lattice property test, 49–52
Lattice structures, 42, 61, 71
Least squares solution of equations, 195–96
Library of elements, 5–6
Limit analysis, 257–66
beam plastic moment capability, 259–61
collapse load lower bound, 263
incremental load analysis method, 262–66
limit analysis models, 257–62
number of analysis steps, 263
proportional loading, 258
Linear buckling analysis, 270–76
analysis method, 273–74
discretization error for beam, 275
induced stiffness matrix, 271–72, 273
lattice models, 274
mathematical model error, 275
rod and beam buckling models, 270–73

Macroscopic equilibrium equations, 3–4, 17–18, 58, 140–41
Material properties matrix, 9, 137–40
anisotropic, 137
isotropic, 139
orthotropic, 139
plane strain, 140
plane stress, 140
Maxwell-Betti theorem, 24, 46
Membrane element models, 131, 133, 142–43, 153
Mesh accuracy measures, 228–30
element discretization residual, 229
optimality measure, 229
Mesh refinement strategies:
adaptive refinement, 230–33
hereditary grids, 225–27
h refinement, 231–37
p refinement, 237–39
regular grids, 240–41
Microscopic equations, 4–5, 54–55, 140–41, 174
Monotonic convergence proof, 237
Multinode element tests. (*see* Continuum element model tests)

N-step solution method, 94–95
algorithms, 94–95
number of calculations, 95
NASTRAN, 8
Nodal local reference axes, 61, 64
Nonlinear structural analysis, 249, 270, 276, 282–83
Norm algebra, 108–11
Numerical singularity measure, 50–52, 111, 112
Numerical tests of elements. (*see* Continuum element model tests *and* Lattice element tests)

Optimum computer arithr etic, 107–8
Optimum grids, 224 27
Orthogonal matr , 65
Overflow of e onent, 103

Parame mapping, 169–71
Parent and child grids. (*see* Hereditary grids)
Perfect computer arithmetic, 105
Plastic collapse, 263
Plastic moment capability, 259–60
Plate element models, 131, 135, 144–45, 146–47
Positive definiteness. (*see* Definiteness of a matrix)
Potential energy analysis test, 163
Preferentiality test, 196–99, 204–5
Prestressed structures, 276–77
analysis method, 276–77
Duhamel's principle, 276, 277
with induced stiffness, 277
Problem descriptive data, 34–35
element data 34
nodal data 34–35

Procedural element models, 5, 69–70
Proportional loading, 258
Pseudotime analysis, 251–52, 258, 278–79, 282

Rank of a matrix, 24, 57, 86
Regular grid refinement, 240–41
Rigid modes imply macroscopic equilibrium, 58
Round-off error, 102–24
  actual errors, 115
  bounds, 108–11, 189–91, 199–201
  control, 123
  measurement, 111–15
  minimization, 119–22
  probable error, 111, 114, 120

Scelernomic analysis, 249–83
  analysis checks, 282–83
  definition, 249, 282
  hyperelastic material behavior, 266--70
  limit analysis, 257–66
  linear buckling analysis, 270–76
  proportional loading, 258, 282
  slotted-joint structures, 277–82
  structures with counters, 250–56
  structures with prestresses, 276–77
Self-consistency checks 37, 96, 114–15
Self-qualifying FEA, 214–48
  adaptive local grid refinement, 230–36
  definition, 214
  discretization error measures, 227–30
  optimum grids, 224–27
  refining meshes, 243–44
  solution extrapolation, 239–43
Semi-infinite element models, 44, 131
Shape function, 157, 175
Shell element models, 5, 136, 148–50
Simplex element model, 142, 157
Singular stresses, 244
Singularity of a matrix, 24, 86
  consistency of equations, 88
  interpretation of, 86
Slotted-joint structures, 277–83
  analysis method, 279
  number of analysis steps, 279–80
  rod element model, 278
Solid element models, 5, 136, 148–49
Solution of FEA equations, 8, 81–101
  direct method, 82–89
  iterative method, 82, 89–94
  method selection, 95–96
  N-step method, 82, 94–95
  validation, 96
St. Venant's semi-inverse method, 215–16
Starting FEA grid, 231, 237
Stepwise-linear analysis. (*see* Scelernomic analysis)
Stiffness comparison test. (*see* Comparison test)
Stiffness matrix:
  bandwidth, 26
  definition, 3
  primary, secondary degeneracy, 188–89, 199
  rank, 24, 57, 185–86, 188, 189, 201
  sparsity, 26
  symmetry, 24
Strain energy, 19, 109
Structural analysis checks, 252, 256, 283
Structural equations, 5, 140
Subdivisibility test. (*see* Preferentiality test)
Superconvergence, 234

Three dimensional elasticity theory, 137–40
Torsion analysis, 215–36
  boundary conditions, 217
  FEA, 219–21
  governing differential equation, 216
Torsion element models, 219–20
Torsional rigidity FEA, 215–36
Triangular factorization, 82–89
  back substitution, 83
  for determining definiteness, 85
  forward substitution, 83
  number of calculations, 85, 88
Two-noded element tests. (*see* Continuum element model tests)

Unbraced and braced elements, 57–59, 65–69
Unimodular matrix, 43–44
Uniqueness symmetry, 68

Validation of the computer configuration, 36–37
Vandermode matrix, 154, 155
Von Mises yield criterion, 261

Tools Search Notes Discuss

MyReportLinks.com Books

Go!

PRESIDENTS

THOMAS JEFFERSON

A MyReportLinks.com Book

Chris Reiter

MyReportLinks.com Books
an imprint of

Enslow Publishers, Inc.
Box 398, 40 Industrial Road
Berkeley Heights, NJ 07922
USA

MyReportLinks.com Books, an imprint of Enslow Publishers, Inc.

Library of Congress Cataloging-in-Publication Data

Reiter, Chris.
Thomas Jefferson / Chris Reiter.
p. cm. — (Presidents)
Includes bibliographical references and index.
Summary: A biography of the third President, who was also creator of the Declaration of Independence.
ISBN 0-7660-5071-8
1. Jefferson, Thomas, 1743–1826—Juvenile literature. 2. Presidents—United States—Biography—Juvenile literature. [1. Jefferson, Thomas, 1743–1826. 2. Presidents.] I. Title. II. Series.
E332.79 .R45 2002
973.4'6'092—dc21

2001004311

Printed in the United States of America

10 9 8 7 6 5 4 3 2 1

To Our Readers:
Through the purchase of this book, you and your library gain access to the Report Links that specifically back up this book.
The Publisher will provide access to the Report Links that back up this book and will keep these Report Links up to date on **www.myreportlinks.com** for three years from the book's first publication date.
We have done our best to make sure all Internet addresses in this book were active and appropriate when we went to press. However, the author and the Publisher have no control over, and assume no liability for, the material available on those Internet sites or on other Web sites they may link to.
The usage of the MyReportLinks.com Books Web site is subject to the terms and conditions stated on the Usage Policy Statement on **www.myreportlinks.com**.
In the future, a password may be required to access the Report Links that back up this book. The password is found on the bottom of page 4 of this book.
Any comments or suggestions can be sent by e-mail to comments@myreportlinks.com or to the address on the back cover.

Photo credits: © Corel Corporation, pp. 1 (background), 3; Courtesy of America's Story, The Library of Congress, p. 17; Courtesy of DiscoverySchool.com, pp. 25, 36; Courtesy of Encyclopedia Americana, p. 30; Courtesy of Frontline/PBS, p. 42; Courtesy of Monticello—The Home of Thomas Jefferson, p. 39; Courtesy of MyReportLinks.com Books, p. 4; Courtesy of PBS Online, p. 33; Courtesy of The American President, pp. 16, 24; Courtesy of The Library of Congress, pp. 11, 13, 23; Courtesy of The University of Virginia, p. 40; Courtesy of The White House, p. 21; The Library of Congress, pp. 20, 26, 37; The United States Department of the Interior, pp. 28, 32.

Cover photo: © Corel Corporation; White House Historical Association.

Report Links

The Internet sites described below can be accessed at **http://www.myreportlinks.com**

*EDITOR'S CHOICE

▶Monticello: The Home of Thomas Jefferson

At this Web site about Thomas Jefferson's estate, Monticello, you can take a virtual tour and learn about the plantation and the people who lived and worked there. You will also find biographical information about Jefferson.

Link to this Internet site from http://www.myreportlinks.com

*EDITOR'S CHOICE

▶Thomas Jefferson: Man of the Millennium

This Web site includes interactive time lines, a collection of Jefferson's writings, and summaries of Jefferson's views on public service, architecture, education, science, and religious freedom, among other things.

Link to this Internet site from http://www.myreportlinks.com

*EDITOR'S CHOICE

▶Echoes From the White House

At this Web site, you will learn about Thomas Jefferson's involvement with helping to complete the President's House, as the White House was then called.

Link to this Internet site from http://www.myreportlinks.com

*EDITOR'S CHOICE

▶Lewis and Clark

This PBS site provides a comprehensive look at the explorations of Lewis and Clark. Here you will learn about the places they traveled to and the discoveries they made.

Link to this Internet site from http://www.myreportlinks.com

*EDITOR'S CHOICE

▶Thomas Jefferson: A Film by Ken Burns

This PBS site explores the ideals Thomas Jefferson fought for and their meaning in today's world. This site also features images of important documents written by Jefferson, including an early draft of the Declaration of Independence.

Link to this Internet site from http://www.myreportlinks.com

*EDITOR'S CHOICE

▶The Philosopher President

This site provides a comprehensive biography of Thomas Jefferson. Here you will learn about his life before, during, and after the presidency.

Link to this Internet site from http://www.myreportlinks.com

Report Links

The Internet sites described below can be accessed at **http://www.myreportlinks.com**

▶ Aaron Burr

Aaron Burr became vice president after the contested election of 1800. At this Web site you will find Burr's biography and learn about his vice presidency and his famous duel with Alexander Hamilton.

Link to this Internet site from http://www.myreportlinks.com

▶ The American Presidency: Thomas Jefferson

This site provides a detailed account of Thomas Jefferson's life and political career. In particular you will learn about Jefferson's contributions toward the cause of liberty, including the Declaration of Independence and Virginia's Statute of Religious Freedom.

Link to this Internet site from http://www.myreportlinks.com

▶ American Presidents, Life Portraits: Thomas Jefferson

This site provides "Life Facts" and "Did You Know?" trivia about Thomas Jefferson. You can also read or listen to a letter written by Jefferson.

Link to this Internet site from http://www.myreportlinks.com

▶ Autobiography of Thomas Jefferson

In Thomas Jefferson's autobiography, you will learn about Jefferson's views on the Declaration of Independence, the Articles of Confederation, and the French Revolution.

Link to this Internet site from http://www.myreportlinks.com

▶ The Avalon Project at the Yale Law School: The Papers of Thomas Jefferson

At this site you will find Jefferson's addresses to Congress, his inaugural addresses, major works, and miscellaneous writings.

Link to this Internet site from http://www.myreportlinks.com

▶ Declaring Independence: Drafting the Documents

From the Library of Congress comes a virtual exhibit that tells the story of how the Declaration of Independence came to be. Here you will find Jefferson's original draft of the Declaration and other original documents.

Link to this Internet site from http://www.myreportlinks.com

Report Links

The Internet sites described below can be accessed at **http://www.myreportlinks.com**

▶The House Selects a President

This site provides a brief description of the hotly contested election of 1800, which was ultimately decided by the House of Representatives.

Link to this Internet site from http://www.myreportlinks.com

▶To Form a More Perfect Union

This Library of Congress site explores the work of the Continental Congress between 1774 and 1789. It also examines the Constitutional Convention of 1787.

Link to this Internet site from http://www.myreportlinks.com

▶Jefferson's Blood

This PBS series explores recent articles by scholars that examine Thomas Jefferson's relationship to Sally Hemings.

Link to this Internet site from http://www.myreportlinks.com

▶*Marbury* v. *Madison* (1803)

This Web site explores the *Marbury* v. *Madison* Supreme Court case. Here you will learn about the key players and the historical significance of the case.

Link to this Internet site from http://www.myreportlinks.com

▶New Perspectives on the West: The Louisiana Purchase Treaty

This PBS site contains the text of the Louisiana Purchase Treaty, along with related documents. Here you will learn how, through the treaty, the United States gained new territory.

Link to this Internet site from http://www.myreportlinks.com

▶Notes on the State of Virginia

In 1781, Thomas Jefferson wrote a description of the state of Virginia. At this Web site you can read the full text of Jefferson's "Notes on the State of Virginia."

Link to this Internet site from http://www.myreportlinks.com

Report Links

The Internet sites described below can be accessed at **http://www.myreportlinks.com**

▶ **Objects From the Presidency**

By navigating though this site you will find objects related to all United States presidents. Here you will find Jefferson's polygraph, paperweight, and a desk designed by Jefferson.

Link to this Internet site from http://www.myreportlinks.com

▶ **Thomas Jefferson**

This site provides an outline of Thomas Jefferson's life and political career. You will also find many links to related online resources, including biographies, historical documents, and notable events.

Link to this Internet site from http://www.myreportlinks.com

▶ **Thomas Jefferson**

This site explores the life of Thomas Jefferson. Here you will learn about his life at Monticello, his creation of a Virginia Republic, and the Declaration of Independence.

Link to this Internet site from http://www.myreportlinks.com

▶ **Thomas Jefferson At Home**

America's Story, from America's Library, a Library of Congress Web site, provides interesting facts about Thomas Jefferson and his home life. Learn about some of his favorite things to cook.

Link to this Internet site from http://www.myreportlinks.com

▶ **Thomas Jefferson was born on April 13, 1743**

America's Story, from America's Library, a Library of Congress Web site, provides a brief introduction to Thomas Jefferson. Here you learn about some of Jefferson's major accomplishments over his lifetime.

Link to this Internet site from http://www.myreportlinks.com

▶ **Thomas Jefferson as an Inventor**

At this Web site you will find brief descriptions and images of Jefferson's inventions and his views on the use of patents.

Link to this Internet site from http://www.myreportlinks.com

Report Links

The Internet sites described below can be accessed at **http://www.myreportlinks.com**

▶Thomas Jefferson Digital Archive: Resources at the University of Virginia

This University of Virginia Web site contains 1,700 documents written by or to Thomas Jefferson. You will also find a large collection of quotations from Jefferson on a variety of topics.

Link to this Internet site from http://www.myreportlinks.com

▶The Thomas Jefferson Papers at the Library of Congress

The Library of Congress holds 27,000 Jefferson documents which you can explore. You will also find time lines accompanied by images.

Link to this Internet site from http://www.myreportlinks.com

▶Thomas Jefferson's Poplar Forest

Jefferson's final architectural masterpiece was Poplar Forest, his private retreat near Lynchburg, Virginia. This site provides a history of Poplar Forest and Jefferson's ties to the property.

Link to this Internet site from http://www.myreportlinks.com

▶The White House: Martha Wayles Skelton Jefferson

The official White House biography of Martha Wayles Skelton Jefferson includes details about Martha's relationship with her husband and her untimely death.

Link to this Internet site from http://www.myreportlinks.com

▶The White House: Thomas Jefferson

This official White House biography of Thomas Jefferson offers a brief overview of Jefferson's political career and rise to the presidency.

Link to this Internet site from http://www.myreportlinks.com

▶Thomas Jefferson's William & Mary

Thomas Jefferson attended the College of William and Mary, in Williamsburg, Virginia, from 1760 to 1762. This College site examines William and Mary as it was in Jefferson's time and describes Jefferson's days as a student there.

Link to this Internet site from http://www.myreportlinks.com

Highlights

1743—*April 13:* Thomas Jefferson is born at Shadwell, Albemarle County, Virginia.

1760–1762—Attends College of William and Mary in Williamsburg, Virginia.

1762–1767—Serves law apprenticeship with George Wythe.

1767—Begins law practice.

1769—Begins building Monticello.

1769–1776—Member of the House of Burgesses.

1772—*Jan. 1:* Marries Martha Wayles Skelton.

1775—Attends Continental Congress in Philadelphia.

1776—Writes the Declaration of Independence.

1776—Elected a member of the Virginia Assembly.

1779–1781—Serves as governor of Virginia.

1782—Martha Wayles Jefferson dies.

1784—Appointed trade minister to Paris.

1785—Appointed United States minister to France.

1790—Named secretary of state.

1796—Elected vice president of the United States.

1800—Elected president of the United States.

1803—Directs Louisiana Purchase and launches Lewis and Clark expedition.

1804—Reelected president.

1807—Signs Embargo Act.

1819—Establishes University of Virginia, which opens its doors to students on March 7, 1825.

1826—*July 4:* Dies at Monticello.

Chapter 1 ▶

The Declaration of Independence, 1776

For several warm and muggy evenings in June of 1776, in his rented rooms in Philadelphia, Thomas Jefferson was at work on an important document for the Continental Congress. The tall Virginian was one of nearly sixty lawyers, merchants, and wealthy farmers who were members of the second Continental Congress. They had come to

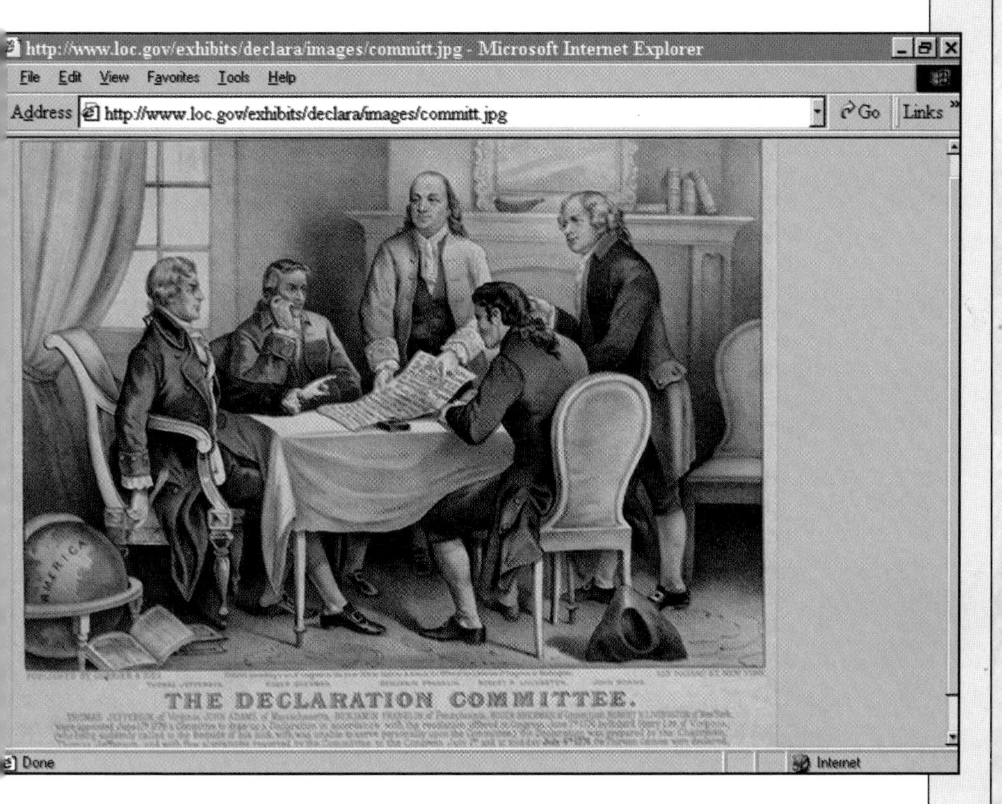

Thomas Jefferson was one of five men who served on a committee to draft a declaration of independence from Great Britain in 1776. Jefferson was the one chosen to write the declaration.

Philadelphia to decide the future of the thirteen colonies, whose citizens were questioning Great Britain's right to govern and tax them.

For ten years, colonial leaders had been challenging the policies of the British Parliament. By 1775, the conflict had led to war, and men were now dying in battle. In the spring of 1776, the delegates of the Continental Congress had met in Philadelphia and decided to declare their independence from Great Britain. A committee was formed to draft a declaration of independence, and Thomas Jefferson was the man chosen to write it.

On the Eve of Independence

In his rooms, Jefferson paused before reviewing the draft. In the morning, he would deliver the document to Congress. There would be a debate over the declaration, and he did not look forward to it. On this warm June night, he picked up the declaration and began to read.

"When in the Course of human events, it becomes necessary for one people to dissolve the political bands which have connected them with another . . . they should declare the causes which impel them to the separation."

In these first lines of the Declaration, Jefferson announced the colonies' intention to sever their ties with Great Britain. Then, in lines that would become "the most quoted statement of human rights in recorded history,"[1] he made clear why the colonies had the right to be free.

"We hold these truths to be self-evident, that all men are created equal, that they are endowed by their Creator with certain unalienable Rights, that among these are Life, Liberty and the pursuit of Happiness.— That to secure these rights, Governments are instituted among Men, deriving their just powers from the consent of the governed,— That whenever any Form of Government becomes destructive of these ends, it is the Right of the People to alter or to abolish it. . . ."

To fortify his case, Jefferson then listed the colonies' grievances. The crown, he wrote, had imposed taxes, influenced trade, suspended state legislatures, and kept British troops in America, all without the consent of the colonies. These and other abuses added up, Jefferson said, to "the establishment of an absolute Tyranny. . . ."

Finally, he declared that "these United Colonies are, and of Right ought to be Free and Independent States. . . ."

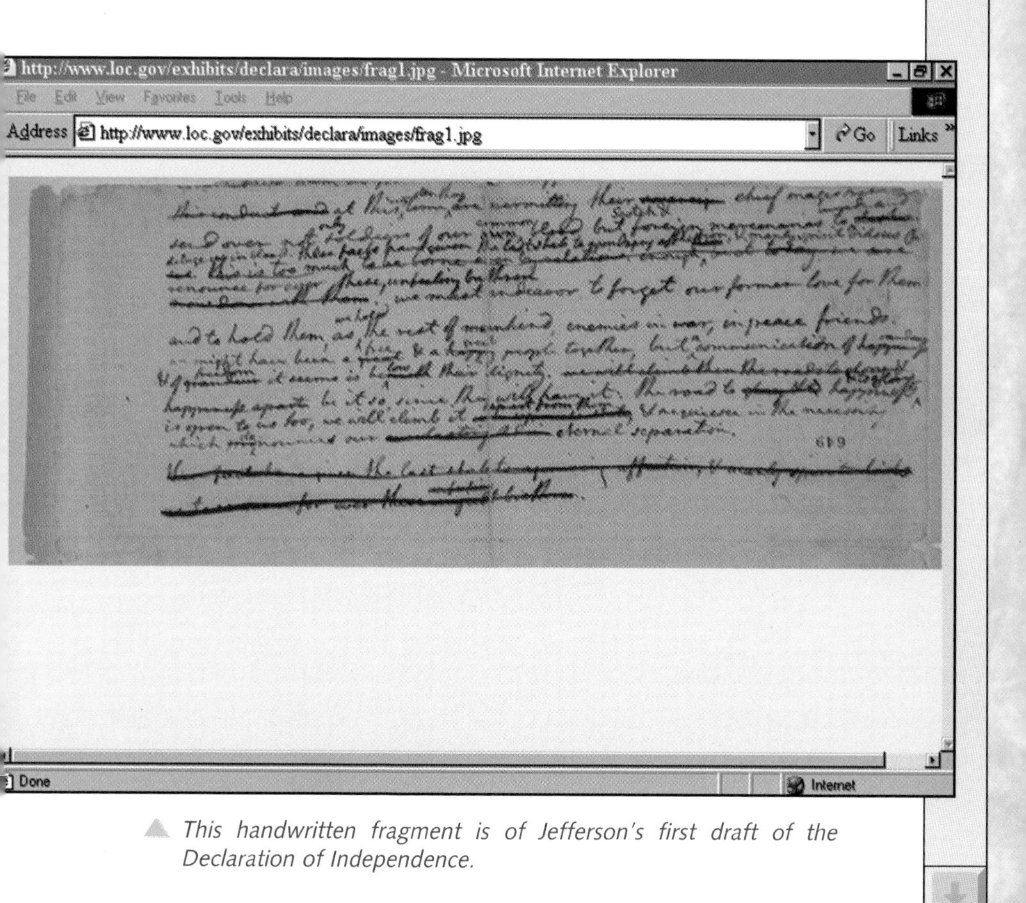

This handwritten fragment is of Jefferson's first draft of the Declaration of Independence.

Independence and Equality

Thomas Jefferson had managed, he thought, to make a commonsense case for independence.[2] He had declared all men equal, endowed with basic human rights. He had declared that the work of government is to secure and protect those rights. He had insisted that governments exist for men, not men for governments.[3] He had constructed an eloquent statement of the principles of a democratic nation.

The Continental Congress agreed. Its members debated the document and made minor changes to it before it was adopted, on July 4, 1776. Four days later, the Declaration of Independence was read to the people of Philadelphia, and before long, copies were circulating throughout the colonies.

As a legislator, diplomat, governor, and president, Jefferson would use the ideas expressed in the Declaration to shape the laws and freedoms of the United States. The new nation, and Jefferson himself, would not always be able to live up to those lofty ideals. It would be many years, for instance, before women, American Indians, and African Americans had the same rights as those of white men. But the Declaration set the democratic ideal in motion, and it continues to light the way to just and humane government. Thomas Jefferson considered it his greatest achievement.[4]

Chapter 2 ▶

The Young Virginian, 1743–1768

Thomas Jefferson was born on April 13, 1743, at Shadwell, a frontier plantation in Albemarle County, Virginia. His father, Peter Jefferson, had been one of the first settlers in the county. His mother, Jane Randolph, was a member of one of the most prominent families in Virginia. Thomas was one of eight children (six girls and two boys) of Peter and Jane Jefferson.

In spite of living in what was then the wilderness, the Jeffersons led a comfortable life. Peter Jefferson was a successful tobacco planter and a well-known explorer and surveyor who had drawn the first map of Virginia. Jane Randolph came from a family that valued education, so young Thomas had access to tutors and colleges at a time when few American settlers attended school. He also had a library within his own home, because books were considered as important as great fortunes. No true gentlemen of the day would be without them.

Young Thomas grew up in a family that valued the ways of the refined gentleman and the self-made farmer. He would always be drawn to both. Many of his journeys would carry him between the world of the frontier and the world of Virginia society.

The World of Virginia Gentlemen

The first journey came when he was only two. Peter Jefferson's dying friend, William Randolph, had asked him to take care of his estate and children. After Randolph

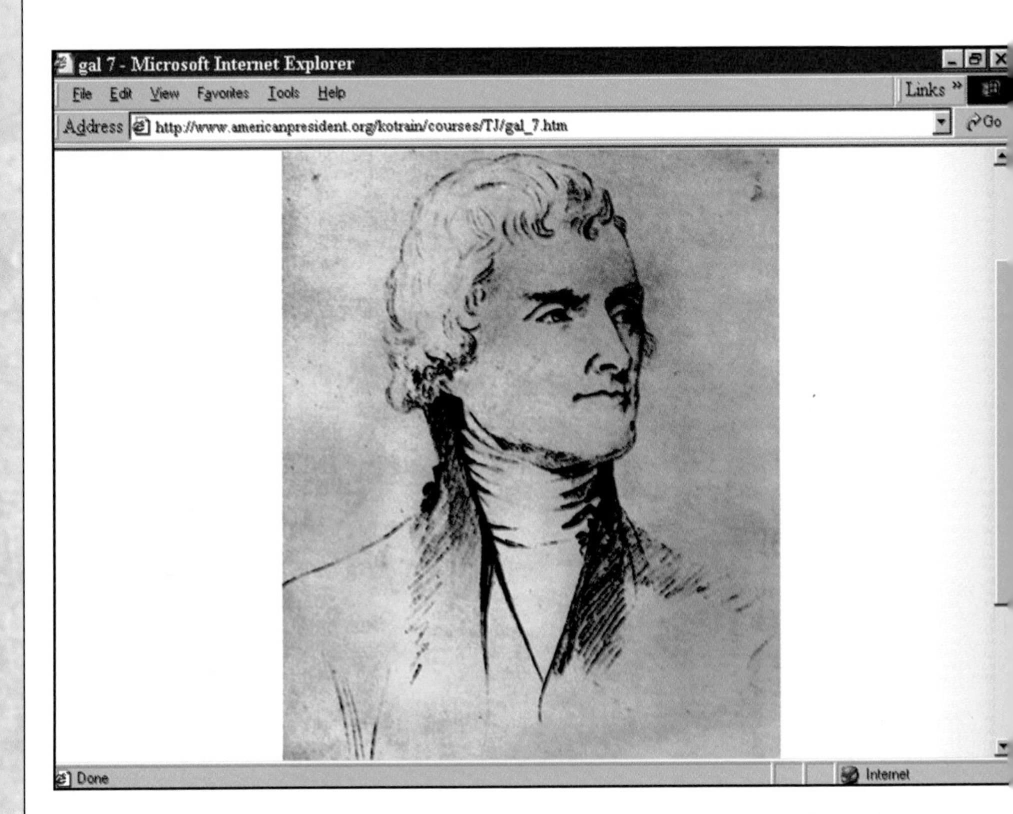

This sketch depicts Thomas Jefferson as a Virginia planter, a legacy he inherited from his father, Peter Jefferson.

died, Peter Jefferson moved the family to the Randolph plantation near Richmond, where they lived for seven years. Thomas was tutored along with the Randolph children and became a voracious reader. He also took an interest in mathematics and music and absorbed the manners of the plantation family. By the time the Jeffersons had left the Randolph estate, nine-year-old Thomas had already learned much about being a Virginia gentleman.

Thomas also learned about slavery at the Randolph estate. More than one hundred black slaves worked on the plantation, in the fields and in the house, and the plantation prospered.[1] But the slaves were not free to leave the plantation and make lives of their own. Though Thomas was allowed to play with black children, they were not allowed to go to school with him. Many years later he would recall how white children learned the rules of slavery by imitating their parents. "Our children," he wrote, "are daily exercised in tyranny."[2]

Once the Jeffersons were back in Albemarle County, they sent Thomas to live for five years with the Reverend William Douglas, who taught Latin to several boys Thomas's age. Then he studied the classics as well as history, geography, and natural science with the Reverend James

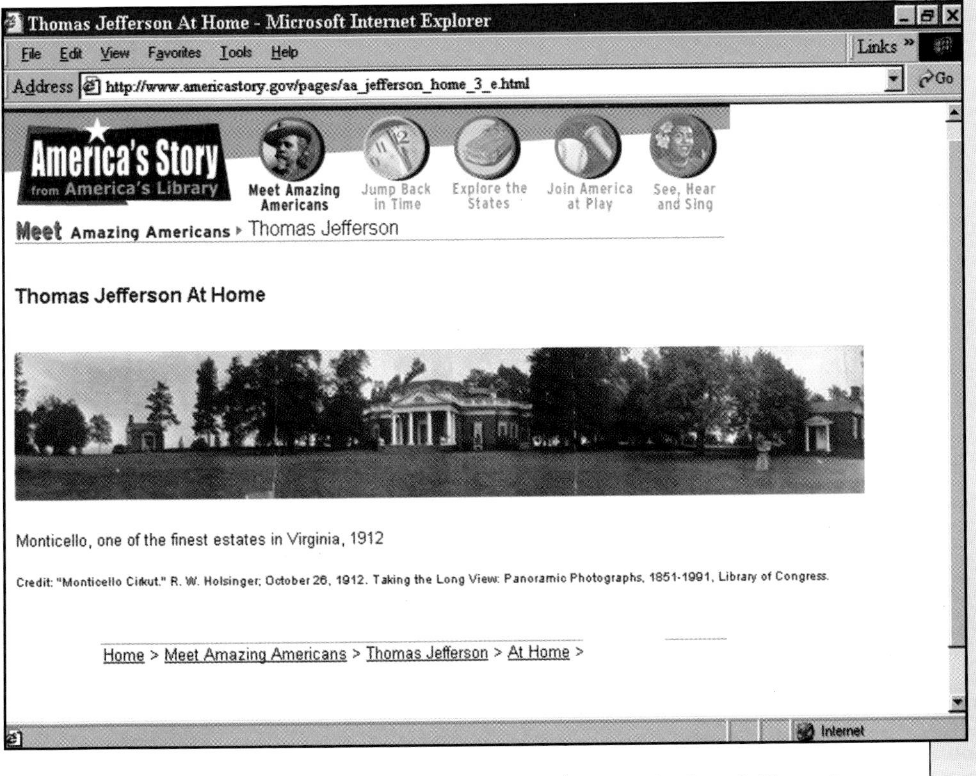

A panoramic photograph captures Monticello, Jefferson's architectural masterpiece, as it looked in 1912. Thomas Jefferson built his home on land inherited from his father.

Maury for two years. He also learned to dance and play the violin.

Thomas also loved the outdoors, a love his father instilled in him. Peter Jefferson believed "it is the strong in body who are both the strong and free in mind."[3] He taught his son to value both aspects of life, and young Thomas tried to emulate him.

But Peter Jefferson died when Thomas was only fourteen. Stunned by the loss, he struggled with his grief for several years. He had never been very close to his mother, so when his father died, he felt alone.

College in Williamsburg

In March 1760, Thomas again left the frontier to continue his education. He traveled to Williamsburg, the capital of Virginia, to attend the College of William and Mary. Williamsburg was a bustling colonial town. It was also the home of the governor and the House of Burgesses, the oldest assembly of elected representatives in the colonies.

Thomas excelled at the College of William and Mary, where he studied science, math, literature, and philosophy. There he was profoundly influenced by the great thinkers of the Enlightenment, particularly the philosophy of John Locke. Locke and other Enlightenment thinkers believed that each person is endowed with reason and as such is able to determine right and wrong individually. But in the eighteenth century, a faith in reason challenged the power of the Church—and of the kings and lords who ruled most nations and who believed their power to rule was derived from God. The Enlightenment ideals of individual freedom were revolutionary for their age. By the time Jefferson left college, he was steeped in them.

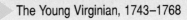

Thomas Jefferson graduated from William and Mary in 1762, where he had concentrated his studies on law. Jefferson then began a five-year apprenticeship with George Wythe, a prominent legal scholar. Wythe was also a member of the House of Burgesses and a friend to the governor. Jefferson accompanied Wythe to meetings of the House of Burgesses and dinners at the governor's mansion, where he conversed with the leaders of Virginia. Thomas Jefferson absorbed everything he saw and heard. He was preparing for a life on the political stage.

Chapter 3 ▶

An Architect of Democracy, 1769–1779

In the spring of 1769, Thomas Jefferson began his political career. Just twenty-six years old, he was elected to the House of Burgesses. Jefferson was already well known and well connected in Williamsburg. He had started a successful law practice after finishing his apprenticeship with Wythe. The leader of the House of Burgesses, Peyton Randolph, was a relative. Soon, Jefferson would befriend another influential Virginian—the representative from Fairfax County, George Washington.[1]

Monticello and Martha

The following year, 1770, was one of great change in the life of Thomas Jefferson. In February, his childhood home, Shadwell, was destroyed by fire. By November, Jefferson was able to move into the home he had been building for the past two years. He situated it on a nearby mountaintop, on land his father had left him. Jefferson designed his home in the style of an Italian villa. He named it Monticello.

While Jefferson was building Monticello, he was also courting a young widow, Martha Wayles Skelton. Martha, the daughter of a

While a member of the House of Burgesses, Thomas Jefferson became friends with a fellow Virginian—and the man who would become the first president of the United States—George Washington.

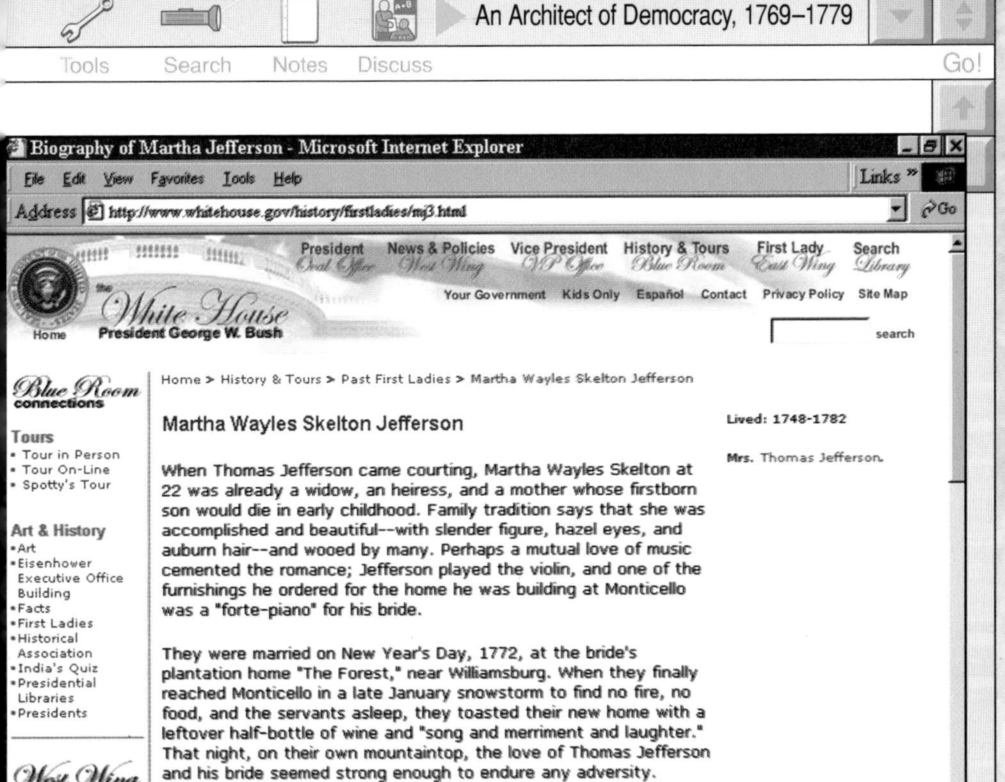
Biography of Martha Jefferson - Microsoft Internet Explorer

http://www.whitehouse.gov/history/firstladies/mj3.html

Home > History & Tours > Past First Ladies > Martha Wayles Skelton Jefferson

Martha Wayles Skelton Jefferson

Lived: 1748-1782

Mrs. Thomas Jefferson.

When Thomas Jefferson came courting, Martha Wayles Skelton at 22 was already a widow, an heiress, and a mother whose firstborn son would die in early childhood. Family tradition says that she was accomplished and beautiful--with slender figure, hazel eyes, and auburn hair--and wooed by many. Perhaps a mutual love of music cemented the romance; Jefferson played the violin, and one of the furnishings he ordered for the home he was building at Monticello was a "forte-piano" for his bride.

They were married on New Year's Day, 1772, at the bride's plantation home "The Forest," near Williamsburg. When they finally reached Monticello in a late January snowstorm to find no fire, no food, and the servants asleep, they toasted their new home with a leftover half-bottle of wine and "song and merriment and laughter." That night, on their own mountaintop, the love of Thomas Jefferson and his bride seemed strong enough to endure any adversity.

The birth of their daughter Martha in September increased their happiness. Within ten years the family gained five more children. Of

Martha Wayles Skelton married Thomas Jefferson on New Year's Day, 1772. She and Jefferson shared a love of music and dancing.

wealthy landowner, was said to be pretty and was an accomplished musician. She also enjoyed riding and dancing and taking long walks. She and Thomas Jefferson shared a love of music. The two often played music together—he on violin and she on harpsichord—when he visited her. For almost two years, he traveled on horseback nearly one hundred miles each way to see her at the Forest, her father's estate. On New Year's Day, 1772, they were married. Soon after the wedding, the young couple began their life together at Monticello. It would not be long, however, before Jefferson would be called away from his wife and his home, as the American colonies began their stirrings toward independence.

The Fight for Independence

By the middle of the eighteenth century, American colonists had begun to resist what they saw as Britain's interference in their lives. Beginning in 1774, a group of elected representatives from each colony met in Philadelphia. As members of the Continental Congress, they were deciding what action to take against Great Britain.

In July of that year, Thomas Jefferson had been asked to write a draft of instructions for the Virginia delegation to the first Continental Congress. In it, Jefferson wrote that Parliament had no rights over the colonies and that, indeed, the colonies had been independent since they were founded. Though the Virginia delegates adopted a less radical position, Jefferson's draft was published as *A Summary View of the Rights of British America* by a group of his friends. It was circulated in London as well as in Philadelphia and New York. By the time Jefferson attended the second Continental Congress, he already had "a reputation for a masterly pen."[2]

Revolution and Reform

The beginnings of the American Revolution again drew Jefferson away from his mountaintop home. In 1775, he was the youngest member of the Virginia delegation to the second Continental Congress. In May of the next year he was appointed to a committee that had been formed to draft a declaration of independence from Great Britain. Also serving on that committee were Benjamin Franklin, John Adams, Robert Livingston, and Roger Sherman. It was decided that Jefferson alone would be entrusted with writing the draft of the declaration. He was selected because of his reputation as a powerful writer and also because he represented Virginia, the most influential colony in the

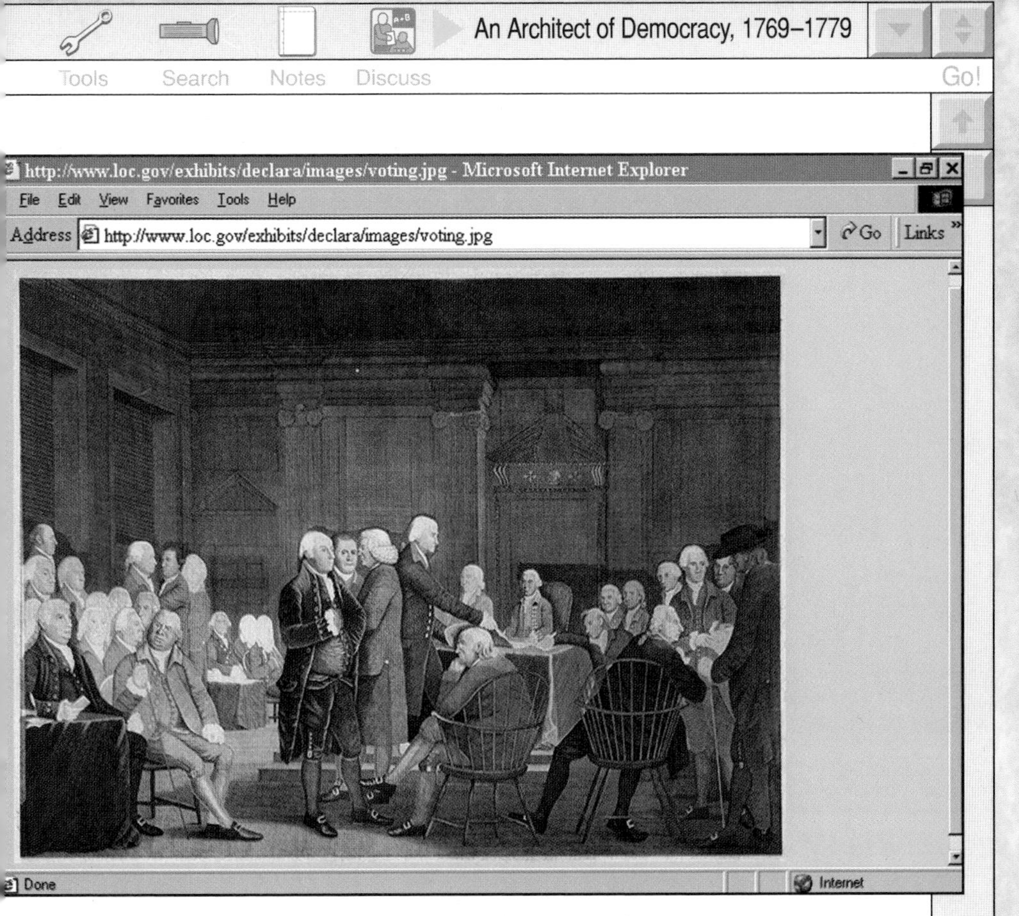

This painting shows members of the Continental Congress in July of 1776 voting to adopt the Declaration of Independence.

South. Jefferson wrote his original draft in less than three days. After changes were made to it by the other members of the committee and then after debate by the other members of Congress, the Declaration was adopted on July 4, 1776. On Congress's orders, a Philadelphia printer began printing copies of it that very day.

After writing the Declaration, Jefferson went to work bringing the democratic ideal to Virginia. As a member of the Virginia Assembly, he wrote bills to make the laws and courts more just. He proposed a plan for a public school system. He wrote a bill that would establish religious freedom and ensure the separation of church and state.

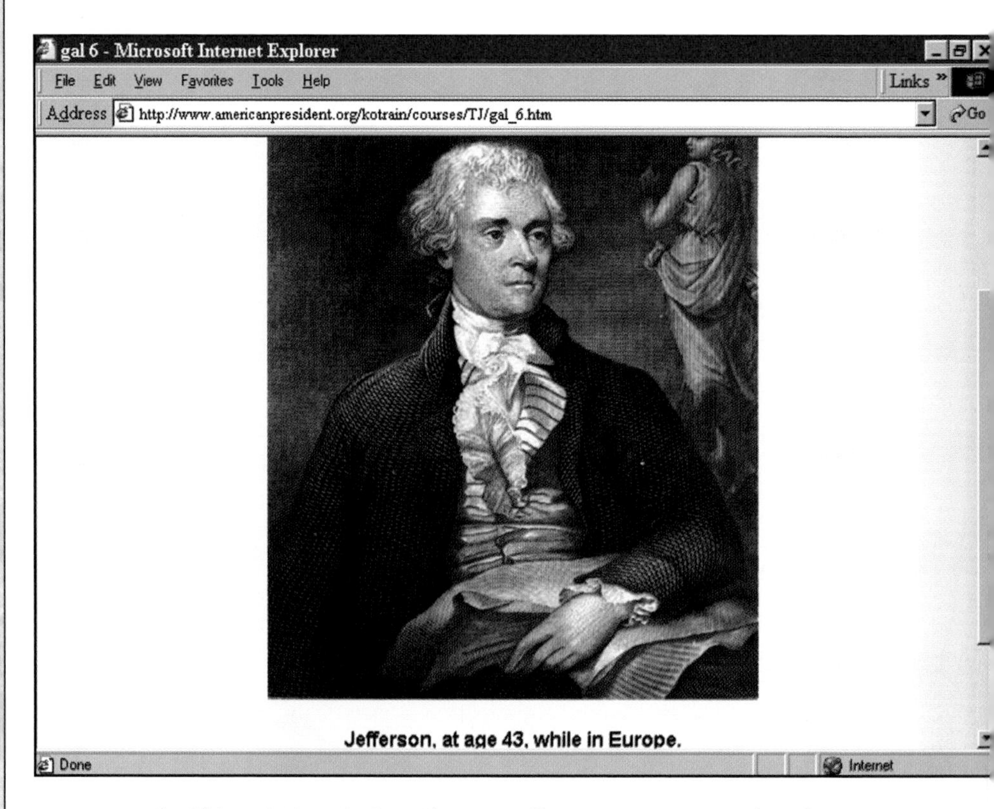

This painting depicts Thomas Jefferson at age 43, when he was serving the new nation as the American minister to France. Jefferson loved the French and their culture, and they admired him as a champion of democracy.

Jefferson then went on to serve two one-year terms as Virginia's governor, from 1779 to 1781.

When his terms as governor had ended, Jefferson returned to Monticello and his family. Sadly, his joy at being reunited with them was not to last. On September 6, 1782, Martha Jefferson died, four months after giving birth to their third daughter. Jefferson was devastated by his wife's death. He did not leave his room for three weeks after Martha's funeral.

Jefferson the Diplomat

But mourning finally gave way to purpose when Jefferson was called back into service for the new nation. The war with Britain had ended in victory for the colonies. The official ending came with the signing of the Treaty of Paris, in 1783. Now the leaders of the new United States of America were trying to establish ties with the countries of Europe so that American farmers and merchants could sell their goods abroad. And expansion at home was on the minds of the new government. In March 1784, Jefferson led a congressional committee in proposing to

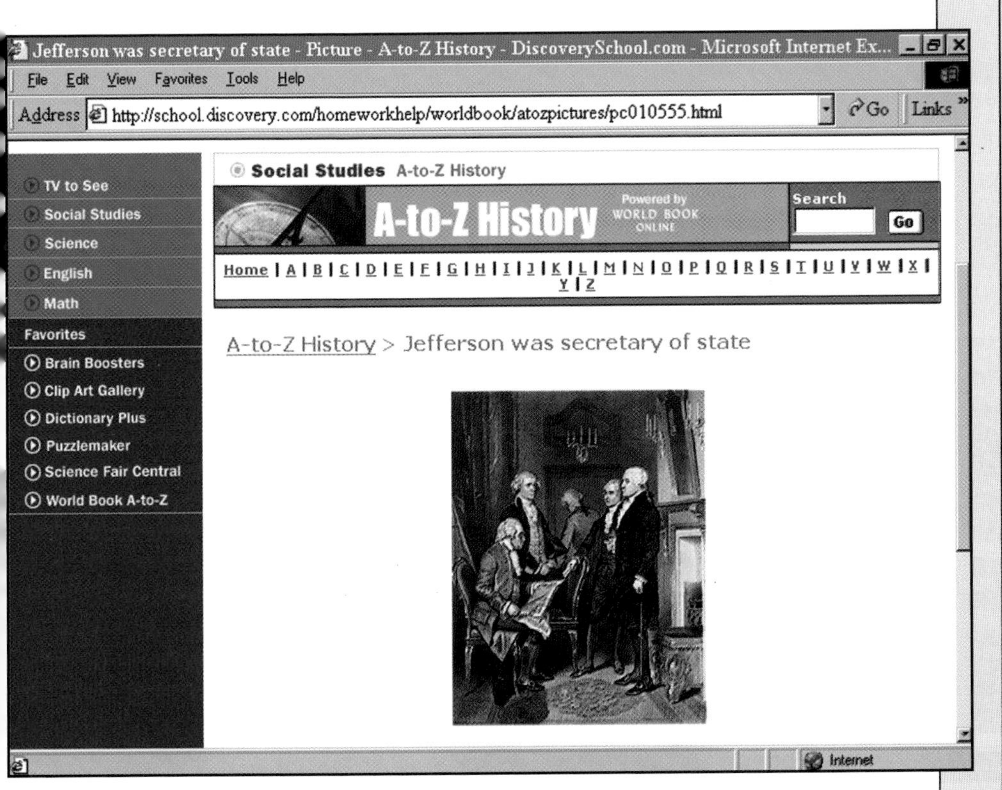

Upon Thomas Jefferson's return from France in 1789, he was asked by President George Washington to serve as his secretary of state. At the prodding of his friend and fellow Virginian James Madison, Jefferson accepted the position.

divide the western territories and make them states. Jefferson also proposed a ban on slavery in all the new states admitted to the Union after 1800. That proposal, however, was not passed by Congress. In May of the same year, Jefferson was asked to join John Adams and Benjamin Franklin in Paris, where the two had been working on trade agreements with the French.

After assisting Franklin for a year, Jefferson became the American minister to France in 1785. For four years he served in Paris, promoting American commerce. The French grew fond of Jefferson, and he of them. They saw him as an American patriot, a champion of democratic ideals.[3]

Jefferson returned to the United States in 1789, where his own countrymen also praised his work in France. President George Washington asked Thomas Jefferson to join his cabinet as secretary of state. In that position, Jefferson would be responsible for overseeing relations between the United States and other countries. His time in Europe as a diplomat had prepared him well for the job. Urged by his friend and fellow Virginian James Madison, Jefferson accepted the appointment. In 1790, Thomas Jefferson became the country's first secretary of state.

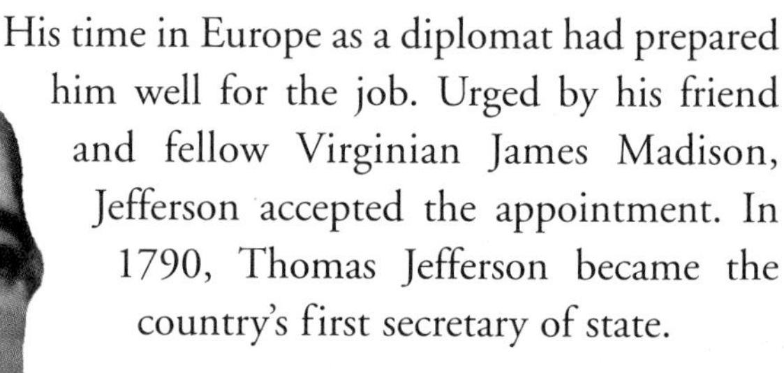

Alexander Hamilton (pictured) and Thomas Jefferson both served in George Washington's cabinet. But the two held opposing views on how powerful the federal branch of the government should be. Hamilton, a Federalist, wanted a strong federal government while Jefferson was committed throughout his political life to limiting the powers of the federal government.

Republicans and Federalists

As secretary of state, Jefferson helped define the future of the new nation. He would soon, however, come into conflict with another cabinet member—Alexander Hamilton, the secretary of the treasury. Hamilton was a Federalist, who believed in a strong federal government. Among the things Hamilton supported were a national bank and the rise of manufacturing. Jefferson was opposed to a national bank and was afraid that an emphasis on manufacturing would hurt farming, which would in turn hurt the South.

Jefferson, who would come to be known as a Republican, was strongly opposed to many of Hamilton's views. He believed that the powers of the federal government needed to be limited if the liberty of American citizens was to be preserved. He feared that an executive branch with too much power would become like the monarchy the Americans had just fought to remove. In 1793, Jefferson resigned his position as secretary of state and once again returned to Virginia and his beloved Monticello. But he would not be out of government for long.

In 1796, John Adams was elected the second president of the United States, and Jefferson was elected its third vice president. As secretary of state and then as vice president, there were times when Jefferson thought the American republic would fail. But in the end, Jefferson rallied support for a government that he had described in the Declaration as "deriving [its] just powers from the consent of the governed." Jefferson would see his hope for the preservation of the republic in the results of the next presidential election.

Leading the New Republic, 1801–1805

In the election of 1800, Thomas Jefferson and Aaron Burr, members of the Republican Party, received more votes than John Adams, the Federalist candidate. Jefferson and Burr, however, each received the same number of electoral votes. Though it had been clear throughout the campaign that Burr was the vice presidential candidate, he refused to

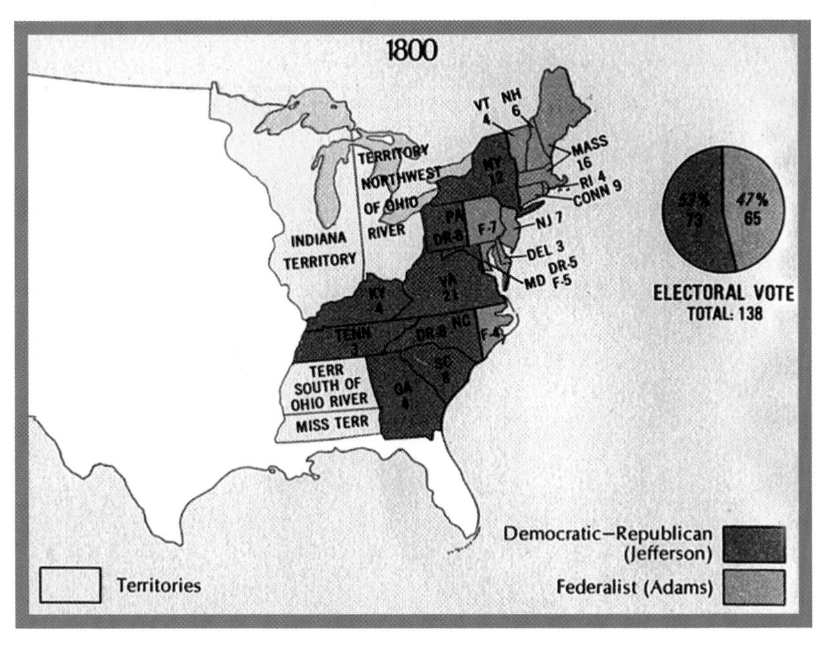

The presidential election of 1800 was one of the most divisive in American history. The outcome was not decided until February of 1801, when it fell to the House of Representatives to choose the president. John Adams, the Federalist candidate, had received fewer votes than Thomas Jefferson and Aaron Burr, members of the Republican Party. But Jefferson and Burr had received the same number of electoral votes. After thirty-six ballots, Jefferson was chosen president and Burr became the vice president.

concede the presidency to Jefferson. The vote total was not made known until December of 1800. And the election was not decided until February of 1801, when it fell to the House of Representatives to choose the president. Though it took six days and thirty-six ballots, Jefferson was finally elected president, and Burr vice president.

On March 4, 1801, Thomas Jefferson became the first president to deliver his inaugural address in Washington, D.C., the new capital of the United States. When Jefferson took office, Washington was a new city, surrounded by farms and forests. The White House had just been built, and the Capitol building was under construction. Only the Senate chamber was complete.[1]

On the morning of March 4, Jefferson walked to Capitol Hill for his inauguration, accompanied by only a few congressmen and dignitaries. Unlike today's inaugural ceremonies, Jefferson's was modest—it was important to Jefferson to have a simple inauguration, to show that he was a man of the people.

Unifying the Nation

Jefferson also wanted to unify the new nation, because the election of 1800 had divided the country. The conflicts between the Federalists and the Republicans had been fierce. They had attacked each other in the newspapers and had argued harshly in the House and Senate. Some people feared that the Federalists would take power by force. Others worried that a federal government under Jefferson would not be strong enough.

Jefferson knew he had to reunite and reassure the nation, and he used his inaugural address to attempt to do so. He wanted to offer the United States a vision of unity

and cooperation. As he walked to Capitol Hill, the crowd in the Senate chamber waited expectantly.

Jefferson reminded his listeners that the "contest" between the Federalists and Republicans had been possible because of the freedoms of the United States. The Constitution had given American citizens the right to "think freely and to speak and write what they think."[2] Citizens might disagree, but they could unite in support of those freedoms.

"Every difference of opinion is not a difference of principle," Jefferson said. "We have called by different names brethren of the same principle. We are all Republicans, we are

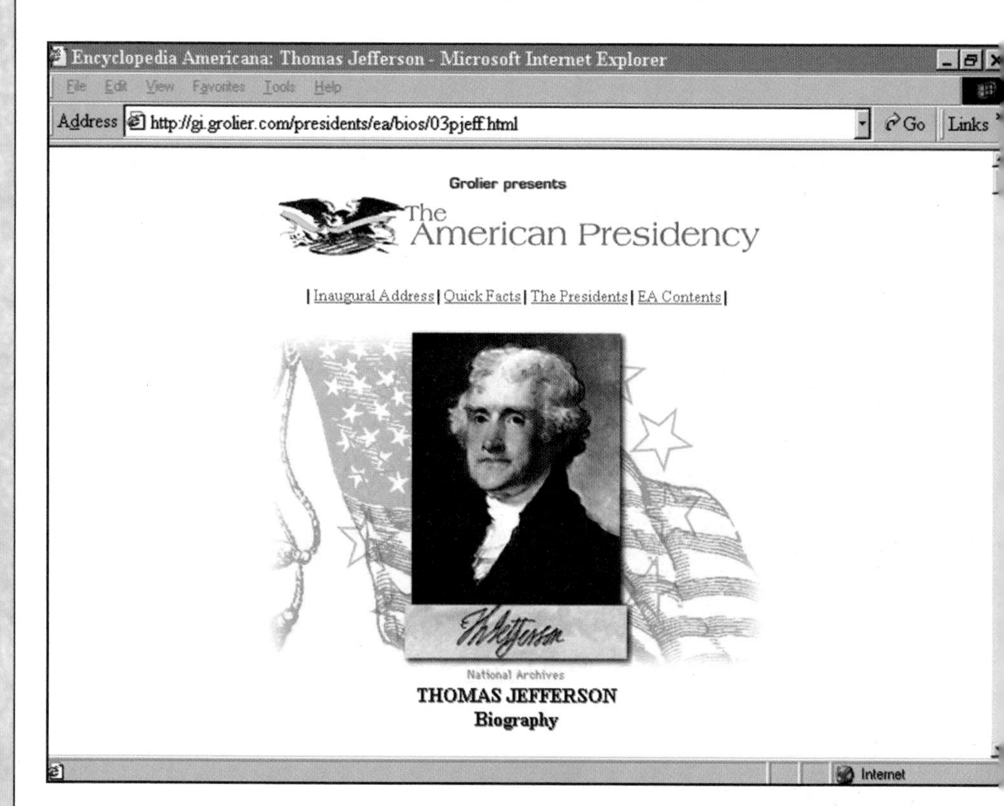

In his First Inaugural Address, Thomas Jefferson as the third president of the United States of America chose words that he hoped would unite the fledgling nation.

all Federalists . . . Let us, then, with courage and confidence pursue our own Federal and Republican principles, our attachment to union and representative government."[3]

Jefferson was trying to suggest to the new nation that their differences should not drive them apart. They could all "arrange themselves" under the laws of the Constitution. They could "unite in common efforts for the common good." This was only possible in the American republic. And this new form of government, Jefferson said, was "the world's best hope."[4]

A Hopeful Time, A Hardworking President

Jefferson's inaugural address set the tone for his presidency. The strife between Republicans and Federalists eased. The government was successful in paying off the nation's debts. Taxes were reduced for many citizens. Farms throughout the nation were productive. Trade with Europe blossomed.

The Louisiana Purchase

In the spring of 1802, Jefferson learned that France was about to take control of the vast province of Louisiana, which included the Mississippi River and the port of New Orleans. (France had first claimed the territory in the seventeenth century, then ceded it to Spain, and Spain was just about to cede it back to France in 1802.) Jefferson feared that the United States would lose access to the West if the Mississippi was controlled by France. He sent a fellow Virginian, James Monroe, to Paris to negotiate with the French. Robert Livingston, the U.S. ambassador to France, also took part in the negotiations. Jefferson hoped they could convince France to sell a portion of the Mississippi Valley, including the port of New Orleans, to the United States.

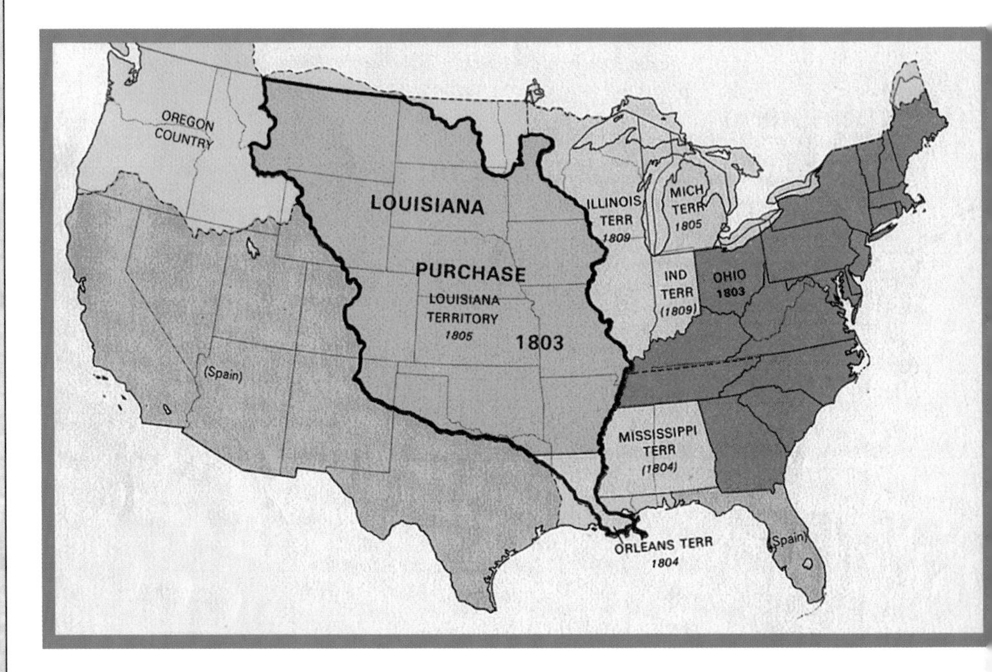

This map shows the growth in territory held by the United States after the Louisiana Purchase.

But Napoléon Bonaparte, the leader of France, surprised the Americans by offering to sell the entire Louisiana Territory to the United States. Napoléon's armies had been fighting battles all over Europe, and they now faced a renewed war with Britain. Napoléon decided he could not support an army in America. On April 30, 1803, he agreed to sell the entire Louisiana Territory to the United States for 15 million dollars.

The Louisiana Purchase was one of the great achievements of Jefferson's presidency. Historians have called it one of the most important presidential actions in American history.[5] The purchase doubled the size of the United States at a cost of only three cents an acre. It kept the Mississippi Valley and New Orleans open to American trade. It also opened the American West to exploration and

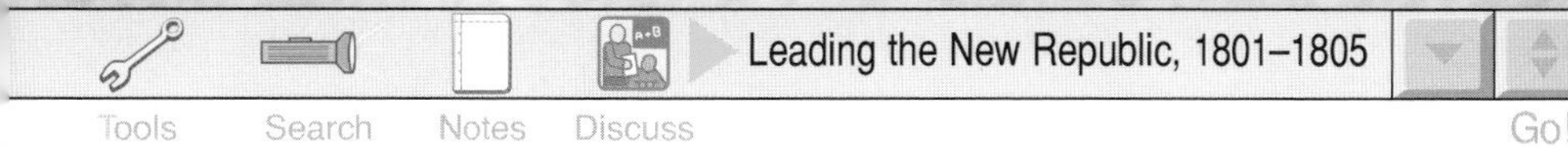

settlement. Jefferson's prompt response to a historic opportunity had changed the course of the nation.

The Lewis and Clark Expedition

Jefferson soon sent an expedition to explore the new territory. He appointed his personal secretary, Meriwether Lewis, and an Army colonel, William Clark, to lead the expedition. Jefferson wanted them to find a route across the continent, to the Pacific, that would encourage trade and settlement. He also asked them to take detailed notes about the plants and wildlife they saw on their travels.

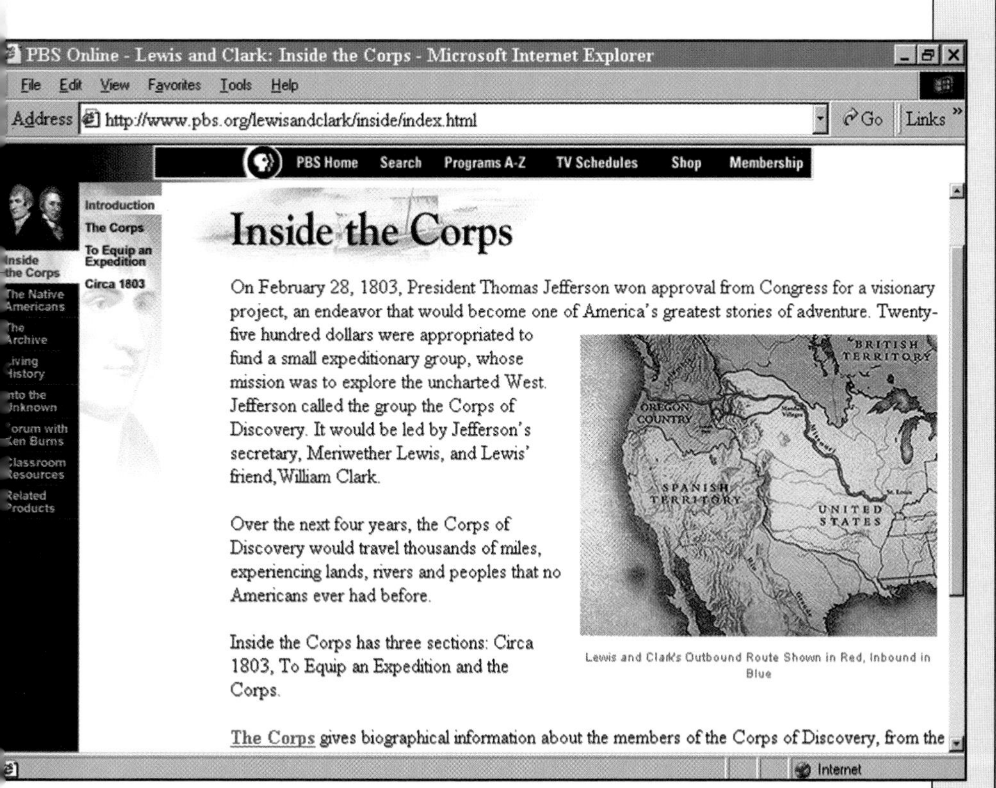

PBS Online - Lewis and Clark: Inside the Corps - Microsoft Internet Explorer
File Edit View Favorites Tools Help
Address http://www.pbs.org/lewisandclark/inside/index.html
Go
Links
PBS Home
Search
Programs A-Z
TV Schedules
Shop
Membership
Introduction
The Corps
To Equip an Expedition
Circa 1803
Inside the Corps
The Native Americans
Inside the Corps
On February 28, 1803, President Thomas Jefferson won approval from Congress for a visionary project, an endeavor that would become one of America's greatest stories of adventure. Twenty-five hundred dollars were appropriated to fund a small expeditionary group, whose mission was to explore the uncharted West. Jefferson called the group the Corps of Discovery. It would be led by Jefferson's secretary, Meriwether Lewis, and Lewis' friend, William Clark.
Over the next four years, the Corps of Discovery would travel thousands of miles, experiencing lands, rivers and peoples that no Americans ever had before.
Inside the Corps has three sections: Circa 1803, To Equip an Expedition and the Corps.
The Corps gives biographical information about the members of the Corps of Discovery, from the
BRITISH TERRITORY
OREGON COUNTRY
SPANISH TERRITORY
UNITED STATES
Lewis and Clark's Outbound Route Shown in Red, Inbound in Blue
Internet

Thomas Jefferson appointed his secretary, Meriwether Lewis, and Lewis's friend, Colonel William Clark, to explore the vast uncharted western territory that became part of the United States after the Louisiana Purchase.

Their maps and journals are a valuable record of America's western wilderness before settlement.[6]

Jefferson and the American West

America's westward expansion would soon sweep away much of the wilderness that Lewis and Clark encountered, however. Jefferson believed that settling new frontiers in what seemed to be a boundless West would invigorate the American republic.[7] But settlement would also displace many of the American Indians who lived in the West, and Jefferson knew there would be conflicts with them. Before he was president, Jefferson had written fondly of American Indians. He had celebrated their culture and praised their leaders. He admired their "love of liberty and independence."[8] Yet Jefferson's push for settlements in the West would ultimately drive American Indians from their land.

Many citizens thought Jefferson's achievements added up to a successful first term. In fact, historians regard his first term as one of the most successful in history. His popularity was high, and in 1804, he was easily reelected. His second term as president, however, would bring far greater challenges.

Chapter 5 ▶

A Troubling Second Term, 1805–1809

As Jefferson's second term began, he was faced with extraordinary difficulties. Great Britain and France were at war, and Jefferson was determined to stay out of it. But British ships began seizing American vessels at sea, because the British wanted to disrupt American trade with France. France, meanwhile, was setting restrictions on trade with the United States.

Trade with Europe was crucial to the American economy. Farmers, shipbuilders, and merchants all depended on the income from goods shipped to Europe. The tobacco plantations in Jefferson's Virginia, for example, made much of their profit from shipping their crops to Great Britain.

A Standoff With Europe

The European powers, however, made it difficult for American ships to reach their ports. In April 1806, Great Britain blockaded the coast of the United States, and British ships harassed American vessels. In November of that year, Napoléon's fleet blockaded Britain. Shipping lanes were stalled and American ports struggled with what to do with all their cargo.

Jefferson tried to solve the problem with diplomacy, knowing that the United States could not afford a war with either Britain or France. He sent diplomats to London to try to negotiate a solution. Jefferson's position was that neutral nations—like the United States—should be free to sail the seas as they pleased. In support of his argument,

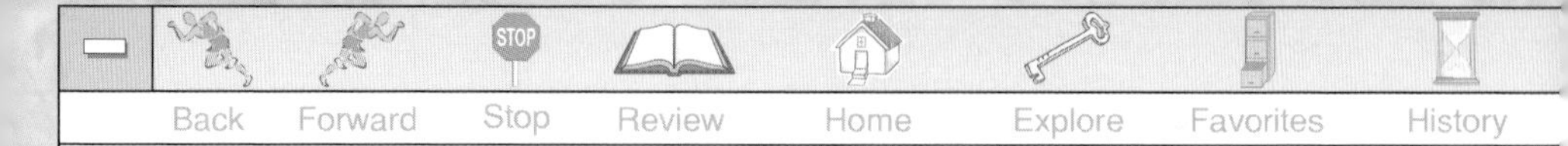

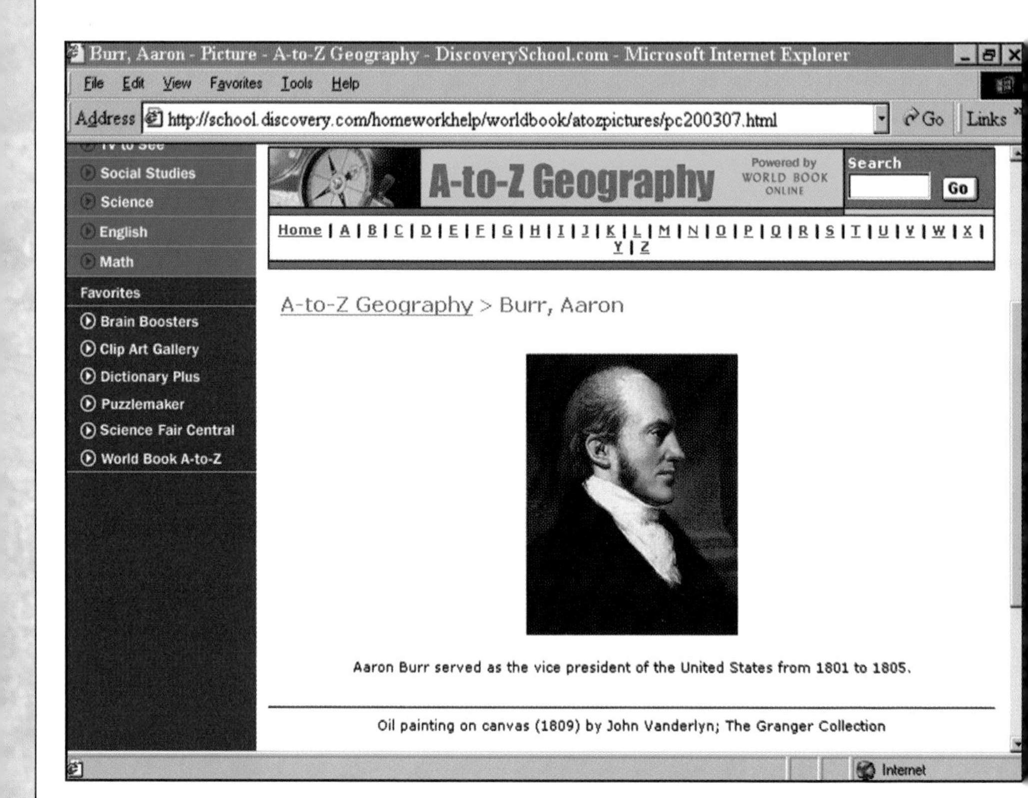

At one time a vice president to Thomas Jefferson, Aaron Burr was tried for conspiracy in 1807. The trial was presided over by Chief Justice William Marshall, an old enemy of Jefferson's. The daily accounts of the trial depicted Jefferson as trying to get even with Burr and Marshall. That publicity and the failure of the embargo of 1807 turned even some of Jefferson's supporters against him.

Congress passed the Non-Importation Act of 1806, which blocked imports from Britain to America. Jefferson hoped this would put economic pressure on Great Britain and persuade the British to stop harassing American ships.

But Britain did not respond to American diplomacy or economic pressure. In June 1807, the British warship *Leopard* fired on the American ship *Chesapeake* off the Virginia coast, and the British used force to board the ship. Americans were outraged by the act, but Jefferson still did not declare war. He decided instead to use an embargo, calling it "the last card we have to play short of war."[1] The Embargo Act of 1807 cut off trade with foreign countries. Jefferson again thought the loss of American goods would have an impact on Britain and France and would force them to respect the neutral rights of American ships on the high seas.

But again, he was wrong. The British and French did nothing. Trade stalled, American ports closed down, and farmers in New England and plantation owners in the South had nowhere to ship their goods. The embargo, said one historian, was a "calamity that virtually wrecked the American economy."[2]

The Trial of Aaron Burr

Meanwhile, another crisis was brewing that year which involved Jefferson's former vice president, Aaron Burr. In 1807, Burr was arrested and charged with treason. He was accused of conspiring to set up an independent country in the American West. The conspiracy cast a shadow on Jefferson's presidency.

Burr's trial was closely watched by the American public. It was presided over by an old enemy of Jefferson's, Chief

Thomas Jefferson's loyal friend and fellow Virginian James Madison succeeded him as president in 1809. These men were responsible for two of the most important documents in American history—Jefferson for writing the Declaration of Independence, and Madison for drafting the Constitution.

Justice John Marshall. In a sense, the proceedings put Jefferson on trial. Justice Marshall used the very public, very political trial to denounce Jefferson. Burr's lawyer attacked Jefferson as well. As Americans read the daily accounts, Jefferson's opponents made the case that Burr was being persecuted by "a vindictive president."[3]

Jefferson paid a high price for the publicity surrounding the Burr affair and the failure of the embargo. The Federalist press stirred up scandal, and some of Jefferson's Republican supporters turned against him. As the economy stalled, more and more merchants and farmers grew impatient with the government. The country as a whole was growing restless.

Many historians have judged that there was little Jefferson could have done.[4] The wars between France and Britain were the cause of the shipping crisis. If the United States had joined the war, the costs would have been great. Jefferson had steered a careful course, but it was also an exhausting course. The political battles had worn him out. After forty years of public service, he was ready to go home. On March 3, 1809, his loyal friend James Madison succeeded him as president. When Jefferson left Washington, D.C., he was relieved to be "shaking off the shackles of power."[5] After Madison's inauguration, Thomas Jefferson mounted his horse and rode home to Monticello.

Chapter 6 ▶

Jefferson at Monticello, 1809–1826

Thomas Jefferson would not leave Virginia again for the rest of his life. Monticello was home to everything that Jefferson held dear. It had always been a place of retreat and study, and it remained so in Jefferson's retirement. He corresponded with friends, spent time with his family, and tended to his farm and gardens.

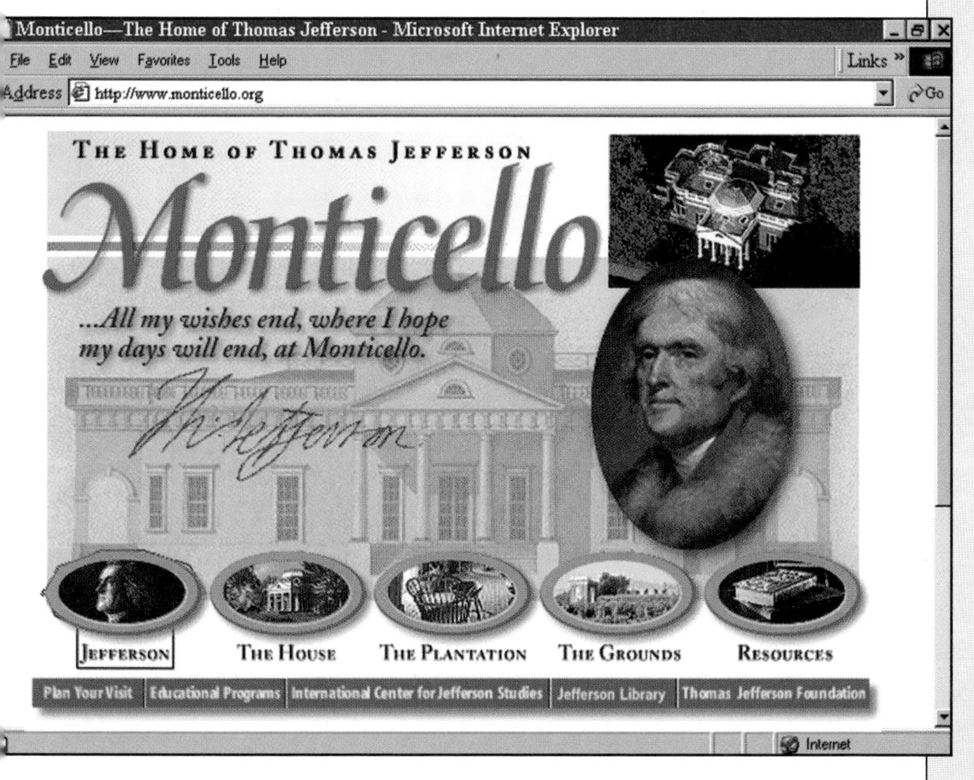

In 1809, Thomas Jefferson left public office and returned home to his beloved Monticello.

STOP
Back
Forward
Stop
Review
Home
Explore
Favorites
History

When Jefferson retired, his daughter Martha, her husband, and their eleven children moved in to Monticello. The family often gathered in the parlor to play music and games or to read together. They walked and played on the grounds of the estate and shared meals with many guests. The steady stream of visitors, according to one guest, enjoyed dinners "served in half Virginian, half French style, in good taste and abundance."[1]

Jefferson's devotion to education did not waver in his final years. To him, the education of America's citizens was

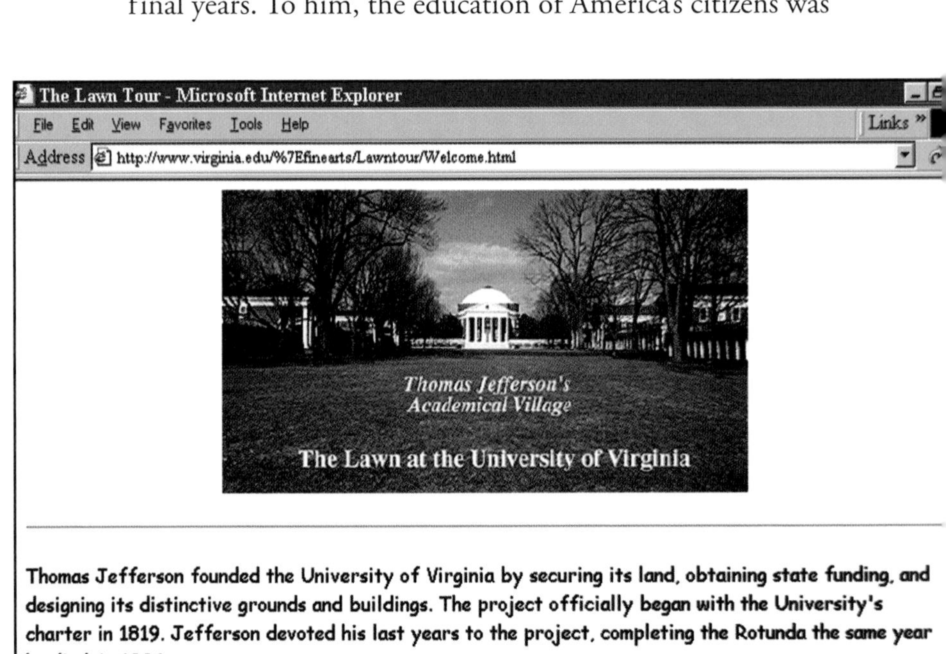
The Lawn Tour - Microsoft Internet Explorer
File Edit View Favorites Tools Help
Links »
Address http://www.virginia.edu/%7Efinearts/Lawntour/Welcome.html
Thomas Jefferson's
Academical Village
The Lawn at the University of Virginia
Thomas Jefferson founded the University of Virginia by securing its land, obtaining state funding, and designing its distinctive grounds and buildings. The project officially began with the University's charter in 1819. Jefferson devoted his last years to the project, completing the Rotunda the same year he died, in 1826.
On his gravestone, Jefferson claimed the University as one of his proudest achievements, along with authoring the Declaration of Independence and Virginia's statute on religious freedom.
Done
Internet

One of Jefferson's proudest achievements was his founding of the University of Virginia. He conceived of the university as an "academical village" where teachers and students would live and learn together. When it opened in 1825, the University of Virginia became the first state university in America.

the key to a strong democracy. "Enlighten the people generally," he wrote, "and tyranny and oppressions of both mind and body will vanish like evil spirits at the dawn of the day."[2] One of Jefferson's proudest achievements was his founding of the University of Virginia, in Charlottesville.

The University of Virginia

Jefferson helped secure land and state funding for the university, which was chartered in 1819. He then designed and supervised much of its construction. The university was conceived as an "academical village," with classrooms and living quarters for students and teachers alike. When its doors opened in 1825, it was the first state university in America. Students who attended the university in its early years read books selected by Jefferson and were taught by professors he recruited. Today, students still study in the buildings he designed and walk the campus he so carefully laid out. Like the Declaration of Independence, the University of Virginia is a living legacy of Jefferson's profound influence on American ideals and institutions.

Jefferson and Slavery

The Jefferson household was the center of a sprawling 5,000-acre plantation. As his father had done before him, Jefferson grew tobacco and wheat. He also raised horses, hogs, cattle, and sheep and cultivated an orchard and vegetable garden. Like most Virginia landowners of the time, Jefferson depended on slaves for the care of his crops and livestock. Slaves also worked as cooks, stable hands, blacksmiths, and carpenters. Nearly one hundred and fifty slaves lived and worked at Monticello at the time of Jefferson's death. [3]

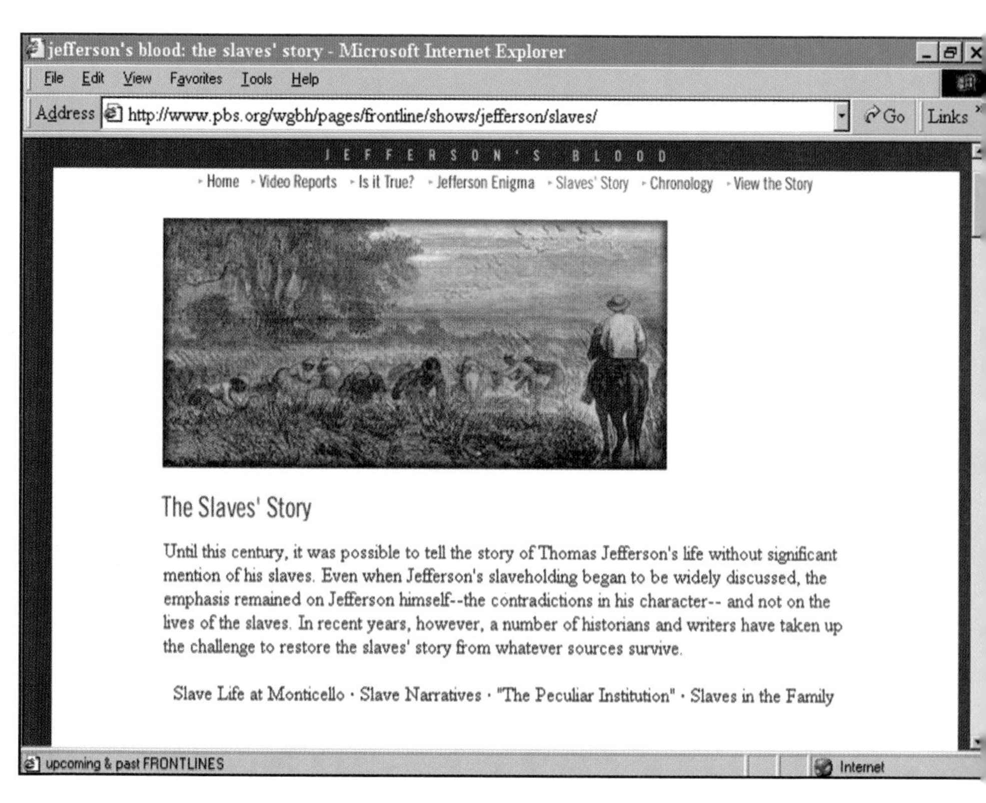
jefferson's blood: the slaves' story - Microsoft Internet Explorer

File Edit View Favorites Tools Help

Address http://www.pbs.org/wgbh/pages/frontline/shows/jefferson/slaves/

JEFFERSON'S BLOOD

Home · Video Reports · Is it True? · Jefferson Enigma · Slaves' Story · Chronology · View the Story

The Slaves' Story

Until this century, it was possible to tell the story of Thomas Jefferson's life without significant mention of his slaves. Even when Jefferson's slaveholding began to be widely discussed, the emphasis remained on Jefferson himself--the contradictions in his character-- and not on the lives of the slaves. In recent years, however, a number of historians and writers have taken up the challenge to restore the slaves' story from whatever sources survive.

Slave Life at Monticello · Slave Narratives · "The Peculiar Institution" · Slaves in the Family

upcoming & past FRONTLINES   Internet

Many historians, as well as ordinary Americans, struggle to reconcile Jefferson's words in the Declaration of Independence—"We hold these truths to be self-evident, that all men are created equal"—with Jefferson the slave owner. Though Jefferson wrote that he believed the institution of slavery to be wrong, he continued to own slaves throughout his adult life.

Jefferson called his slaves "my family."[4] But the slaves who lived and worked at Monticello actually made up many different, unique families. There were husbands and wives, children and grandchildren. Joseph and Edith Fossett, for example, were longtime slaves at Monticello. Edith Fossett worked in Jefferson's kitchen. Joseph Fossett ran the Monticello blacksmith shop. Their workdays stretched as long as fourteen hours. Then Edith cooked and cleaned for her family, Joseph did extra work to earn a small wage, and they both took care of their children. Their son Peter would win his freedom more than twenty years after Jefferson's death. He would become a successful caterer and a Baptist minister in Cincinnati, Ohio.

But there was no path to freedom from Jefferson's Monticello. Though Jefferson called slavery an "abominable crime,"[5] he continued to keep slaves throughout his life. In this, he was a product of the culture of colonial Virginia. The wealthy men of Virginia were all slaveholders. Like them, Jefferson had grown up with the institution of slavery. He was no different from the children he had observed who learned the "tyranny of slavery" from their parents. Yet Jefferson wrote that he believed the institution of slavery was wrong, and he tried as a politician to abolish the African slave trade. The people who have studied Jefferson's life, as well as the people who have been influenced by his ideals, still struggle with the contradictions in Jefferson's attitude toward slavery.

Jefferson's Legacy

Though scholars have long debated the contradictions in the words and deeds of Thomas Jefferson, they are in agreement on the power and durability of those words. They are words which, according to the historian Garry Wills, "continue to express what is deepest and best in America's struggle toward equality for all."[6]

"We hold these truths to be self-evident, that all men are created equal. . . ." Those words have inspired people all over the world. The eventual end of slavery, the rights of citizenship finally extended to African Americans and women, the protection of individual freedoms—all flowed from the powerful ideas of the Declaration.[7]

Thomas Jefferson and the other founding fathers seem to have had a sense of the lasting power of those words. On July 4, 1826, exactly fifty years after the signing of the Declaration, Thomas Jefferson, its author and the nation's third president, died. What Jefferson could not know was that a few hours later that same day, John Adams, the nation's second president and Jefferson's friend and fellow patriot, would also die. Adams, who did not know that Jefferson had died, spoke these words as his last: "Thomas Jefferson survives."[8] And he still survives—in the words of the Declaration, in the strength of the American republic, and in the ideals of democracy that he gave voice to.

Chapter Notes

Chapter 1. The Declaration of Independence, 1776

1. Joseph J. Ellis, *American Sphinx: The Character of Thomas Jefferson* (New York: Alfred A. Knopf, 1997), p. 53.

2. Willard Sterne Randall, *Thomas Jefferson: A Life* (New York: Henry Holt and Company, 1993), p. 273.

3. Carl L. Becker, *The Declaration of Independence: A Study in the History of Political Ideas* (New York: Random House, 1958), p. 72.

4. Randall, p. 279.

Chapter 2. The Young Virginian, 1743–1768

1. Willard Sterne Randall, *Thomas Jefferson: A Life* (New York: Henry Holt and Company, 1993), p. 12.

2. Randall, p. 12.

3. Ibid., p. 13.

Chapter 3. An Architect of Democracy, 1769–1779

1. Willard Sterne Randall, *Thomas Jefferson: A Life* (New York: Henry Holt and Company, 1993), pp. 124–125.

2. Dumas Malone, *Thomas Jefferson: A Brief Biography* (Charlottesville, Va.: Thomas Jefferson Memorial Foundation, 1992), p. 15.

3. Joseph J. Ellis, *American Sphinx: The Character of Thomas Jefferson* (New York: Alfred A. Knopf, 1997), p. 82.

Chapter 4. Leading the New Republic, 1801–1805

1. Joseph J. Ellis, *American Sphinx: The Character of Thomas Jefferson* (New York: Alfred A. Knopf, 1997), pp. 172–173.

2. Merrill D. Peterson, ed., *The Political Writings of Thomas Jefferson* (Charlottesville, Va.: Thomas Jefferson Memorial Foundation, 1993), p. 139.

3. Peterson, p. 140.

4. Ibid., pp. 139–140.

5. Ellis, p. 204.

6. Daniel B. Botkin, *Passage of Discovery* (New York: Perigee, 1999), pp. 2–5.

7. Ellis, p. 212.

8. Ibid., p. 201.

Chapter 5. A Troubling Second Term, 1805–1809

1. Adrienne Koch and William Peden, *The Life and Selected Writings of Thomas Jefferson* (New York: The Modern Library, 1998), p. xxxvii.

2. Joseph J. Ellis, *American Sphinx: The Character of Thomas Jefferson* (New York: Alfred A. Knopf, 1997), p. 237.

3. Dumas Malone, *Thomas Jefferson: A Brief Biography* (Charlottesville, Va.: Thomas Jefferson Memorial Foundation, 1992), p. 39.

4. Malone, p. 39.

5. Ellis, p. 238.

Chapter 6. Jefferson at Monticello, 1809–1826

1. Lucia Stanton, *Slavery at Monticello* (Charlottesville, Va.: Thomas Jefferson Memorial Foundation, 1996), p. 36.

2. Dumas Malone, *Thomas Jefferson: A Brief Biography* (Charlottesville, Va.: Thomas Jefferson Memorial Foundation, 1992), p. 43.

3. Stanton, p. 11.

4. Ibid., p. 13.

5. Thomas Jefferson Memorial Foundation, *Monticello: The Home of Thomas Jefferson* (Charlottesville, Va.: Thomas Jefferson Memorial Foundation, 1999), p. 1.

6. Garry Wills, introduction to *Thomas Jefferson, Genius of Liberty,* Joseph J. Ellis, ed. (New York: Viking Studio, 2000), p. xxii.

7. Joseph J. Ellis, *American Sphinx: The Character of Thomas Jefferson* (New York: Alfred A. Knopf, 1997), p. 54.

8. James Bishop Peabody, ed., *The Founding Fathers: John Adams, Biography in His Own Words* (New York: Newsweek, Inc., 1973), p. 404.

Further Reading

Boorstin, Daniel J. *The Lost World of Thomas Jefferson.* Chicago: University of Chicago Press, 1993.

Brown, David S. *Thomas Jefferson: A Biographical Companion.* Santa Barbara, Calif.: ABC–CLIO, 1998.

Burstein, Andrew. *The Inner Jefferson: Portrait of a Grieving Optimist.* Charlottesville: The University Press of Virginia, 1995.

Koch, Adrienne and William Peden. *The Life and Selected Writings of Thomas Jefferson.* New York: The Modern Library, 1998.

Malone, Dumas. *Thomas Jefferson: A Brief Biography.* Charlottesville, Va.: Thomas Jefferson Memorial Foundation, 1986.

Old, Wendie C. *Thomas Jefferson.* Springfield, N.J.: Enslow Publishers, Inc., 1997.

Peterson, Merrill D., ed. *The Political Writings of Thomas Jefferson.* Charlottesville, Va.: Thomas Jefferson Memorial Foundation, 1993.

Stanton, Lucia. *Slavery at Monticello.* Charlottesville, Va.: Thomas Jefferson Memorial Foundation, 1996.

Whitelaw, Nancy. *Thomas Jefferson: Philosopher & President.* Greensboro, N.C.: Morgan Reynolds, 2002.

Zinn, Howard. *A People's History of the United States: 1492–Present.* New York: HarperCollins, 1995.

Index

A
Adams, John, 22, 26, 27, 28, 44
American Revolution, 12, 22, 25

B
Burr, Aaron, 28–29, 37–38

C
Clark, William, 33–34
College of William and Mary, 18, 19
Continental Congress, 11, 12, 14, 22, 23

D
Declaration of Independence, 11–14, 22–23, 41, 43–44
Douglas, Rev. William, 16

E
Embargo Act of 1807, 37

F
Fossett, Edith, 42
Fossett, Joseph, 42
Fossett, Peter, 43
Franklin, Benjamin, 22, 26

H
Hamilton, Alexander, 26, 27
House of Burgesses, 18, 19

J
Jefferson, Jane Randolph (mother), 15
Jefferson, Martha Wayles Skelton (wife), 20–21, 24
Jefferson, Peter (father), 15, 16, 18

L
Lewis, Meriwether, 33–34
Livingston, Robert, 22, 31
Locke, John, 18
Louisiana Purchase, 31–32

M
Madison, James, 25, 26, 38
Marshall, John, 38
Maury, Rev. James, 17–18
Monroe, James, 32
Monticello, 20, 21, 24, 27, 38, 39–43

N
Napoléon Bonaparte, 32, 35

R
Randolph, Peyton, 20
Randolph, William, 15, 16

S
Shadwell, 15, 20
Sherman, Roger, 22
slavery, 16–17, 41–43

U
University of Virginia, 41

W
Washington, George, 20, 26
Wythe, George, 19